STUDENT SOLUTIONS MANUAL

Julia K. Sawyer
Northeastern State University, Tahlequah

INTRODUCTORY STATISTICS

Sixth Edition

Prem S. Mann
Eastern Connecticut State University

JOHN WILEY & SONS, INC.

Cover Photo: Paolo Curto/Getty Images

To order books or for customer service, please call 1-800-CALL-WILEY (225-5945).

ISBN-13 978-0-471-75531-9
ISBN-10 0-471-75531-1

Printed in the United States of America.

10 9 8 7 6 5 4 3 2 1

Printed and bound by Bind-Rite Graphics, Inc.

Table of Contents

Chapter One

Sections 1.1 - 1.2

1.1
1) **Statistics** refers to numerical facts such as the age of a student or the income of a family.
2) Statistics refers to the field or discipline of study. Statistics is a group of methods used to collect, analyze, present, and interpret data and to make decisions.

Section 1.3

1.3 A **population** is the collection of all elements whose characteristics are being studied. A **sample** is a portion of the population selected for study. A **representative sample** is a sample that represents the characteristics of the population as closely as possible, and a **random sample** is a sample drawn in such a way that each element of the population has a chance of being included in the sample. **Sampling with replacement** refers to a sampling procedure in which the item selected at each selection is put back in the population before the next item is drawn; **sampling without replacement** is a sampling procedure in which the item selected at each selection is not replaced in the population.

1.5 A **census** is a survey that includes every member of the population. A survey based on a portion of the population is called a **sample survey**. A sample survey is preferred over a census for the following reasons:
1) Conducting a census is very expensive because the size of the population is often very large.
2) Conducting a census is very time consuming.
3) In many cases it is impossible to identify each element of the target population.

1.7 a. Population b. Sample c. Population d. Sample e. Population

Section 1.4

1.9 With reference to this table, we have the following definitions:
- Member: Each city included in the table

- Variable: Number of dog bites reported
- Measurement: Number of dog bites in a specific city
- Data set: Collection of dog bite numbers for the six cities listed in the table.

1.11 a. Number of dog bites b. Six c. Six (cities)

Section 1.5

1.13 a. A **quantitative variable** is a variable that can be measured numerically.

b. A variable that cannot assume a numeric value but can be classified into two or more nonnumeric categories is called a **qualitative variable**.

c. A **discrete variable** is a variable whose values are countable.

d. A variable that can assume any numerical value over a certain interval or intervals is called a **continuous variable**.

e. Data collected on a quantitative variable is called **quantitative data**.

f. **Qualitative data** is data collected on a qualitative variable.

1.15 a. Quantitative b. Quantitative c. Qualitative d. Quantitative e. Quantitative

1.17 a. Discrete b. Continuous d. Discrete e. Continuous

Sections 1.6 - 1.7

1.19 **Internal sources** of data are sources inside the organization conducting the study which make needed data available. **External sources** of data are the sources outside the organization from which data is available.

1.21 a. Cross-section data b. Cross-section data c. Time-series data d. Time-series data

Section 1.8

1.23

m	f	m^2	mf	m^2f
3	16	9	48	144
6	11	36	66	396
25	16	625	400	10,000
12	8	144	96	1152
15	4	225	60	900
18	14	324	252	4536
$\Sigma m = 79$	$\Sigma f = 69$	$\Sigma m^2 = 1363$	$\Sigma mf = 922$	$\Sigma m^2f = 17,128$

a. $\sum f = 69$ b. $\sum m^2 = 1363$ c. $\sum mf = 922$ d. $\sum m^2 f = 17{,}128$

1.25

x	y	xy	x^2	y^2
4	12	48	16	144
18	5	90	324	25
25	14	350	625	196
9	7	63	81	49
12	12	144	144	144
20	8	160	400	64
$\sum x = 88$	$\sum y = 58$	$\sum xy = 855$	$\sum x^2 = 1590$	$\sum y^2 = 622$

a. $\sum x = 88$ b. $\sum y = 58$ c. $\sum xy = 855$ d. $\sum x^2 = 1590$ e. $\sum y^2 = 622$

1.27 a. $\sum y = 83 + 205 + 87 + 154 = \$529$ b. $(\sum y)^2 = (529)^2 = 279{,}841$

c. $\sum y^2 = (83)^2 + (205)^2 + (87)^2 + (154)^2 = 80{,}199$

1.29 a. $\sum x = 7 + 39 + 21 + 16 + 3 + 43 + 19 = 148$ students b. $(\sum x)^2 = (148)^2 = 21{,}904$

c. $\sum x^2 = (7)^2 + (39)^2 + (21)^2 + (16)^2 + (3)^2 + (43)^2 + (19)^2 = 4486$

Supplementary Exercises

1.31 With reference to this table, we have the following definitions

- Variable: Number of Americans who took cruises
- Measurement: Number of Americans who took cruises for a specific year
- Data Set: Collection of the number of Americans who took cruises for the 10 years listed in the table

1.33 a. Sample b. Population c. Sample d. Population

1.35 a. This is an example of sampling without replacement because once a patient is selected, he/she will not be replaced before the next patient is selected.

b. This is an example of sampling with replacement because both times the selection is made from the same group of professors.

1.37 a. $\sum x = 8 + 14 + 3 + 7 + 10 + 5 = 47$ shoe pairs b. $(\sum x)^2 = (47)^2 = 2209$

c. $\sum x^2 = (8)^2 + (14)^2 + (3)^2 + (7)^2 + (10)^2 + (5)^2 = 443$

1.39

m	f	f^2	mf	m^2f	m^2
3	7	49	21	63	9
16	32	1024	512	8192	256
11	17	289	187	2057	121
9	12	144	108	972	81
20	34	1156	680	13,600	400
$\sum m = 59$	$\sum f = 102$	$\sum f^2 = 2662$	$\sum mf = 1508$	$\sum m^2f = 24{,}884$	$\sum m^2 = 867$

a. $\sum m = 59$ b. $\sum f^2 = 2662$ c. $\sum mf = 1508$ d. $\sum m^2f = 24{,}884$ e. $\sum m^2 = 867$

Self-Review Test

1. b

2. c

3. a. Sample without replacement b. Sample with replacement

4. a. Qualitative b. Quantitative; continuous c. Quantitative; discrete d. Qualitative

5. With reference to this table, we have the following definitions:

- Member: Each player listed in the table
- Variable: Total number of career yards rushed
- Measurement: Total number of yards rushed for a specific player
- Data set: Collection of total number of yards rushed for the 10 players listed in the table

6. a. $\sum x = 2 + 5 + 3 + 12 + 7 = 29$ credit cards b. $(\sum x)^2 = (29)^2 = 841$

c. $\sum x^2 = (2)^2 + (5)^2 + (3)^2 + (12)^2 + (7)^2 = 231$

7.

m	f	m^2	mf	m^2f	f^2
3	15	9	45	135	225
6	25	36	150	900	625
9	40	81	360	3240	1600
12	20	144	240	2880	400
15	12	225	180	2700	144
$\sum m = 45$	$\sum f = 112$	$\sum m^2 = 495$	$\sum mf = 975$	$\sum m^2f = 9855$	$\sum f^2 = 2994$

a. $\sum m = 45$ b. $\sum f = 112$ c. $\sum m^2 = 495$ d. $\sum mf = 975$ e. $\sum m^2f = 9855$ f. $\sum f^2 = 2994$

Chapter Two

Section 2.1 – 2.2

2.1 Data in their original form are often too large and unmanageable. It is easier to make sense of grouped data than ungrouped data and easier to make decisions and draw conclusions using grouped data.

2.3 a. & b.

Category	Frequency	Relative Frequency	Percentage
A	8	8/30 = .267	26.7
B	8	8/30 = .267	26.7
C	14	14/30 = .467	46.7

c. 26.7 % of the elements in this sample belong to category B.

d. 26.7 + 46.7 = 73.4% of the elements in this sample belong to category A or C.

e.

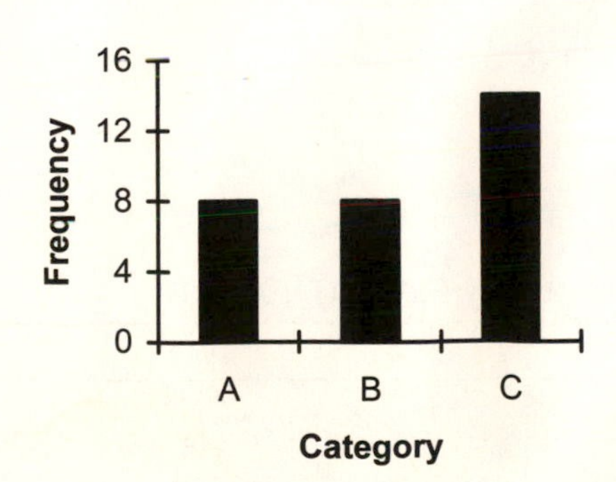

2.5 a. & b.

Category	Frequency	Relative Frequency	Percentage
F	12	12/50 = .24	24
SO	12	12/50 = .24	24
J	15	15/50 = .30	30
SE	11	11/50 = .22	22

c. 30 + 22 = 52% of the students are juniors or seniors.

d.

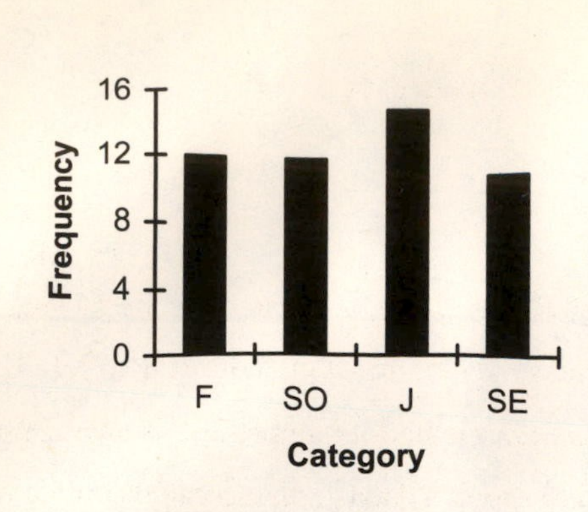

2.7 a. & b.

Category	Frequency	Relative Frequency	Percentage
H	10	10/20 = .50	50
C	6	6/20 = .30	30
O	4	4/20 = .20	20

c. 50% of the dieticians gave *Health* as the major reason for people to lose weight.

d.

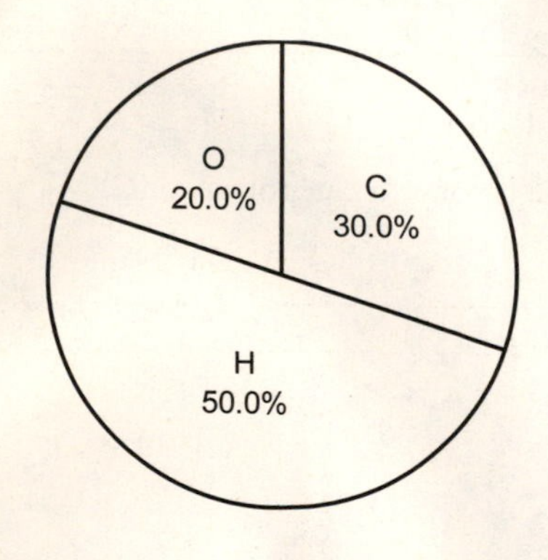

2.9 Let the five categories listed in the table be denoted by E, VG, G, F, and P.

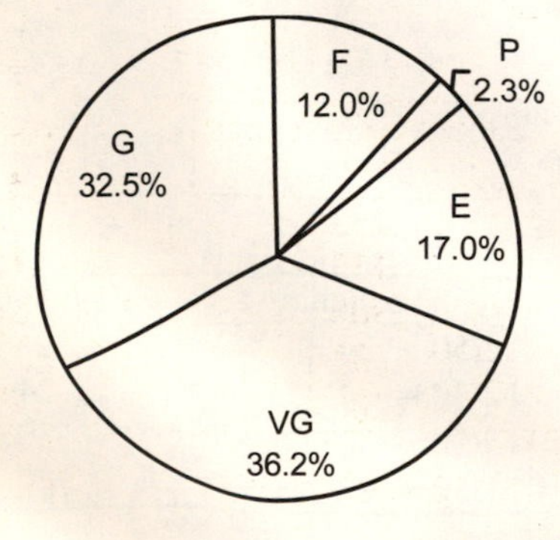

Section 2.3 – 2.4

2.11
1. The number of classes to be used to group the given data.
2. The width of each class.
3. The lower limit of the first class.

2.13 A data set that does not contain fractional values is usually grouped by using classes with limits. Example 2–4 is an example of the writing classes using limits method. A data set that contains fractional values is grouped by using the *less than* method. Example 2–5 is an example of the *less than* method. Single-valued classes are used to group a data set that contains only a few distinct values. Example 2–6 is an example of the single-valued classes method.

2.15 a. & c.

Class Boundaries	Class Midpoint	Relative Frequency	Percentage
17.5 to less than 30.5	24	.24	24
30.5 to less than 43.5	37	.38	38
43.5 to less than 56.5	50	.28	28
56.5 to less than 69.5	63	.10	10

b. Yes, each class has a width of 13.

d. 24 + 38 = 62% of the employees are 43 years old or younger.

2.17 a., b., & c.

Class Limits	Class Boundaries	Class Width	Class Midpoint
1 to 25	.5 to less than 25.5	25	13
26 to 50	25.5 to less than 50.5	25	38
51 to 75	50.5 to less than 75.5	25	63
76 to 100	75.5 to less than 100.5	25	88
101 to 125	100.5 to less than 125.5	25	113
126 to 150	125.5 to less than 150.5	25	138

2.19 a. & b.

Number of Computer Monitors Manufactured	Frequency	Relative Frequency	Percentage
21 to 23	7	.233	23.3
24 to 26	4	.133	13.3
27 to 29	9	.300	30.0
30 to 32	4	.133	13.3
33 to 35	6	.200	20.0

c.

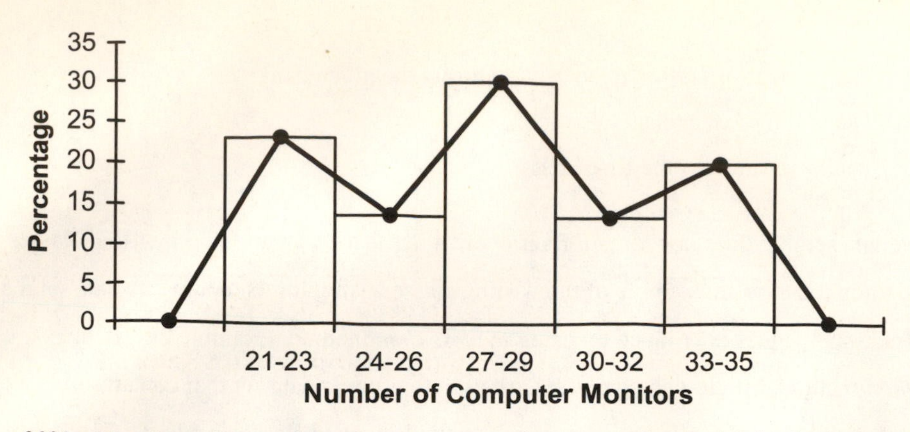

d. For 30% of the days, the number of computer monitors manufactured is in the interval 27 to 29.

2.21 a. & b.

Charitable Contributions (millions of dollars)	Frequency	Relative Frequency	Percentage
2 to less than 12	2	.133	13.3
12 to less than 22	6	.400	40.0
22 to less than 32	4	.267	26.7
32 to less than 42	0	.000	0.0
42 to less than 52	3	.200	20.0

2.23 a., b., & c. The lowest burglary rate is 379.4, and the highest is 1196.3. The following table groups the data into five classes of equal width (165) with a starting point of 375.

Burglary Rate	Frequency	Relative Frequency	Percentage
375 to less than 540	10	.385	38.5
540 to less than 705	6	.231	23.1
705 to less than 870	4	.154	15.4
870 to less than 1035	2	.077	7.7
1035 to less than 1200	4	.154	15.4

2.25 a. & b. The lowest aggravated assault rate is 56.8, and the highest rate is 626.5. The following table groups the data into six classes of equal width (100) with a starting point of 50.

Aggravated Assault Rate	Frequency	Relative Frequency	Percentage
50 to less than 150	5	.192	19.2
150 to less than 250	9	.346	34.6
250 to less than 350	5	.192	19.2
350 to less than 450	3	.115	11.5
450 to less than 550	3	.115	11.5
550 to less than 650	1	.038	3.8

c.

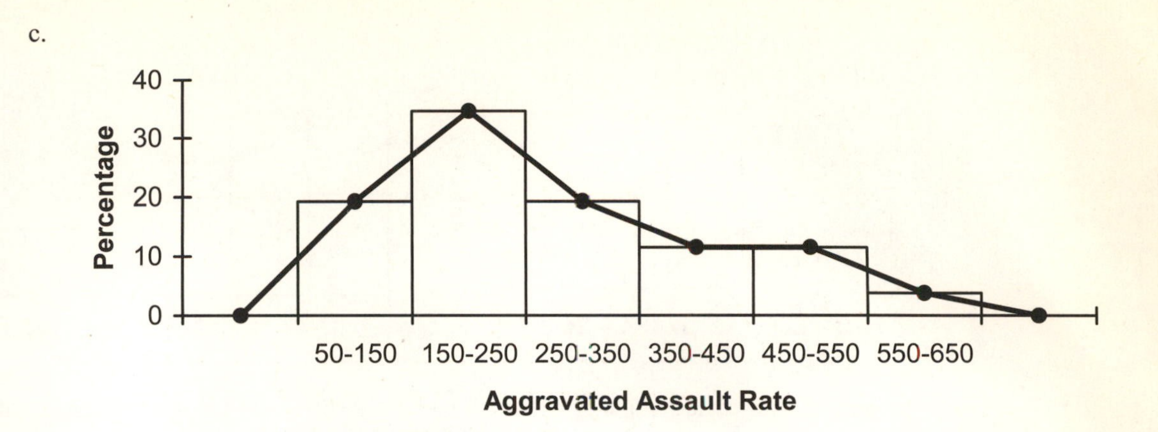

2.27 a. & b

ERA	Frequency	Relative Frequency	Percentage
3.50 to less than 4.00	3	.188	18.8
4.00 to less than 4.50	10	.625	62.5
4.50 to less than 5.00	1	.063	6.3
5.00 to less than 5.50	1	.063	6.3
5.50 to less than 6.00	1	.063	6.3

2.29 a. & b.

Number of Errors in Credit Reports	Frequency	Relative Frequency	Percentage
0	7	.28	28
1	7	.28	28
2	5	.20	20
3	3	.12	12
4	2	.08	8
5	1	.04	4

c. $5 + 3 + 2 + 1 = 11$ credit reports have two or more errors.

d.

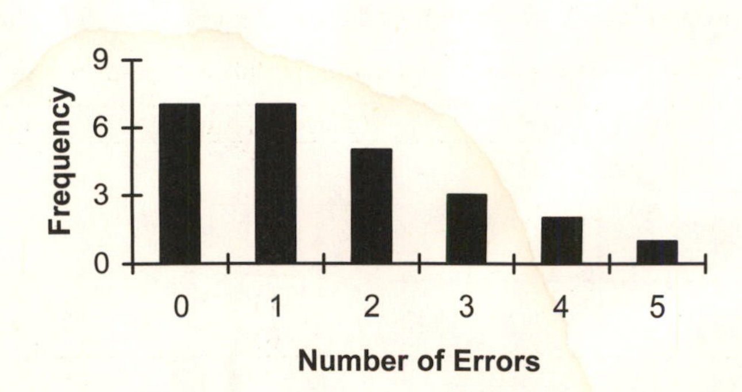

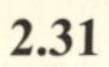
2.31

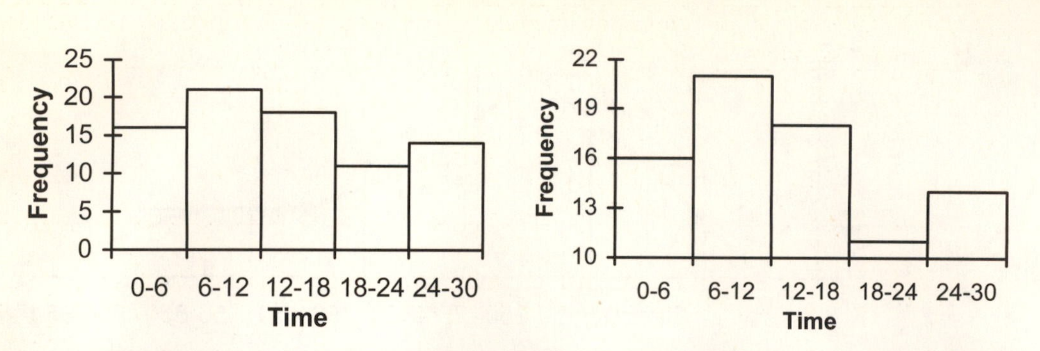

The graph with the truncated frequency axis exaggerates the differences in the frequencies of the various time intervals.

Section 2.5

2.33 An **ogive** is drawn for a cumulative frequency distribution, a cumulative relative frequency distribution, or a cumulative percentage distribution. An ogive can be used to find the approximate cumulative frequency (cumulative relative frequency or cumulative percentage) for any class interval.

2.35 a. & b.

Age (years)	Cumulative Frequency	Cumulative Relative Frequency	Cumulative Percentage
18 to 30	12	.24	24
18 to 43	31	.62	62
18 to 56	45	.90	90
18 to 69	50	1.00	100

c. 100 – 62 = 38% of the employees are 44 years of age or older.

d.

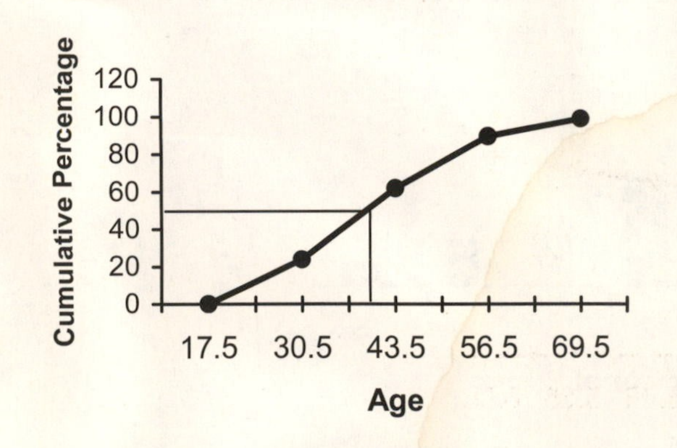

e. Approximately 52% of the employees are 40 years of age or younger as indicated on the ogive in part d.

2.37

Number of Computer Monitors	Cumulative Frequency	Cumulative Relative Frequency	Cumulative Percentage
21 to 23	7	.233	23.3
21 to 26	11	.367	36.7
21 to 29	20	.667	66.7
21 to 32	24	.800	80.0
21 to 35	30	1.000	100.0

2.39

Burglary Rate	Cumulative Frequency	Cumulative Relative Frequency	Cumulative Percentage
375 to less than 540	10	.385	38.5
375 to less than 705	16	.615	61.5
375 to less than 870	20	.769	76.9
375 to less than 1035	22	.846	84.6
375 to less than 1200	26	1.000	100.0

2.41

Motor Vehicle Theft Rate	Cumulative Frequency	Cumulative Relative Frequency	Cumulative Percentage
100 to less than 200	3	.115	11.5
100 to less than 300	10	.385	38.5
100 to less than 400	17	.654	65.4
100 to less than 500	24	.923	92.3
100 to less than 600	25	.962	96.2
100 to less than 700	26	1.000	100.0

2.43

ERA	Cumulative Frequency	Cumulative Relative Frequency	Cumulative Percentage
3.50 to less than 4.00	3	.188	18.8
3.50 to less than 4.50	13	.813	81.3
3.50 to less than 5.00	14	.875	87.5
3.50 to less than 5.50	15	.938	93.8
3.50 to less than 6.00	16	1.000	100.0

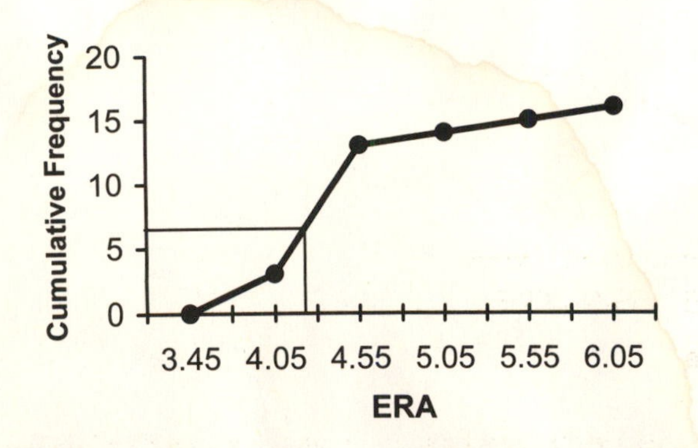

Approximately 7 of the teams had an ERA of less than 4.20.

Section 2.6

2.45 The advantage of a stem-and-leaf display over a frequency distribution is that by preparing a stem-and-leaf display we do not lose information on individual observations. From a stem-and-leaf display we can obtain the original data. However, we cannot obtain the original data from a frequency distribution table. Consider the stem-and-leaf display from Example 2–8:

5	2 0 7
6	5 9 1 8 4
7	5 9 1 2 6 9 7 1 2
8	0 7 1 6 3 4 7
9	6 3 5 2 2 8

The data that were used to make this stem-and-leaf display are: 52, 50, 57, 65, 69, 61, 68, 64, 75, 79, 71, 72, 76, 79, 77, 71, 72, 80, 87, 81, 86, 83, 84, 87, 96, 93, 95, 92, 92, 98

2.47 The data that were used to make this stem-and-leaf display are: 218, 245, 256, 329, 367, 383, 397, 404, 427, 433, 471, 523, 537, 551, 563, 581, 592, 622, 636, 647, 655, 678, 689, 810, 841

2.49

7	45 75
8	48 00 57
9	21 33 67 95
10	24 09
11	33 45
12	75

7	45 75
8	00 48 57
9	21 33 67 95
10	09 24
11	33 45
12	75

2.51

4	5 8 1 6 4 2 8 8 6 3 7 4 7 9
5	2 6 3 1 3 1 2 0 4 0 2

4	1 2 3 4 4 5 6 6 7 7 8 8 8 9
5	0 0 1 1 2 2 2 3 3 4 6

2.53

0	5 7
1	0 1 7 5 9
2	3 6 6 9 1 2
3	3 9 2
4	8 3
5	0
6	5

0	5 7
1	0 1 5 7 9
2	1 2 3 6 6 9
3	2 3 9
4	3 8
5	0
6	5

2.55 a.

Stem	Leaves							
1	58							
2	10	20	45	65	68	70		
3	20	45	50	68	90			
4	30	38	57	60	75	87	90	
5	05	28	30	38	40	60	65	70
6	17	35	38					
7	02	05	06	20	21			

b.

Stem	Leaves															
1-3	58	*	10	20	45	65	68	70	*	20	45	50	68	90		
4-5	30	38	57	60	75	87	90	*	05	28	30	38	40	60	65	70
6-7	17	35	38	*	02	05	06	20	21							

Section 2.7

2.57 A **stacked dotplot** is used to compare two or more data sets by creating a dotplot for each data set with numbers lines for all data sets on the same scale. The data sets are placed on top of each other.

2.59

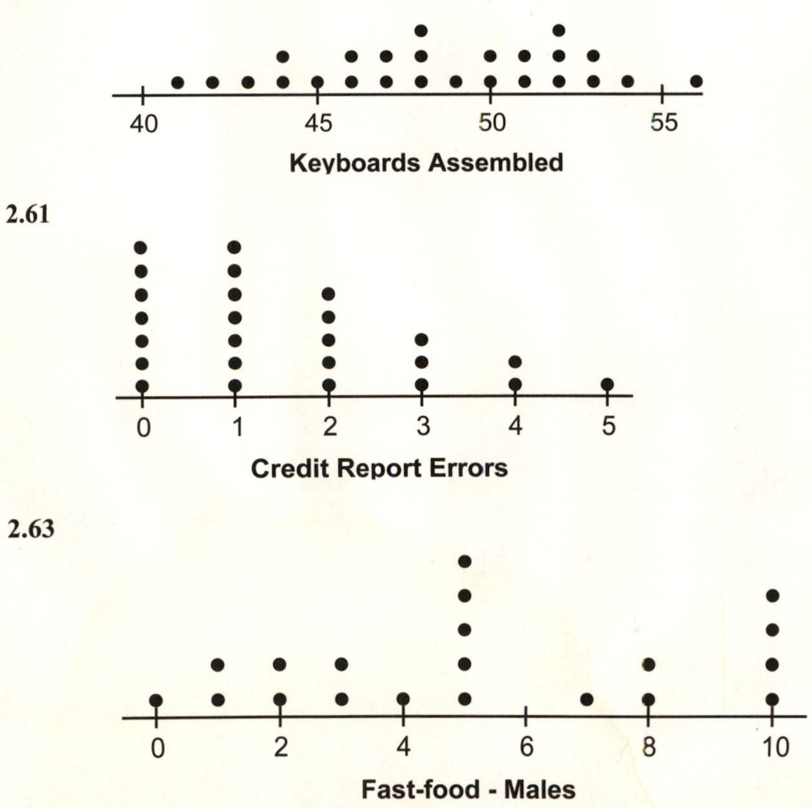

The data for males is clustered in two groups with the first group having values from zero to five, and the second having values from seven to 10.

2.65

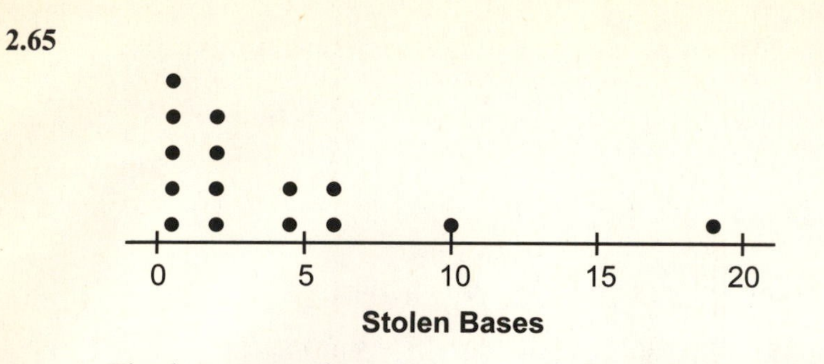

The data set contains two clusters – the first from one to two and the second from four to six. The value 10 appears to be a mild outlier while the value 19 is an extreme value for number of stolen bases.

Supplementary Exercises

2.67 a. & b.

Coping Strategy	Frequency	Relative Frequency	Percentage
C	13	.433	43.3
N	7	.233	23.3
W	5	.167	16.7
S	5	.167	16.7

c.

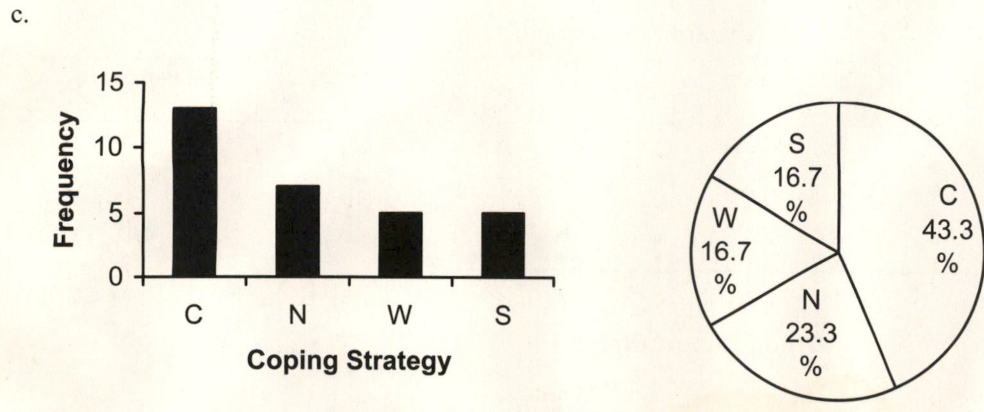

d. 23.3% of these respondents preferred to cope with afternoon drowsiness by taking a nap.

2.69 a. & b.

Correct Names	Frequency	Relative Frequency	Percentage
0	1	.042	4.2
1	3	.125	12.5
2	4	.167	16.7
3	6	.250	25.0
4	4	.167	16.7
5	6	.250	25.0

c. 4.2 + 12.5 = 16.7% of the students named fewer than two of the representatives correctly.

d.

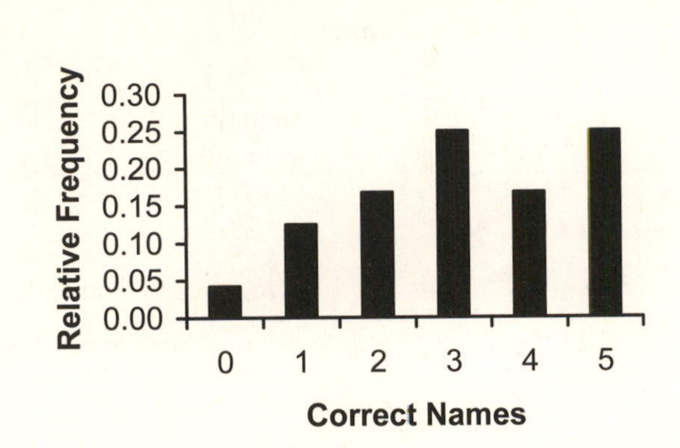

2.71 a. & b.

Number of Orders	Frequency	Relative Frequency	Percentage
23 – 29	4	.133	13.3
30 – 36	9	.300	30.0
37 – 43	6	.200	20.0
44 – 50	8	.267	26.7
51 – 57	3	.100	10.0

c. For 20.0 + 26.7 + 10.0 = 56.7% of the hours in this sample, the number of orders was more than 36.

2.73 a. & b.

Car Repair Costs (dollars)	Frequency	Relative Frequency	Percentage
1 – 1400	11	.367	36.7
1401 – 2800	10	.333	33.3
2801 – 4200	3	.100	10.0
4201 – 5600	2	.067	6.7
5601 – 7000	4	.133	13.3

c.

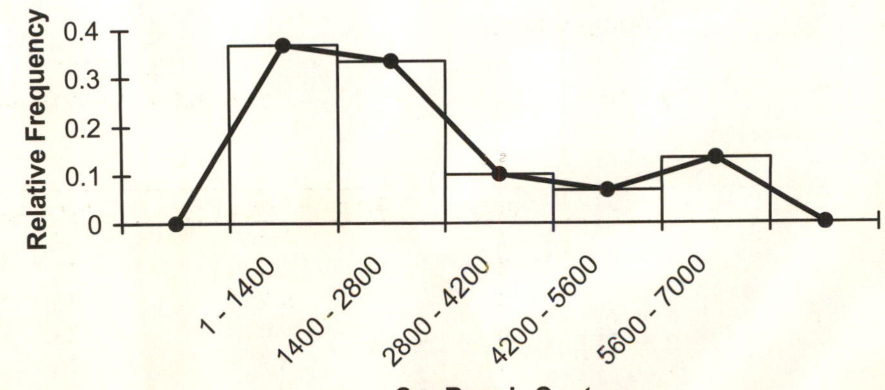

d. The class boundaries of the fourth class are \$4200.50 and \$5600.50. The width of this class is \$1400.

2.75

Number of Orders	Cumulative Frequency	Cumulative Relative Frequency	Cumulative Percentage
23 – 29	4	.133	13.3
23 – 36	13	.433	43.3
23 – 43	19	.633	63.3
23 – 50	27	.900	90.0
23 – 57	30	1.000	100.0

2.77

Car Repair Costs (dollars)	Cumulative Frequency	Cumulative Relative Frequency	Cumulative Percentage
1 – 1400	11	.367	36.7
1 – 2800	21	.700	70.0
1 – 4200	24	.800	80.0
1 – 5600	26	.867	86.7
1 – 7000	30	1.000	100.0

2.79

```
2 | 8 4 7 7
3 | 4 1 8 5 2 9 3 7 0 8 4 6 0
4 | 4 1 7 6 1 9 5 6 7
5 | 2 3 7 0
```

2.81

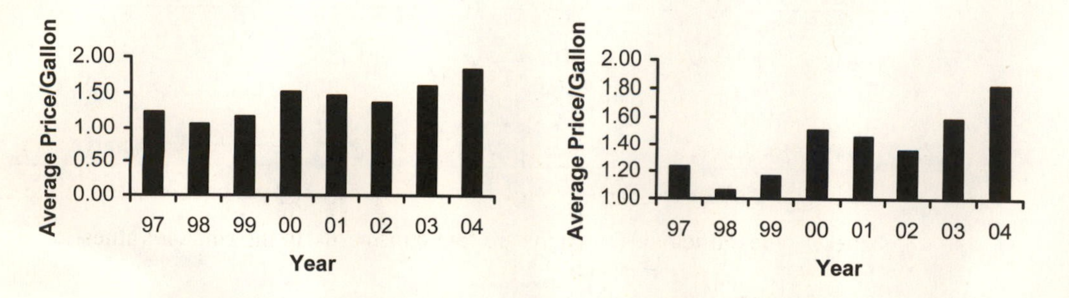

The truncated graph exaggerates the differences in average price per gallon for the eight year period.

2.83

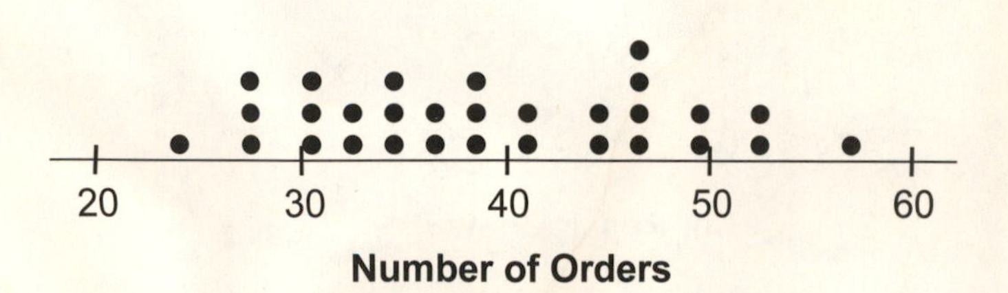

2.85

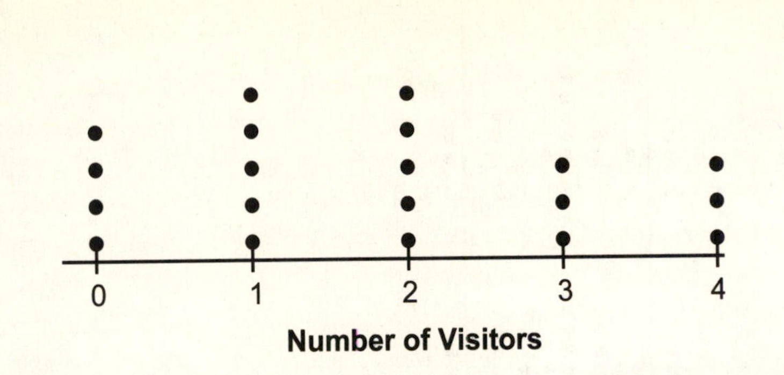

2.87 The greater relative frequency of accidents in the older age group does not imply that they are more accident-prone than the younger group. For instance, the older group may drive more miles during a week than the younger group.

2.89 a. Answers will vary.

b. i.

9	9					
10	2	8	8			
11	0	4	5	5	6	9
12	3	3	3	5	8	8
13	2	3	8			
14	6	7	7	8		
15	5	9				
16	1	2	4	8		
17	4	4	5	9	9	9
18	0	2	3	9		
19	3	3	5			
20	2	4				

ii. The display shows a bimodal distribution, due to the presence of both females and males in the sample. The males tend to be heavier, so their weights are concentrated in the larger values, while the females' weights are found primarily in the smaller values.

c.

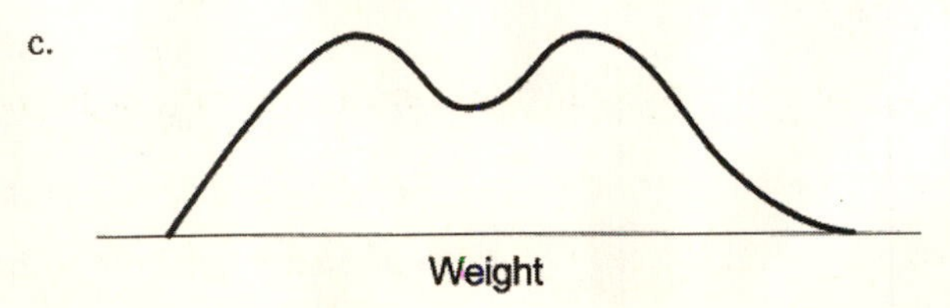

2.91

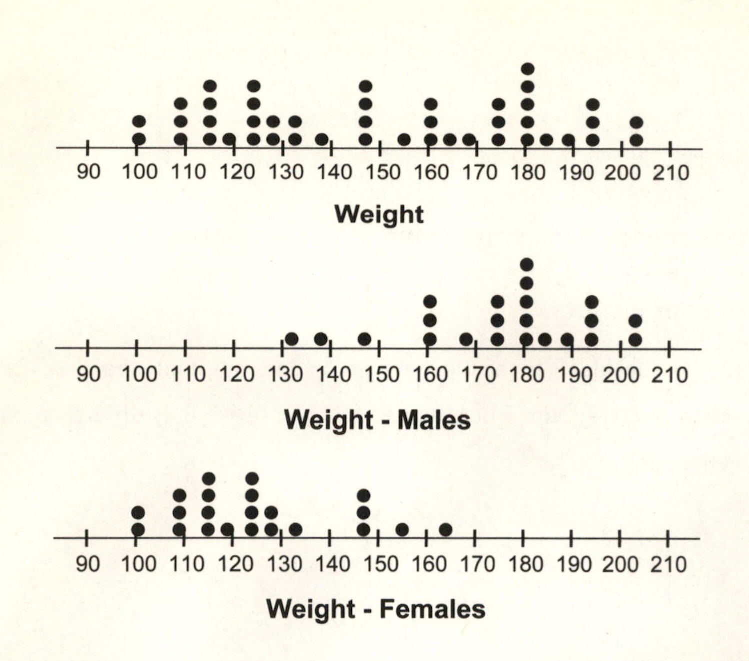

The distribution of all weights is bimodal. The distribution of weights for males is skewed to the left while the distribution for females is skewed to the right. You cannot distinguish between the lightest males and heaviest females in the dotplot of all weights as the distributions overlap in the area between 130 and 170 pounds.

Self-Review Test

1. An **ungrouped data set** contains information on each member of a sample or population individually. The first part of Example 2-1 in the text, listing the responses of each of the 30 employees, is an example of ungrouped data. Data presented in the form of a frequency table are called **grouped data**. Table 2.4 in the solution of Example 2-1 is an example of grouped data.

2. a. 5 b. 7 c. 17 d. 6.5 e. 13 f. 90 g. .30

3. A histogram that is identical on both sides of its central point is called a symmetric histogram. A histogram that is skewed to the right has a longer tail on the right side, and a histogram that is skewed to the left has a longer tail on the left side. The following three histograms present these three cases. Figure 2.8 in the text provides graphs of symmetric histograms, Figure 2.9a displays a histogram skewed to the right, and Figure 2.9b displays a histogram that is skewed to the left.

4. a. & b.

Category	Frequency	Relative Frequency	Percentage
B	8	.40	40
F	4	.20	20
M	7	.35	35
S	1	.05	5

c. 35% of the children live with their mothers only.

d.

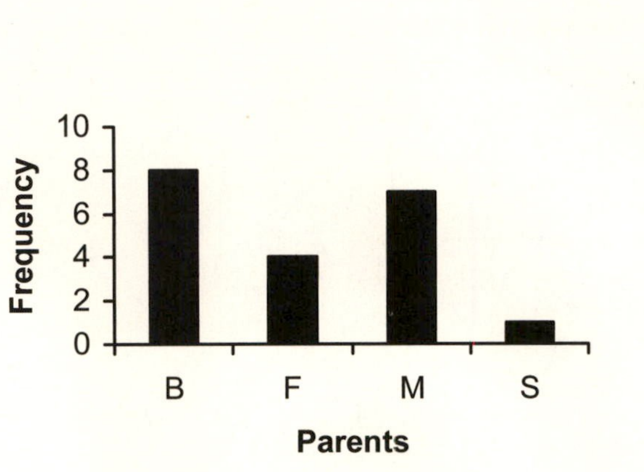

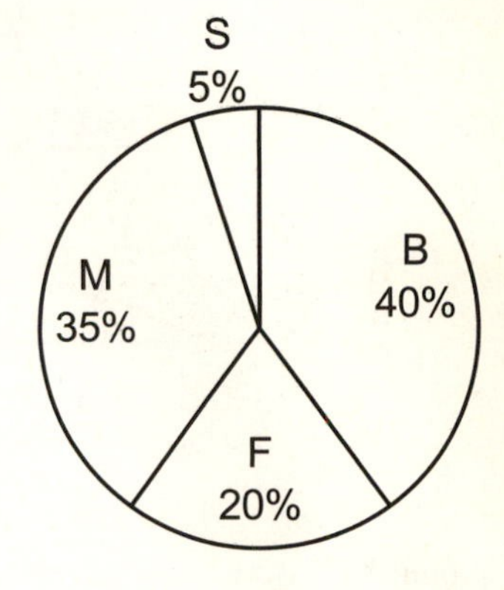

5. a. & b.

Number of False Alarms	Frequency	Relative Frequency	Percentage
1 – 3	5	.208	20.8
4 – 6	6	.250	25.0
7 – 9	6	.250	25.0
10 – 12	4	.167	16.7
13 – 15	3	.125	12.5

c. 20.8 + 25.0 + 25.0 = 70.8% of the weeks had 9 or fewer false alarms.

d.

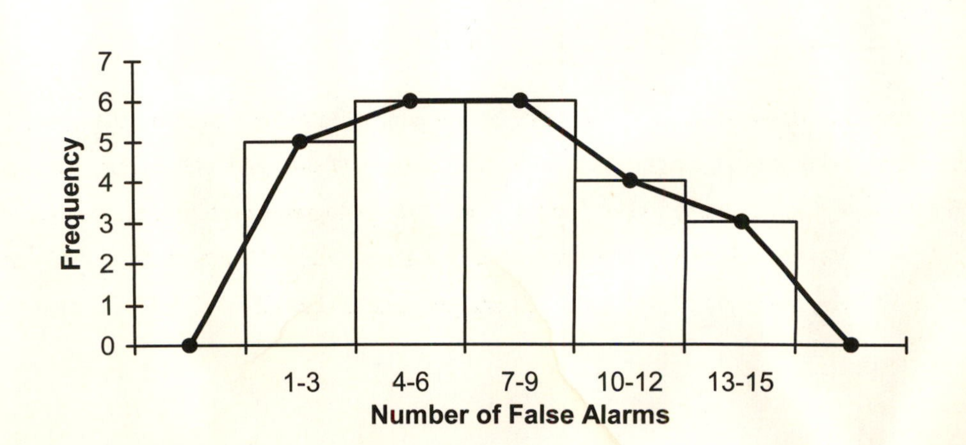

6.

Number of False Alarms	Cumulative Frequency	Cumulative Relative Frequency	Cumulative Percentage
1 – 3	5	.208	20.8
1 – 6	11	.458	45.8
1 – 9	17	.708	70.8
1 – 12	21	.875	87.5
1 – 15	24	1.000	100.0

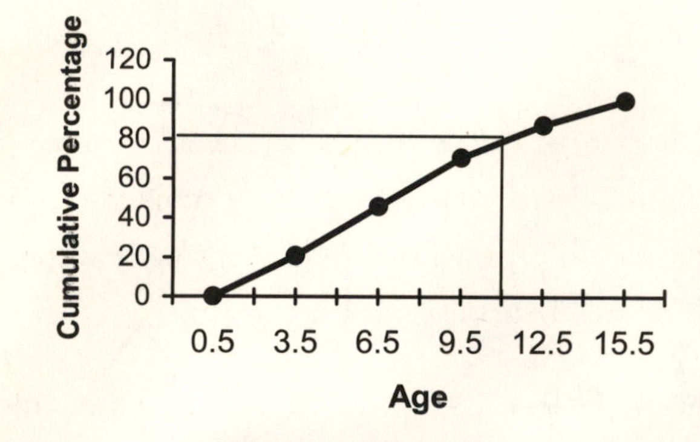

About 83% of the employees have been with their current employer for 11 or fewer years as indicated in the above ogive.

7.

```
0 | 4 6 7 8
1 | 0 2 2 3 4 4 5 6 6 6 7 8 9
2 | 0 1 2 2 5 9
3 | 2
```

8. 30 33 37 42 44 46 47 49 51 53 53 56 60 67 67 71 79

9.

0 3 6 9 12 15

Chapter Three

Section 3.1

3.1 For a data set with an odd number of observations, first we rank the data set in increasing (or decreasing) order and then find the value of the middle term. This value is the median. For a data set with an even number of observations, first we rank the data set in increasing (or decreasing) order and then find the average of the two middle terms. The average gives the median.

3.3 Suppose the exam scores for seven students are 73, 82, 95, 79, 22, 86, and 91 points. Then, Mean = (73 + 82 + 95 + 79 + 22 + 86 + 91)/7 = 75.43 points. If we drop the outlier (22), Mean = (73 + 82 + 95 + 79 + 86 + 91)/6 = 84.33 points. This shows how an outlier can affect the value of the mean.

3.5 The mode can assume more than one value for a data set. Examples 3–8 and 3–9 of the text present such cases.

3.7 For a symmetric histogram (with one peak), the values of the mean, median, and mode are all equal. Figure 3.2 of the text shows this case. For a histogram that is skewed to the right, the value of the mode is the smallest and the value of the mean is the largest. The median lies between the mode and the mean. Such a case is presented in Figure 3.3 of the text. For a histogram that is skewed to the left, the value of the mean is the smallest, the value of the mode is the largest, and the value of the median lies between the mean and the mode. Figure 3.4 of the text exhibits this case.

3.9 $\sum x = 5 + (-7) + 2 + 0 + (-9) + 16 + 10 + 7 = 24$

$\mu = (\sum x)/N = 24/8 = 3$

Median = value of the 4.5^{th} term in ranked data = (2 + 5)/2 = 3.50

This data set has no mode.

3.11 $\bar{x} = (\sum x)/n = 164{,}348/10 = \$16{,}434.80$

Median = value of the 5.5^{th} term in ranked data set = (15,820 + 16,449)/2 = $16,134.50

3.13 a. $\mu = (\sum x)/N = 56{,}446/50 = 1128.92$ thousand workers

Median = value of the 25.5th term in ranked data = (739 + 759)/2 = 749 thousand workers

b. The mode is 1200 since this value occurs three times and no other value occurs more than twice.

3.15 $\bar{x} = (\sum x)/n = 1{,}422{,}432/13 = \$109{,}417.85$

Median = value of the 7th term in ranked data = \$98,500

3.17 $\bar{x} = (\sum x)/n = 1967/6 = \$327.83$ million

Median = value of the 3.5th term in ranked data = (158 + 192)/2 = \$175 million

This data set has no mode because no value appears more than once.

3.19 $\bar{x} = (\sum x)/n = 35/12 = 2.92$ outages

Median = value of the 6.5th term in ranked data = (2 + 3)/2 = 2.5 outages

Mode = 2 outages

3.21 $\bar{x} = (\sum x)/n = 294/10 = 29.4$ computer monitors

Median = value of the 5.5th term in ranked data = (28 + 29)/2 = 28.5 computer monitors

Mode = 23 computer monitors

3.23 a. $\bar{x} = (\sum x)/n = 443/11 = 40.27$ casinos

Median = value of the 6th term in ranked data = 13 casinos

b. Yes, 256 is an outlier. When we drop this value,

Mean = 187/10 = 18.7 casinos

Median = value of the 5.5th term in ranked data = (12 + 13)/2 = 12.5 casinos

As we observe, the mean is affected more by the outlier.

c. The median is a better measure because it is not as sensitive to outliers as the mean.

3.25 $n_1 = 10,\ n_2 = 8,\ \bar{x}_1 = \$95,\ \bar{x}_2 = \$104$

$$\bar{x} = \frac{n_1\bar{x}_1 + n_2\bar{x}_2}{n_1 + n_2} = \frac{(10)(95) + (8)(104)}{10 + 8} = \frac{1782}{18} = \$99$$

3.27 Total money spent by 10 persons = $\sum x = n\bar{x} = 10(105.50) = \$1055$

3.29 Sum of the ages of six persons = (6)(46) = 276 years, so the age of sixth person = 276 – (57 + 39 + 44 + 51 + 37) = 48 years.

3.31 For Data Set I: Mean = 123/5 = 24.6 For Data Set II: Mean = 158/5 = 31.6

The mean of the second data set is greater than the mean of the first data set by 7.

3.33 The ranked data are: 19 23 26 31 38 39 47 49 53 67

By dropping 19 and 67, we obtain $\sum x = 23 + 26 + 31 + 38 + 39 + 47 + 49 + 53 = 306$

10% Trimmed Mean = $(\sum x)/n = 306/8 = 38.25$ years

3.35 From the given information: $x_1 = 73$, $x_2 = 67$, $x_3 = 85$, $w_1 = w_2 = 1$, $w_3 = 2$

$$\text{Weighted mean} = \frac{\sum xw}{\sum w} = \frac{(73)(1)+(67)(1)+(85)(2)}{4} = \frac{310}{4} = 77.5$$

Section 3.2

3.37 Suppose the exam scores for seven students are 73, 82, 95, 79, 22, 86, and 91.

Then, Range = Largest value – Smallest value = 95 – 22 = 73 points.

If we drop the outlier (22) and calculate the range,

Range = Largest value – Smallest value = 95 – 73 = 22 points.

Thus, when we drop the outlier, the range decreases from 73 to 22 points.

3.39 The value of the standard deviation is zero when all values in a data are the same. For example, suppose the exam scores of a sample of seven students are 82, 82, 82, 82, 82, 82, and 82. As this data set has no variation, the value of the standard deviation is zero for these observations. This is shown below:

$\sum x = 574$ and $\sum x^2 = 47{,}068$

$$s = \sqrt{\frac{\sum x^2 - \frac{(\sum x)^2}{n}}{n-1}} = \sqrt{\frac{47{,}068 - \frac{(574)^2}{7}}{7-1}} = \sqrt{\frac{47{,}068 - 47{,}068}{6}} = 0$$

3.41 Range = Largest value – Smallest value = 16 – (–9) = 25, $\sum x = 24$, $\sum x^2 = 564$ and $N = 8$

$$\sigma^2 = \frac{\sum x^2 - \frac{(\sum x)^2}{N}}{N} = \frac{564 - \frac{(24)^2}{8}}{8} = \frac{564 - 72}{8} = 61.5 \qquad \sigma = \sqrt{61.5} = 7.84$$

3.43 a. $\bar{x} = (\sum x)/n = 72/8 = 9$ shoplifters caught

Shoplifters caught	Deviations from the Mean
7	$7 - 9 = -2$
10	$10 - 9 = 1$
8	$8 - 9 = -1$
3	$3 - 9 = -6$
15	$15 - 9 = 6$
12	$12 - 9 = 3$
6	$6 - 9 = -3$
11	$11 - 9 = 2$
	Sum = 0

Yes, the sum of the deviations from the mean is zero.

b. Range = Largest value – Smallest value = 15 – 3 = 12, $\sum x = 72$, $\sum x^2 = 748$, and $n = 8$

$$s^2 = \frac{\sum x^2 - \frac{(\sum x)^2}{n}}{n-1} = \frac{748 - \frac{(72)^2}{8}}{8-1} = 14.2857 \qquad s = \sqrt{14.2857} = 3.78$$

3.45 $\sum x = 81$, $\sum x^2 = 699$, and $n = 12$

Range = Largest value – Smallest value = 15 – 2 = 13 thefts

$$s^2 = \frac{\sum x^2 - \frac{(\sum x)^2}{n}}{n-1} = \frac{699 - \frac{(81)^2}{12}}{12-1} = 13.8409 \qquad s = \sqrt{13.8409} = 3.72 \text{ thefts}$$

3.47 $\sum x = 291$, $\sum x^2 = 9171$, and $n = 10$

Range = Largest value – Smallest value = 41 – 14 = 27 pieces

$$s^2 = \frac{\sum x^2 - \frac{(\sum x)^2}{n}}{n-1} = \frac{9171 - \frac{(291)^2}{10}}{10-1} = 78.1 \qquad s = \sqrt{78.1} = 8.84 \text{ pieces}$$

3.49 $\sum x = 27$, $\sum x^2 = 111$, and $n = 13$

Range = Largest value – Smallest value = 7 – 0 = 7 patients

$$s^2 = \frac{\sum x^2 - \frac{(\sum x)^2}{n}}{n-1} = \frac{111 - \frac{(27)^2}{13}}{13-1} = 4.5769 \qquad s = \sqrt{4.5769} = 2.14 \text{ patients}$$

3.51 $\sum x = 80$, $\sum x^2 = 1552$, and $n = 8$

Range = Largest value – Smallest value = 23 – (–7) = 30° F

$$s^2 = \frac{\Sigma x^2 - \frac{(\Sigma x)^2}{n}}{n-1} = \frac{1552 - \frac{(80)^2}{8}}{8-1} = 107.4286 \qquad s = \sqrt{107.4286} = 10.36° \text{ F}$$

3.53 $\Sigma x = 262.2$, $\Sigma x^2 = 7837.08$, and $n = 10$

Range = Largest value – Smallest value = 46.5 – 18.3 = $28.2 billion

$$s^2 = \frac{\Sigma x^2 - \frac{(\Sigma x)^2}{n}}{n-1} = \frac{7837.08 - \frac{(262.2)^2}{10}}{10-1} = 106.9107 \qquad s = \sqrt{106.9107} = \$10.34 \text{ billion}$$

3.55 $\Sigma x = 96$, $\Sigma x^2 = 1152$, and $n = 8$

$$s = \sqrt{\frac{\Sigma x^2 - \frac{(\Sigma x)^2}{n}}{n-1}} = \sqrt{\frac{1152 - \frac{(96)^2}{8}}{8-1}} = \sqrt{\frac{1152-1152}{7}} = 0$$

The standard deviation is zero because all these data values are the same and there is no variation among them.

3.57 For the yearly salaries of all employees, CV = $(\sigma/\mu) \times 100\%$ = (6,820/62,350) × 100 = 10.94%

For the years of experience of these employees, CV = $(\sigma/\mu) \times 100\%$ = (2/15) × 100 = 13.33%

The relative variation in salaries is lower than that in years of experience.

3.59 For Data Set I: $\Sigma x = 123$, $\Sigma x^2 = 3883$, and $n = 5$

$$s = \sqrt{\frac{\Sigma x^2 - \frac{(\Sigma x)^2}{n}}{n-1}} = \sqrt{\frac{3883 - \frac{(123)^2}{5}}{5-1}} = \sqrt{214.300} = 14.64$$

For Data Set II: $\Sigma x = 158$, $\Sigma x^2 = 5850$, and $n = 5$

$$s = \sqrt{\frac{\Sigma x^2 - \frac{(\Sigma x)^2}{n}}{n-1}} = \sqrt{\frac{5850 - \frac{(158)^2}{5}}{5-1}} = \sqrt{214.300} = 14.64$$

The standard deviations of the two data sets are equal.

Section 3.3

3.61 The values of the mean and standard deviation for a grouped data set are the approximate values of the mean and standard deviation. The exact values of the mean and standard deviation are obtained only when ungrouped data are used.

3.63

x	f	m	mf	m^2f
0 to less than 4	17	2	34	68
4 to less than 8	23	6	138	828
8 to less than 12	15	10	150	1500
12 to less than 16	11	14	154	2156
16 to less than 20	8	18	144	2592
20 to less than 24	6	22	132	2904
	$N = \sum f = 80$		$\sum mf = 752$	$\sum m^2f = 10{,}048$

$\bar{x} = (\sum mf)/n = 752/80 = 9.40$

$$s^2 = \frac{\sum m^2 f - \frac{(\sum mf)^2}{n}}{n-1} = \frac{10{,}048 - \frac{(752)^2}{80}}{80-1} = 37.7114 \qquad s = \sqrt{37.7114} = 6.14$$

3.65

Hours Per Week	Number of Students	m	mf	m^2f
0 to less than 5	7	2.5	17.5	43.75
5 to less than 10	12	7.5	90.0	675.00
10 to less than 15	15	12.5	187.5	2343.75
15 to less than 20	13	17.5	227.5	3981.25
20 to less than 25	8	22.5	180.0	4050.00
25 to less than 30	5	27.5	137.5	3781.25
	$N = \sum f = 60$		$\sum mf = 840$	$\sum m^2f = 14{,}875$

$\mu = (\sum mf)/N = 840/60 = 14$ hours

$$\sigma^2 = \frac{\sum m^2 f - \frac{(\sum mf)^2}{n}}{n-1} = \frac{14{,}875 - \frac{(840)^2}{60}}{60} = 51.9167 \qquad \sigma = \sqrt{51.9167} = 7.21 \text{ hours}$$

3.67

Miles Driven in 2002 (in thousands)	Number of Car Owners	m	mf	m^2f
0 to less than 5	7	2.5	17.5	43.75
5 to less than 10	26	7.5	19.5	1462.50
10 to less than 15	59	12.5	737.5	9218.75
15 to less than 20	71	17.5	1242.5	21,743.75
20 to less than 25	62	22.5	1395.0	31,387.50
25 to less than 30	39	27.5	1072.5	29,493.75
30 to less than 35	22	32.5	715.0	23,237.50
35 to less than 40	14	37.5	525.0	19,687.50
	$N = \sum f = 300$		$\sum mf = 5900$	$\sum m^2f = 136{,}275$

$\bar{x} = (\sum mf)/n = 5900/300 = 19.67$ or 19,670 miles

$$s^2 = \frac{\sum m^2 f - \frac{(\sum mf)^2}{n}}{n-1} = \frac{136{,}275 - \frac{(5900)^2}{300}}{300-1} = 67.6979 \qquad s = \sqrt{67.6979} = 8.23 \text{ or } 8230 \text{ miles}$$

Each value in the column labeled *mf* gives the approximate total mileage for the car owners in the corresponding class. For example, the value of $mf = 17.5$ for the first class indicates that the seven car owners in this class drove a total of approximately 17,500 miles. The value $\sum mf = 5900$ indicates that the total mileage for all 300 car owners was approximately 5,900,000 miles.

3.69

x	f	m	mf	m^2f
0 to less than 20	14	10	140	1400
20 to less than 40	18	30	540	16,200
40 to less than 60	9	50	450	22,500
60 to less than 80	5	70	350	24,500
80 to less than 100	4	90	360	32,400
	$N = \sum f = 50$		$\sum mf = 1840$	$\sum m^2f = 97{,}000$

$\bar{x} = (\sum mf)/n = 1840/50 = 36.80$ minutes

$$s^2 = \frac{\sum m^2 f - \frac{(\sum mf)^2}{n}}{n-1} = \frac{97{,}000 - \frac{(1840)^2}{50}}{50-1} = 597.7143 \qquad s = \sqrt{597.7143} = 24.45 \text{ minutes}$$

3.71 a. $\bar{x} = \sum x/n = 776.41/15 =$ \$51.76 per barrel

b.

Price of Oil per Barrel	f	m	mf
47 to less than 49	2	48	96
49 to less than 51	6	50	300
51 to less than 53	2	52	104
53 to less than 55	2	54	108
55 to less than 57	3	56	168
	$N = \sum f = 15$		$\sum mf = 776$

c. $\bar{x} = (\sum mf)/n = 776/15 =$ \$51.73 per barrel

d. The two means are not equal because the second method uses approximations (mid points of the range) and the first one does not. This leads to very slightly different results.

Section 3.4

3.73 The empirical rule is applied to a bell–shaped distribution. According to this rule, approximately

1) 68% of the observations lie within one standard deviation of the mean.
2) 95% of the observations lie within two standard deviations of the mean.

3) 99.7% of the observations lie within three standard deviations of the mean.

3.75 For the interval $\mu \pm 2\sigma$: $k = 2$, and $1 - \frac{1}{k^2} = 1 - \frac{1}{(2)^2} = 1 - .25 = .75$ or 75%. Thus, at least 75% of the observations fall in the interval $\mu \pm 2\sigma$.

For the interval $\mu \pm 2.5\sigma$: $k = 2.5$, and $1 - \frac{1}{k^2} = 1 - \frac{1}{(2.5)^2} = 1 - .16 = .84$ or 84%. Thus, at least 84% of the observations fall in the interval $\mu \pm 2.5\sigma$.

For the interval $\mu \pm 3\sigma$: $k = 3$, and $1 - \frac{1}{k^2} = 1 - \frac{1}{(3)^2} = 1 - .11 = .89$ or 89%. Thus, at least 89% of the observations fall in the interval $\mu \pm 3\sigma$.

3.77 Approximately 68% of the observations fall in the interval $\bar{x} \pm s$, approximately 95% fall in the interval $\bar{x} \pm 2s$, and about 99.7% fall in the interval $\bar{x} \pm 3s$.

3.79 a. Each of the two values is \$1.2 million from μ = \$2.3 million. Hence,

$$k = 1.2/.6 = 2 \text{ and } 1 - \frac{1}{k^2} = 1 - \frac{1}{(2)^2} = 1 - .25 = .75 \text{ or } 75\%.$$

Thus, at least 75% of all firms had 2005 gross sales of \$1.1 to \$3.5 million.

b. Each of the two values is \$1.5 million from μ = \$2.3 million. Hence,

$$k = 1.5/.6 = 2.5 \text{ and } 1 - \frac{1}{k^2} = 1 - \frac{1}{(2.5)^2} = 1 - .16 = .84 \text{ or } 84\%.$$

Thus, at least 84% of all firms had 2005 gross sales of \$.8 to \$3.8 million.

c. Each of the two values is \$1.8 million from μ = \$2.3 million. Hence,

$$k = 1.8/.6 = 3 \text{ and } 1 - \frac{1}{k^2} = 1 - \frac{1}{(3)^2} = 1 - .11 = .89 \text{ or } 89\%.$$

Thus, at least 89% of all firms had 2005 gross sales of \$.5 to \$ 4.1 million.

3.81 a. i. Each of the two values is \$680 from μ = \$2365. Hence,

$$k = 680/340 = 2 \text{ and } 1 - \frac{1}{k^2} = 1 - \frac{1}{(2)^2} = 1 - .25 = .75 \text{ or } 75\%.$$

Thus, at least 75% of all homeowners pay a monthly mortgage of \$1685 to \$3045.

ii. Each of the two values is \$1020 from μ = \$2365. Hence,

$k = 1020/340 = 3$ and $1 - \frac{1}{k^2} = 1 - \frac{1}{(3)^2} = 1 - .11 = .89$ or 89%.

Thus, at least 89% of all homeowners pay a monthly mortgage of \$1345 to \$5385.

b. $1 - \frac{1}{k^2} = .84$ gives $\frac{1}{k^2} = 1 - .84 = .16$ or $k^2 = \frac{1}{.16}$ so $k = 2.5$.

$\mu - 2.5\sigma = 2365 - 2.5(340) = \$1515$ and $\mu + 2.5\sigma = 2365 + 2.5(340) = \$3215$

Thus, the required interval is \$1515 to \$3215.

3.83 $\mu = \$1481$ and $\sigma = \$355$

a. The interval \$771 to \$2191 is $\mu - 2\sigma$ to $\mu + 2\sigma$. Hence, approximately 95% of employees have annual premium payments between \$771 and \$2191.

b. The interval \$1126 to \$1836 is $\mu - \sigma$ to $\mu + \sigma$. Hence, approximately 68% of employees have annual premium payments between \$1126 and \$1836.

c. The interval \$416 to \$2546 is $\mu - 3\sigma$ to $\mu + 3\sigma$. Hence, approximately 99.7% of employees have annual premium payments between \$416 and \$2546.

3.85 $\mu = 72$ mph and $\sigma = 3$ mph

a. i. The interval 63 to 81 mph is $\mu - 3\sigma$ to $\mu + 3\sigma$. Hence, about 99.7% of speeds of all vehicles are between 63 and 81 mph.

ii. The interval 69 to 75 mph is $\mu - \sigma$ to $\mu + \sigma$. Hence, about 68% of the speeds of all vehicles are between 69 and 75 mph.

a. $\mu - 2\sigma = 72 - 2(3) = 66$ mph and $\mu + 2\sigma = 72 + 2(3) = 78$ mph. The interval that contains the speeds of 95% of the vehicles is 66 to 78 mph.

Section 3.5

3.87 The **interquartile range** (IQR) is given by $Q_3 - Q_1$, where Q_1 and Q_3 are the first and third quartiles, respectively. Examples 3–20 and 3–21 of the text show how to find the IQR for a data set.

3.89 If x_i is a particular observation in the data set, the **percentile rank of x_i** is the percentage of the values in the data set that are less than x_i. Thus,

$$\text{Percentile rank of } x_i = \frac{\text{Number of values less than } x_i}{\text{Total number of values in the data set}} \times 100$$

3.91 The ranked data are: 68 68 69 69 71 72 73 74 75 76 77 78 79

a. The three quartiles are $Q_1 = (69 + 69)/2 = 69$, $Q_2 = 73$, and $Q_3 = (76 + 77)/2 = 76.5$

$IQR = Q_3 - Q_1 = 76.5 - 69 = 7.5$

b. $kn/100 = 35(13)/100 = 4.55 \approx 5$

Thus, the 35th percentile can be approximated by the value of the 5th term in the ranked data, which is 71. Therefore, $P_{35} = 71$.

c. Four values in the given data set are smaller than 71. Hence, the percentile rank of $71 = (4/13) \times 100 = 30.77\%$.

3.93 The ranked data are: 3 4 4 5 5 5 5 5 6 6 6 6 6 6 6
7 7 8 8 8 8 8 9 9 10 10 11 13 18

a. The quartiles are $Q_1 = 5$, $Q_2 = (6 + 7)/2 = 6.5$, and $Q_3 = 8$

$IQR = Q_3 - Q_1 = 8 - 5 = 3$

b. $kn/100 = 63(30)/100 = 18.9 \approx 19$

Thus, the 63rd percentile can be approximated by the value of the 19th term in the ranked data, which is 8. Therefore, $P_{63} = 8$.

c. Twenty-five values in the given data are smaller than 10. Hence, the percentile rank of $10 = (25/30) \times 100 = 83.33\%$.

3.95 The ranked data are: 20 22 23 23 23 23 24 25 26 26 27 27 27 28 28
29 29 31 31 31 32 33 33 33 34 35 35 36 37 43

a. The three quartiles are $Q_1 = 25$, $Q_2 = (28+ 29)/2 = 28.5$, and $Q_3 = 33$

$IQR = Q_3 - Q_1 = 33 - 25 = 8$

The value 31 lies between Q_2 and Q_3, which means that it is in the third 25% group from the bottom in the (ranked) data set.

b. $kn/ 100 = 65(30)/100 \approx 19.5$

Thus, the 65th percentile may be approximated by the average of the nineteenth and twentieth terms in the ranked data. Therefore, $P_{65} = (31 + 31)/2 = 31$. Thus, we can state that the number of computer monitors produced by Nixon Corporation is less than 31 for approximately 65% of the days in this sample.

c. Twenty values in the given data are less than 32. Hence, the percentile rank of $32 = (20/30) \times 100 = 66.67\%$. Thus, on 66.67% of the days in this sample, fewer than 32 monitors were produced. Hence, for $100 - 66.67 = 33.33\%$ of the days, the company produced 32 or more monitors.

3.97 The ranked data are: 397 495 499 509 543 544 550 605 649 675
698 700 710 752 789 808 881 910 929 1016

a. The three quartiles are

$Q_1 = (543 + 544)/2 = 543.5$, $Q_2 = (675 + 698)/2 = 686.5$, and $Q_3 = (789 + 808)/2 = 798.5$

$IQR = Q_3 - Q_1 = 798.5 - 543.5 = 255$

The value 649 lies between Q_1 and Q_2, which indicates that it is in the second 25% group from the bottom in the (ranked) data set.

b. $kn/100 = 77(20)/100 = 15.4 \approx 15.5$

Thus, the 77th percentile may be approximated by the average of the 15th and 16th terms in the ranked data. Therefore, $P_{77} = (789 + 808)/2 = 798.5$. Thus, approximately 77% of the sale prices of homes in San Diego are below $798.5 thousand.

c. Eleven values in the given data are smaller than 700. Hence, the percentile rank of $700 = (11/20) \times 100 = 55\%$. Thus, 55% of the sale prices of homes in San Diego are below $700 thousand.

Section 3.6

3.99 The ranked data are: 22 24 25 28 31 32 34 35 36 41 42 43
47 49 52 55 58 59 61 61 63 65 73 98

Median $= (43 + 47)/2 = 45$, $Q_1 = (32 + 34)/2 = 33$, and $Q_3 = (59 + 61)/2 = 60$,

$IQR = Q_3 - Q_1 = 60 - 33 = 27$, $1.5 \times IQR = 1.5 \times 27 = 40.5$,

Lower inner fence $= Q_1 - 40.5 = 33 - 40.5 = -7.5$,

Upper inner fence $= Q_3 + 40.5 = 60 + 40.5 = 100.5$

The smallest and largest values within the two inner fences are 22 and 98, respectively. The data set has no outliers. The box–and–whisker plot is shown below.

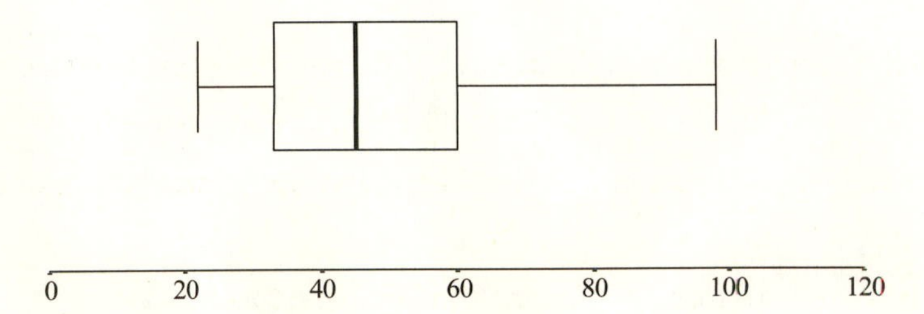

3.101 The ranked data are: 3 5 5 6 8 10 14 15 16 17 17 19 21 22 23 25 30 31 31 34

Median $= (17 + 17)/2 = 17$, $Q_1 = (8 + 10)/2 = 9$, and $Q_3 = (23 + 25)/2 = 24$,

$IQR = Q_3 - Q_1 = 24 - 9 = 15$, $1.5 \times IQR = 1.5 \times 15 = 22.5$,

Lower inner fence = $Q_1 - 22.5 = 9 - 22.5 = -13.5$,

Upper inner fence = $Q_3 + 22.5 = 24 + 22.5 = 46.5$

The smallest and the largest values within the two inner fences are 3 and 34, respectively. The data set contains no outliers.

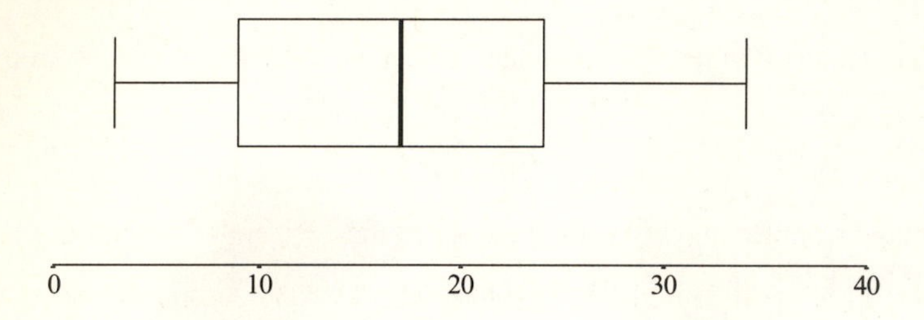

The data are nearly symmetric.

3.103 The ranked data are: 22.5 25.0 26.8 27.0 29.8 51.2 64.0 80.3 94.4 97.7 112.0 261.7

Median = $(51.2 + 64.0)/2 = 57.6$, $Q_1 = (26.8 + 27.0)/2 = 26.9$, and $Q_3 = (94.4 + 97.7)/2 = 96.05$,

$IQR = Q_3 - Q_1 = 96.05 - 26.9 = 69.15$, $1.5 \times IQR = 1.5 \times 69.15 = 103.725$,

Lower inner fence = $Q_1 - 103.725 = 26.9 - 103.725 = -76.825$,

Upper inner fence = $Q_3 + 103.725 = 96.05 + 103.725 = 199.775$

The smallest and largest values within the two inner fences are 22.5 and 112, respectively. The value 261.7 is an outlier.

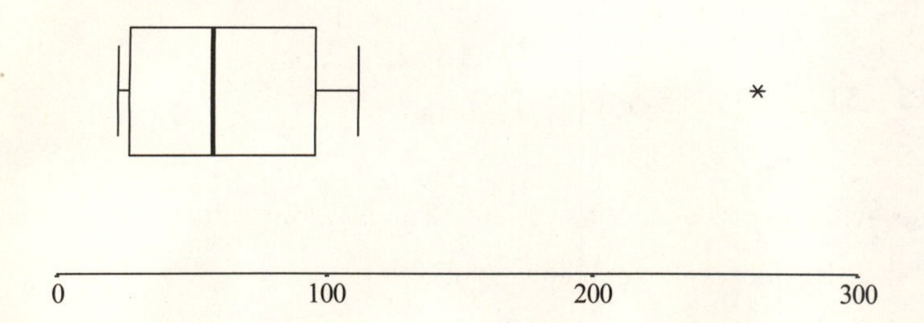

The data are skewed to the right.

3.105 The ranked data are:

3	4	4	5	5	5	5	5	6	6	6	6	6	6	6
7	7	8	8	8	8	8	8	9	9	10	10	11	13	18

Median = $(6 + 7)/2 = 6.5$, $Q_1 = 5$, and $Q_3 = 8$,

$IQR = Q_3 - Q_1 = 8 - 5 = 3$, $1.5 \times IQR = 1.5 \times 3 = 4.5$,

Lower inner fence = $Q_1 - 4.5 = 5 - 4.5 = .5$,

Upper inner fence = $Q_3 + 4.5 = 8 + 4.5 = 12.5$

The smallest and largest values within the two inner fences are 3 and 11, respectively. The values 13 and 18 are outliers.

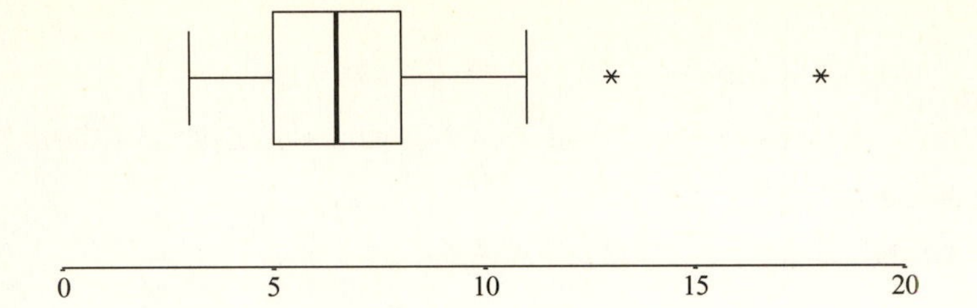

The data are skewed to the right.

3.107 The ranked data are: 20 21 22 23 23 23 23 24 25 26 26 27 27 27 27
28 28 28 29 29 31 31 31 32 33 33 33 34 35 35

Median = (27 + 28)/2 = 27.5, Q_1 = 24, and Q_3 = 31,

IQR = $Q_3 - Q_1$ = 31 – 24 = 7, 1.5 × IQR = 1.5 × 7 = 10.5,

Lower inner fence = Q_1 – 10.5 = 24 – 10.5 = 13.5,

Upper inner fence = Q_3 + 10.5 = 31 + 10.5 = 41.5

The smallest and largest values within the two inner fences are 20 and 35, respectively. There are no outliers.

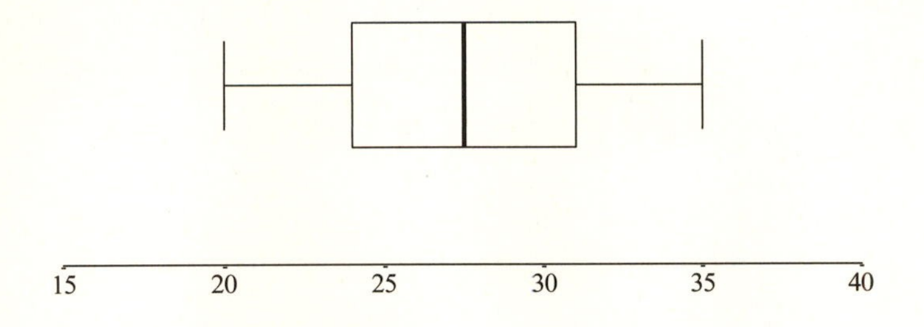

The data are nearly symmetric.

Supplementary Exercises

3.109 a. $\bar{x} = (\sum x)/n$ = 587/10 = \$58.7 thousand

Median = value of the 5.5[th] term in ranked data = (56 + 57)/2= \$56.5 thousand

b. Yes, 104 is an outlier. After dropping this value,

$\bar{x} = (\sum x)/n$ = 483/9 = \$53.67 thousand

Median = \$56 thousand

The value of the mean changes by a larger amount.

c. The median is a better summary measure for these data since it is influenced less by outliers.

3.111 a. $\bar{x} = (\sum x)/n = 1849/10 = 184.9$ yards

Median = value of the 5.5th term in ranked data = (171 + 179)/2 = 175 yards

This data set has two modes at 167 and 170 yards since these two values each occur twice and no other value occurs more than once.

b. Range = Largest value – Smallest value = 233 – 167 = 66 yards

$$s^2 = \frac{\sum x^2 - \frac{(\sum x)^2}{n}}{n-1} = \frac{346{,}269 - \frac{(1849)^2}{10}}{10-1} = 487.6556 \qquad s = \sqrt{487.6556} = 22.08 \text{ yards}$$

3.113

Rainfall	Number of Cities	m	mf	m^2f
0 to less than 2	6	1	6	6
2 to less than 4	10	3	30	90
4 to less than 6	20	5	100	500
6 to less than 8	7	7	49	343
8 to less than 10	4	9	36	324
10 to less than 12	3	11	33	363
	$N = \sum f = 50$		$\sum mf = 254$	$\sum m^2 f = 1626$

$\bar{x} = (\sum mf)/n = 254/50 = 5.08$ inches

$$s^2 = \frac{\sum m^2 f - \frac{(\sum mf)^2}{n}}{n-1} = \frac{1626 - \frac{(254)^2}{50}}{50-1} = 6.8506 \qquad s = \sqrt{6.8506} = 2.62 \text{ inches}$$

The values of these summary measures are sample statistics since they are based on a sample of 50 cities.

3.115 a i. Each of the two values is 40 minutes from $\mu = 200$. Hence,

$k = 40/20 = 2$ and $1 - \frac{1}{k^2} = 1 - \frac{1}{(2)^2} = 1 - .25 = .75$ or 75%.

Thus, at least 75% of the students will learn the basics in 160 to 240 minutes.

ii. Each of the two values is 60 minutes from $\mu = 200$. Hence,

$k = 60/20 = 3$ and $1 - \frac{1}{k^2} = 1 - \frac{1}{(3)^2} = 1 - .11 = .89$ or 89%.

Thus, at least 89% of the students will learn the basics in 140 to 260 minutes.

b. $1 - \frac{1}{k^2} = .75$ gives $\frac{1}{k^2} = 1 - .75 = .25$ or $k^2 = \frac{1}{.25}$, so $k = 2$

$\mu - 2\sigma = 200 - 2(20) = 160$ minutes and $\mu + 2\sigma = 200 + 2(20) = 240$ minutes

Thus, the required interval is 160 to 240 minutes.

3.117 $\mu = 200$ minutes and $\sigma = 20$ minutes

a. i. The interval 180 to 220 minutes is $\mu - \sigma$ to $\mu + \sigma$. Thus, approximately 68% of the students will learn the basics in 180 to 220 minutes.

ii. The interval 160 to 240 minutes is $\mu - 2\sigma$ to $\mu + 2\sigma$. Hence, approximately 95% of the students will learn the basics in 160 to 240 minutes.

b. $\mu - 3\sigma = 200 - 3(20) = 140$ minutes and $\mu + 3\sigma = 200 + 3(20) = 260$ minutes. The interval that contains the learning time of 99.7% of the students is 140 to 260 minutes.

3.119 The ranked data are: 40 44 44 47 56 57 59 68 68 104

a. The three quartiles are $Q_1 = 44$, $Q_2 = (56 + 57)/2 = 56.5$, and $Q_3 = 68$

$IQR = Q_3 - Q_1 = 68 - 44 = 24$

The value 40 lies below Q_1, which indicates that it is in the bottom 25% group in the (ranked) data set.

b. $kn/100 = 70(10)/100 = 7$

Thus, the 70th percentile occurs at the seventh term in the ranked data, which is 59. Therefore, $P_{70} = 59$. This means that about 70% of the values in the data set are smaller than 59.

c. Three values in the given data are smaller than 47. Hence, the percentile rank of $47 = (3/10) \times 100 = 30\%$. This means approximately 30% of the values in the data set are less than 47.

3.121 The ranked data are: 62 67 72 73 75 77 81 83 84 85 90 93 107 112 135

Median = 83, $Q_1 = 73$, and $Q_3 = 93$,

$IQR = Q_3 - Q_1 = 93 - 73 = 20$, $1.5 \times IQR = 1.5 \times 20 = 30$,

Lower inner fence $= Q_1 - 30 = 73 - 30 = 43$,

Upper inner fence $= Q_3 + 30 = 93 + 30 = 123$

The smallest and largest values within the two inner fences are 62 and 112, respectively. The value 135 is an outlier.

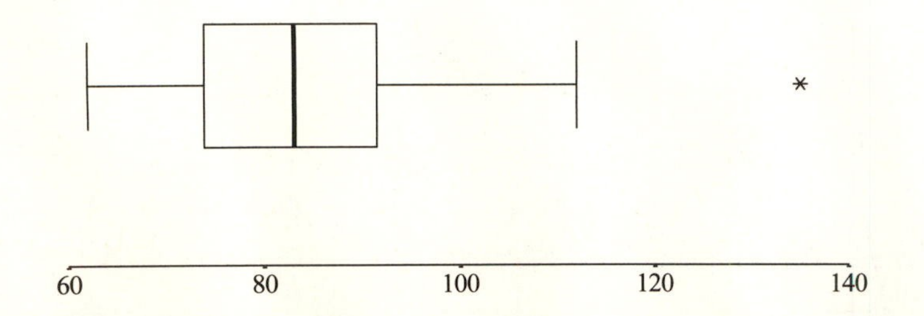

The data are skewed to the right.

3.123 Let y = Melissa's score on the final exam. Then, her grade is $\frac{75+69+87+y}{5}$. To get a B, she needs this to be at least 80. So we solve,

$$80 = \frac{75+69+87+y}{5}$$

$5(80) = 75 + 69 + 87 + y$

$400 = 231 + y$

$y = 169$

Thus, the minimum score that Melissa needs on the final exam in order to get a B grade is 169 out of 200 points.

3.125 a. Since $\bar{x} = (\sum x)/n$, we have $76 = (\sum x)/5$, so $\sum x = 5(76) = 380$ inches. If we replace the tallest player by a substitute who is two inches taller, the sum of the new heights is 380 + 2 = 382 inches. Thus, the new mean is $\bar{x} = 382/5 = 76.4$ inches. Since Range = Largest value – Smallest value, and the largest value has increased by two while the smallest value is unchanged, the range has increased by two. Thus, the new range is 11 + 2 = 13 inches. The median is the height of the third player (if their heights are ranked) and this does not change. So, the median remains 78 inches.

b. If we replace the tallest player by a substitute who is four inches shorter, then by reasoning similar to that in part a, we have a new mean of $\bar{x} = 376/5 = 75.2$ inches. You cannot determine the new median or range with only the information given. We do not know how the new player's height compares to the rest of the players on the team; we have no knowledge of whether the substitute is now the tallest player or not.

3.127 The mean price per barrel of oil purchased in that week is

Mean = [(1000)(51) + (200)(64) + (100)(70)]/1300 = 70,800/1300 = $54.46 per barrel

3.129 a. Mean = (9.4 + 9.5 + 9.5 + 9.5 + 9.6)/5 = 9.5

b. The percentage of trimmed mean is 2/7 × 100 ≈ 28.6/2 = 14.3 % since we dropped two of the seven values.

c. Suppose gymnast B has the following scores: 9.4, 9.4, 9.5, 9.5, 9.5, 9.5, and 9.9. Then, the mean for the gymnast B is μ = (9.4 + 9.4 + 9.5 + 9.5 + 9.5 + 9.5 + 9.9)/7 = 9.5286, and the mean for gymnast A is μ = (9.4 + 9.7 + 9.5 + 9.5 + 9.4 + 9.6 + 9.5)/7 = 9.5143. So, Gymnast B would win if all seven scores were counted. The trimmed mean of B is (9.4 + 9.5 + 9.5 + 9.5 + 9.5)/5 = 9.4800. This is less than the trimmed mean for A (9.500), so gymnast A would win using the trimmed mean.

3.131 a. For people age 30 and under, we have the following death rates from heart attack:

Country A: $\frac{\text{number of deaths}}{\text{number of patients}} \times 1000 = \frac{1}{40} \times 1000 = 25$

Country B: $\frac{\text{number of deaths}}{\text{number of patients}} \times 1000 = \frac{.5}{25} \times 1000 = 20$

So the death rate for people 30 and under is lower in Country B.

b. For people age 31 and older, the death rates from heart attack are as follows:

Country A: $\frac{\text{number of deaths}}{\text{number of patients}} \times 1000 = \frac{2}{20} \times 1000 = 100$

Country B: $\frac{\text{number of deaths}}{\text{number of patients}} \times 1000 = \frac{3}{35} \times 1000 = 85.7$

Thus, the death rate for Country A is greater than that for Country B for people age 31 and older.

c. The overall death rates are as follows:

Country A: $\frac{\text{number of deaths}}{\text{number of patients}} \times 1000 = \frac{3}{60} \times 1000 = 50$

Country B: $\frac{\text{number of deaths}}{\text{number of patients}} \times 1000 = \frac{3.5}{60} \times 1000 = 58.3$

Thus, overall the death rate for country A is *lower* than the death rate for Country B.

d. In both countries people age 30 and under have a lower percentage of death due to heart attack than people age 31 and over. Country A has 2/3 of its population age 30 and under while more than 1/2 of the people in Country B are age 31 and over. Thus, more people in Country B than in County A fall into the higher risk group which drives up Country B's overall death rate from heart attacks.

3.133 $\mu = 70$ minutes and $\sigma = 10$

a. Using Chebyshev's theorem, we need to find k so that

$1 - \frac{1}{k^2} = .50$ gives $\frac{1}{k^2} = 1 - .50 = .50$ or $k^2 = \frac{1}{.50} = 2$, so $k = \sqrt{2} \approx 1.4$.

Thus, at least 50% of the scores are within 1.4 standard deviations of the mean.

b. Using Chebyshev's theorem, we first find k so that at least 1 –.20 = .80 of the scores are within k standard deviations of the mean.

$1 - \frac{1}{k^2} = .80$ gives $\frac{1}{k^2} = 1 - .80 = .20$ or $k^2 = \frac{1}{.20} = 5$, so $k = \sqrt{5} \approx 2.2$.

Thus, at least 80% of the scores are within 2.2 standard deviations of the mean, but this means that at most 10% of the scores are greater than 2.2 standard deviations above the mean.

3.135 a. $\bar{x} = \$13,872$ $Q_1 = \$50$ $s^2 = 4,110,408,767$

Median = \$500 $Q_3 = \$1400$ $s = \$64,112.47$

Mode = \$0 IQR = \$1350 Lowest = \$0

Range = \$321,500 Highest = \$321,500

Below is the histogram for the given data.

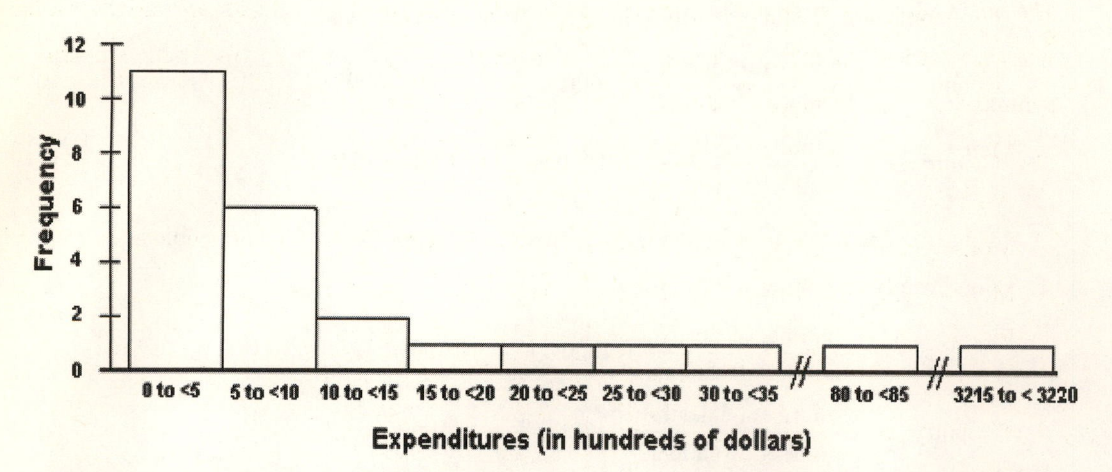

The vacation expenditures are strongly skewed to the right. Most of the expenditures are relatively small (\$3400 or less) but there are two extreme outliers (\$8200 and \$321,500).

b. Neither the mode (\$0) nor the mean (\$13,872) are typical of these expenditures. Thus, the median (\$500) is the best indicator of the average family's vacation expenditures.

3.137 a.

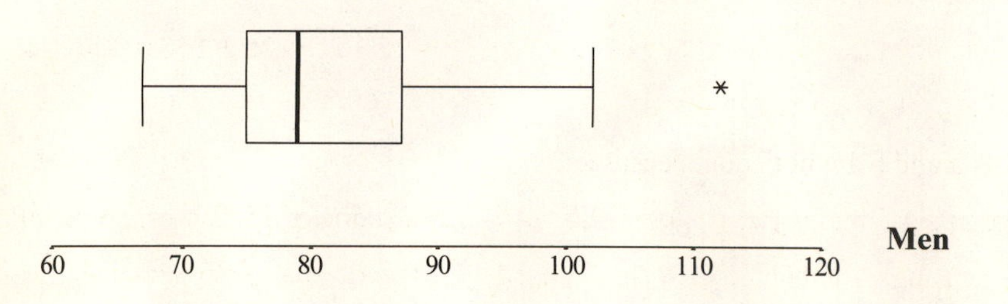

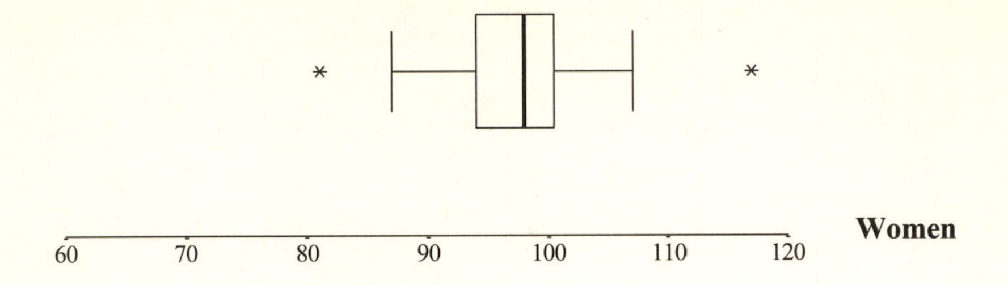

The box–and–whisker plots show that the men's scores tend to be lower and more varied than the women's scores. The men's scores are skewed to the right, while the women's are more nearly symmetric.

b. **Men** | **Women**

Men	Women
$\bar{x} = 82$	$\bar{x} = 97.53$
Median = 79	Median = 98
Modes = 75, 79, and 92	Modes = 94 and 100
Range = 45	Range = 36
$s^2 = 145.8750$	$s^2 = 71.2667$
$s = 12.08$	$s = 8.44$
$Q_1 = 73.5$	$Q_1 = 94$
$Q_3 = 89.5$	$Q_3 = 101$
IQR = 16	IQR = 7

These numerical measures confirm the observations based on the box–and–whisker plots.

3.139 a. The total enrollment in the 25 freshman engineering classes is (24)(25) + 150 = 750. Then, the mean size of these 25 classes is 750/25 = 30.

b. Each student attends five classes with total enrollment of 25 + 25 + 25 + 25 + 150 = 250. Then, the mean size of the class is 250/5 = 50.

The means in parts a and b are not equal because:

1) From the college's point of view, the large class of 150 is just one of 25 classes, so its influence on the mean is strongly offset by the 24 small classes. This leads to a relatively small mean of 30 students per class.
2) From the point of view of each student, the larger class is one of just five, so it has a stronger influence on the mean. This results in a larger mean of 50.

3.141 $\mu = 6$ inches and $\sigma = 2$ inches

a. Each of the two values is 3 inches from $\mu = 6$ inches. Hence,

$k = 3/2 = 1.5$ and $1 - \frac{1}{k^2} = 1 - \frac{1}{(1.5)^2} = 1 - .444 = .556$ or 55.6%.

Thus, at least 55.6% of the fish are between 3 and 9 inches in length.

b. $1 - \frac{1}{k^2} = .84$ gives $\frac{1}{k^2} = 1 - .84 = .16$ or $k^2 = \frac{1}{.16}$, so $k = 2.5$

$\mu - 2.5\sigma = 6 - 2.5(2) = 1$ inch and $\mu + 2.5\sigma = 6 + 2.5(2) = 11$ inches

Thus, the required interval is 1 to 11 inches.

c. 100 – 36 = 64% of the fish have lengths *inside* the required interval. Then

$1 - \frac{1}{k^2} = .64$ gives $\frac{1}{k^2} = 1 - .64 = .36$ or $k^2 = \frac{1}{.36}$, so $k = 1.67$

$\mu - 1.67\sigma = 6 - 1.67(2) = 2.66$ inches and $\mu + 1.67\sigma = 6 + 1.67(2) = 9.34$ inches

Thus, the required interval is 2.66 to 9.34 inches.

3.143 a. The mean, median, and standard deviations of the weights of males and females in grams are:

Men	**Women**
$\bar{x} = 76{,}188.65$	$\bar{x} = 54{,}428.95$
Median = 77,970.61	Median = 53,577.57
$s = 8326.7209$	$s = 7613.1618$

The mean, median, and standard deviations of the weights of males and females in stones are:

Men	**Women**
$\bar{x} = 12.49$	$\bar{x} = 8.93$
Median = 12.79	Median = 8.79
$s = 1.3654$	$s = 1.2484$

b. Converting the answers from Problem 3.140 to grams yields:

Men	**Women**
$\bar{x} = 76{,}189.05$	$\bar{x} = 54{,}426.97$
Median = 77,970.61	Median = 53,577.57
$s = 8326.7384$	$s = 7613.1549$

Converting the answers from Problem 3.140 to stones yields:

Men	**Women**
$\bar{x} = 12.49$	$\bar{x} = 8.93$
Median = 12.79	Median = 8.79
$s = 1.3686$	$s = 1.2484$

The answers are the same as those from part a with the exception of rounding error.

c. When converting from a larger unit to a smaller unit, the summary measures get larger by the conversion factor. When converting from a smaller unit to a larger unit, the summary measures get smaller by the conversion factor.

d. The distribution in units of pounds has more variability than that of stones. This is so because we converted from a smaller unit to a larger unit; hence, the standard deviation was reduced by the amount of the conversion factor (s for stones = s for pounds/14).

e.

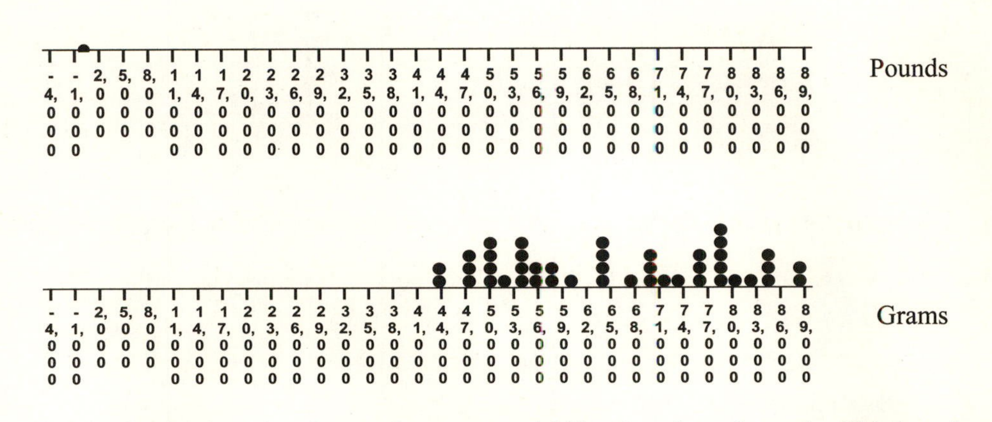

The distribution in units of grams has more variability than that of pounds. This is so because we converted from a larger unit to a smaller unit; hence, the standard deviation was increased by the amount of the conversion factor (s for grams = s for pounds × 435.59). Because of the large difference in units, retaining the same scale for the stacked plots displays all of the data values for pounds in one location with no variability visible.

3.145 This golfer's score was not an outlier; therefore, her score must be less than the value of the upper inner fence. From Exercise 3.137, $Q_3 = 101$ and IQR = 7. Then, $1.5 \times \text{IQR} = 1.5 \times 7 = 10.5$ and $Q_3 + 10.5 = 111.5$. Since this golfer had the uniquely highest score, and the next highest score was 107, she shot between 108 and 111.

Self–Review Test

1. b	**2.** a and d	**3.** c	**4.** c	**5.** b	**6.** b	**7.** a
8. a	**9.** b	**10.** a	**11.**b	**12.** c	**13.** a	**14.** a

15. $n = 10$, $\sum x = 109$, and $\sum x^2 = 1775$

$\bar{x} = (\sum x)/\ n = 109/10 = 10.9$ times

Median = value of the 5.5^{th} term in ranked data = $(7 + 9)/2 = 8$ times

Mode = 6 times

Range = Largest value – Smallest value = 28 – 2 =26 times

$$s^2 = \frac{\sum x^2 - \frac{(\sum x)^2}{n}}{n-1} = \frac{1775 - \frac{(109)^2}{10}}{10-1} = 65.2111 \qquad s = \sqrt{65.2111} = 8.08 \text{ times}$$

16. Suppose the exam scores for seven students are 73, 82, 95, 79, 22, 86, and 91 points. Then, Mean = (73 + 82 + 95 + 79 + 22 + 86 + 91)/7 = 75.43 points. If we drop the outlier (22), Mean = (73 + 82 + 95 + 79 + 86 + 91)/6 = 84.33 points.

17. Suppose the exam scores for seven students are 73, 82, 95, 79, 22, 86, and 91 points. Then,

Range = Largest value – Smallest value = 95 – 22 = 73 points.

If we drop the outlier (22) and calculate the range,

Range = Largest value – Smallest value = 95 – 73 = 22 points.

Thus, when we drop the outlier, the range decreases from 73 to 22 points.

18. The value of the standard deviation is zero when all the values in a data set are the same. For example, suppose the heights (in inches) of five women are: 67 67 67 67 67

This data set has no variation. As shown below the value of the standard deviation is zero for this data set. For these data: $n = 5$, $\sum x = 335$, and $\sum x^2 = 22{,}445$.

$$s = \sqrt{\frac{\sum x^2 - \frac{(\sum x)^2}{n}}{n-1}} = \sqrt{\frac{22{,}445 - \frac{(335)^2}{5}}{5-1}} = \sqrt{\frac{22{,}445 - 22{,}445}{4}} = 0$$

19.

a. The frequency column gives the number of weeks for which the number of computers sold was in the corresponding class.

b. For the given data: $n = 25$, $\sum mf = 486.50$, and $\sum m^2 f = 10{,}524.25$

$\bar{x} = (\sum mf)/n = 486.50/25 = 19.46$ computers

$$s^2 = \frac{\sum m^2 f - \frac{(\sum mf)^2}{n}}{n-1} = \frac{10{,}524.25 - \frac{(486.50)^2}{25}}{25-1} = 44.0400, \; s = \sqrt{44.0400} = 6.64 \text{ computers}$$

20. a. i. Each of the two values is 5.5 years from $\mu = 7.3$ years. Hence,

$$k = 5.5/2.2 = 2.5 \text{ and } 1 - \frac{1}{k^2} = 1 - \frac{1}{(2.5)^2} = 1 - .16 = .84 \text{ or } 84\%$$

Thus, at least 84% of the cars are 1.8 to 12.8 years old.

ii. Each of the two values is 6.6 years from $\mu = 7.3$ years. Hence

$$k = 6.6/2.2 = 3 \text{ and } 1 - \frac{1}{k^2} = 1 - \frac{1}{(3)^2} = 1 - .11 = .89 \text{ or } 89\%$$

Thus, at least 89% of the cars are .7 to 13.9 years old.

b. $1 - \frac{1}{k^2} = .75$ gives $\frac{1}{k^2} = 1 - .75 = .25$ or $k^2 = \frac{1}{2.5}$, so $k = 2$

$\mu - 2\sigma = 7.3 - 2(2.2) = 2.9$ hours and $\mu + 2\sigma = 7.3 + 2(2.2) = 11.7$ hours

Thus, the required interval is 2.9 to 11.7 years.

21. $\mu = 7.3$ years and $\sigma = 2.2$ years

a. i. The intervals 5.1 to 9.5 years is $\mu - \sigma$ to $\mu + \sigma$. Hence, approximately 68% of the cars are 5.1 to 9.5 years old.

ii. The interval .7 to 13.9 years is $\mu - 3\sigma$ to $\mu + 3\sigma$. Hence, approximately 99.7% of the cars are .7 to 13.9 years.

b. $\mu - 2\sigma = 7.3 - 2(2.2) = 2.9$ hours and $\mu + 2\sigma = 7.3 + 2(2.2) = 11.7$ hours. The interval that contains ages of 95% of the cars will be 2.9 to 11.7 years old.

22. The ranked data are: 0 1 2 3 4 5 7 8 10 11 12 13 14 15 20

a. The three quartiles are $Q_1 = 3$, $Q_2 = 8$, and $Q_3 = 13$.

$IQR = Q_3 - Q_1 = 13 - 3 = 10$.

The value 4 lies between Q_1 and Q_2, which indicates that this value is in the second from the bottom 25% group in the ranked data.

b. $kn/100 = 60(15)/100 = 9$. Thus, the 60th percentile may be represented by the value of the ninth term in the ranked data, which is 10. Therefore, $P_{60} = 10$. Thus, approximately 60% of the half hour time periods had fewer than 10 passengers set off the metal detectors during this day.

c. Ten values in the given data are less than 12. Hence, the percentile rank of

12 is (10/15) × 100 = 66.67%. Thus, 66.67% of the half hour time periods had fewer than 12 passengers set off the metal detectors during this day.

23. The ranked data are: 0 1 2 3 4 5 7 8 10 11 12 13 14 15 20

Median = 8, Q_1 = 3, and Q_3 = 13,

IQR = $Q_3 - Q_1$ = 13 – 3 = 10, 1.5 × IQR = 1.5 × 10 =15,

Lower inner fence = Q_1 – 15 = 3 – 15 = –12,

Upper inter fence = Q_3 + 15 = 13 + 15 = 28

The smallest and largest values in the data set within the two inner fences are 0 and 20, respectively. The data does not contain any outliers.

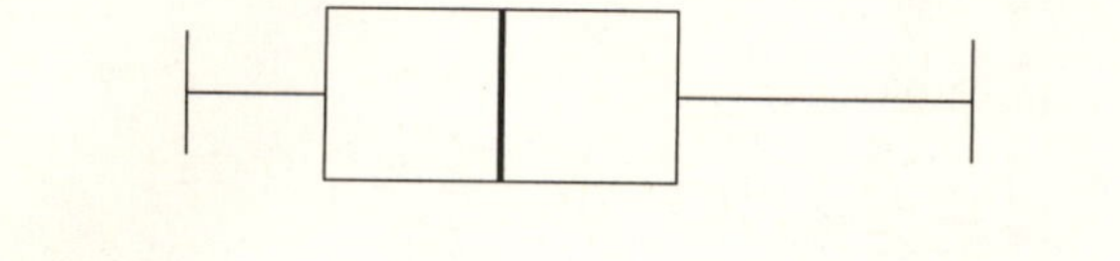

The data are skewed slightly to the right.

24. From the given information: $n_1 = 15$, $n_2 = 20$, $\bar{x}_1$ = \$435, $\bar{x}_2$ = \$490

$$\bar{x} = \frac{n_1\bar{x}_1 + n_1\bar{x}_2}{n_1 + n_2} = \frac{(15)(435) + (20)(490)}{15 + 20} = \frac{16{,}325}{35} = \$466.43$$

25. Sum of the GPAs of five students = (5)(3.21) = 16.05

Sum of the GPAs of four students = 3.85 + 2.67 + 3.45 + 2.91 = 12.88

GPA of the fifth student = 16.05 – 12.88 = 3.17

26. The ranked data are: 58 149 163 166 179 193 207 238 287 2534

Thus, to find the 10% trimmed mean, we drop the smallest value and the largest value (10% of 10 is 1) and find the mean of the remaining 8 values. For these 8 values,

$\sum x$ = 149 + 163 + 166 + 179 + 193 + 207 + 238 + 287 = 1582

10% trimmed mean = $(\sum x)/8$ = 1582/8 = \$197.75 thousand = \$197,750. The 10% trimmed mean is a better summary measure for these data than the mean of all 10 values because it eliminates the effect of the outliers, 58 and 2534.

27. a. For Data Set I: $\bar{x} = (\sum x)/n = 79/4 = 19.75$

For Data Set II: $\bar{x} = (\sum x)/n = 67/4 = 16.75$

The mean of Data Set II is smaller than the mean of Data Set I by 3.

b. For Data Set I: $\sum x = 79$, $\sum x^2 = 1945$, and $n = 4$

$$s = \sqrt{\frac{\sum x^2 - \frac{(\sum x)^2}{n}}{n-1}} = \sqrt{\frac{1945 - \frac{(79)^2}{4}}{4-1}} = 11.32$$

c. For Data Set II: $\sum x = 67$, $\sum x^2 = 1507$, and $n = 4$

$$s = \sqrt{\frac{\sum x^2 - \frac{(\sum x)^2}{n}}{n-1}} = \sqrt{\frac{1507 - \frac{(67)^2}{4}}{4-1}} = 11.32$$

The standard deviations of the two data sets are equal.

Chapter Four

Section 4.1

4.1 An **experiment** is a process that, when performed, results in one and only one of many observations. An **outcome** is the result of the performance of an experiment. The collection of all outcomes for an experiment is called a **sample space**. A **simple event** is an event that includes one and only one of the final outcomes of an experiment. A **compound event** is a collection of more than one outcome of an experiment.

4.3 The experiment of selecting two items from the box without replacement has the following six possible outcomes: *AB*, *AC*, *BA*, *BC*, *CA*, *CB*. The sample space is written as $S = \{AB, AC, BA, BC, CA, CB\}$.

4.5 Let L = person is computer literate and I = person is computer illiterate. The experiment has four outcomes: *LL, LI, IL,* and *II.*

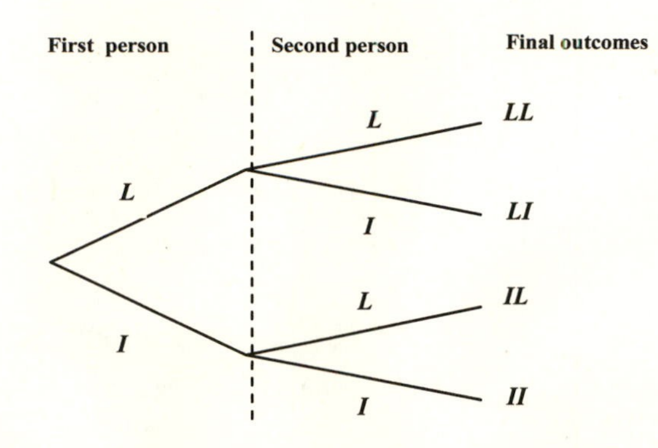

4.7 Let G = the selected part is good and D = the selected part is defective. The four outcomes for this experiment are: *GG, GD, DG,* and *DD.*

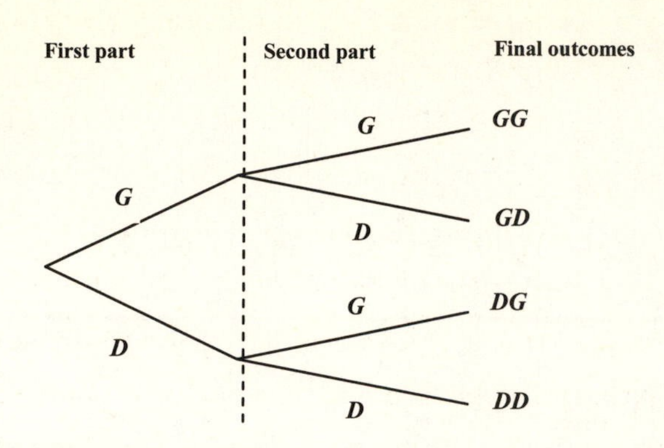

4.9 Let H = a toss results in a head and T = a toss results in a tail. The sample space is written as $S = \{HHH, HHT, HTH, HTT, THH, THT, TTH, TTT\}$.

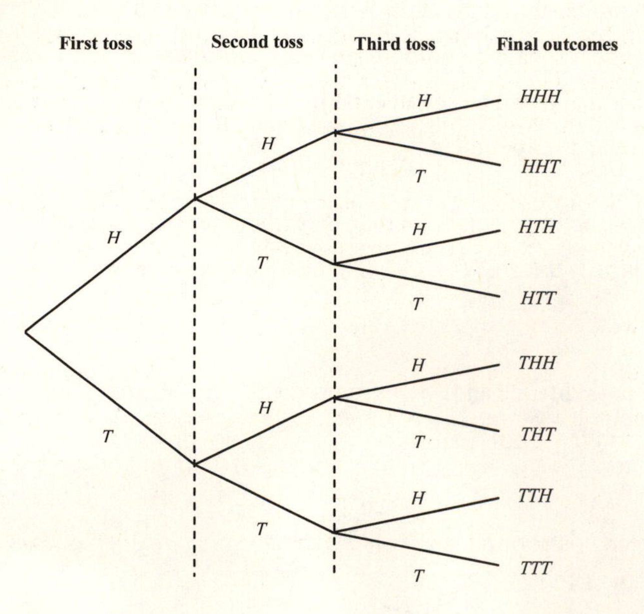

4.11
a. $\{LI, IL\}$; a compound event
b. $\{LL, LI, IL\}$; a compound event
c. $\{II, IL, LI\}$; a compound event
d. $\{LI\}$; a simple event

4.13
a. $\{DG, GD, GG\}$; a compound event
b. $\{DG, GD\}$; a compound event
c. $\{GD\}$; a simple event
d. $\{DD, DG, GD\}$; a compound event

Section 4.2

4.15 1. The probability of an event always lies in the range zero to 1, that is:

$$0 \le P(E_i) \le 1 \text{ and } 0 \le P(A) \le 1$$

2. The sum of the probabilities of all simple events for an experiment is always 1, that is:

$$\sum P(E_i) = P(E_1) + P(E_2) + P(E_3) + \cdots = 1$$

4.17 1. **Classical probability approach:** When all outcomes are equally likely, the probability of an event A is given by: $P(A) = \dfrac{\text{Number of outcomes favorable to A}}{\text{Total number of outcomes for the experiment}}$.

For example, the probability of observing a head when a fair coin is tossed once is 1/2.

2. **Relative frequency approach:** If an event A occurs f times in n repetitions of an experiment, then $P(A)$ is approximately f/n. As the experiment is repeated more and more times, f/n approaches $P(A)$. For example, if 50 of the last 5000 cars off the assembly line are lemons, the probability that the next car is a lemon is approximately $P(\text{lemon}) = f/n = 50/5000 = .01$.
3. **Subjective probability approach:** Probabilities are assigned based on subjective judgment, experience, information and belief. For example, a teacher might estimate the probability of a student earning an A on a statistics test to be 1/6 based on previous classes.

4.19 The values −.55, 1.56, 5/3, and −2/7 cannot be probabilities of events because the probability of an event can never be less than zero or greater than one.

4.21 These two outcomes would not be equally likely unless exactly half of the passengers entering the metal detectors set it off, which is unlikely. We would have to obtain a random sample of passengers going through New York's JFK airport, collect information on whether they set off the metal detector or not, and use the relative frequency approach to find the probabilities.

4.23 This is a case of subjective probability because the given probability is based on the president's judgment.

4.25
a. P(marble selected is red) = 18/40 = .45
b. P(marble selected is green) = 22/40 = .55

4.27 P(adult selected has shopped on the internet) = 1120/2000 = .56

4.29 P(executive selected has a type A personality) = 29/50 = .58

4.31
a. P(her answer is correct) = 1/5 = .2
b. P(her answer is wrong) = 4/5 = .8

Yes, these probabilities add up to 1.0 because this experiment has two and only two outcomes, and according to the second property of probability, the sum of their probabilities must be equal to 1.0.

4.33 *P*(person selected is a woman) = 4/6 = .6667

P(person selected is a man) = 2/6 = .3333

Yes, the sum of these probabilities is 1.0 because of the second property of probability.

4.35 *P*(company selected offers free health fitness center) = 130/400 = .325

Number of companies that do not offer free health fitness center = 400 − 130 = 270

P(company selected does not offer free health fitness center) = 270/400 = .675

Yes, the sum of the probabilities is 1.0 because of the second property of probability.

4.37

Credit Cards	Frequency	Relative Frequency
0	80	.098
1	116	.141
2	94	.115
3	77	.094
4	43	.052
5 or more	410	.500

a. *P*(person selected has three credit cards) = .094

b. *P*(person selected has five or more cards) = .500

4.39 Take a random sample of families from Los Angeles and determine how many of them earn more than $125,000 per year. Then use the relative frequency approach.

Sections 4.3 - 4.7

4.41 **Marginal probability** is the probability of a single event without consideration of any other event. **Conditional probability** is the probability that an event will occur given that another event has already occurred. For example, when a single die is rolled, the marginal probability of a number less than 4 is 1/2; the conditional probability of an odd number given that a number less than 4 has occurred is 2/3.

4.43 Two events are **independent** if the occurrence of one does not affect the probability of the occurrence of the other. Two events are **dependent** if the occurrence of one affects the probability of the occurrence of the other. If two events A and B satisfy the condition $P(A|B) = P(A)$, or $P(B|A) = P(B)$, they are independent; otherwise they are dependent.

4.45 Total outcomes for four rolls of a die = $6 \times 6 \times 6 \times 6 = 1296$

4.47
a. Events A and B are not mutually exclusive since they have the element "2" in common.
b. $P(A) = 3/8$ and $P(A|B) = 1/3$. Since these probabilities are not equal, A and B are dependent.
c. $\overline{A} = \{1, 3, 4, 6, 8\}$; $P(\overline{A}) = 5/8 = .625$ and

$\overline{B} = \{1, 3, 5, 6, 7\}$; $P(\overline{B}) = 5/8 = .625$

4.49 Total selections = $10 \times 5 = 50$

4.51 Total outcomes = $4 \times 8 \times 5 \times 6 = 960$

4.53
a. i. P(selected adult has never shopped on the internet) = 1200/2000 = .600

ii. P(selected adult is a male) = 1200/2000 = .600

iii. P(selected adult has shopped on the internet given that this adult is a female) = 300/800 = .375

iv. P(selected adult is a male given that this adult has never shopped on the internet) = 700/1200 = .583

b. The events "male" and "female" are mutually exclusive because they cannot occur together. The events "have shopped" and "male" are not mutually exclusive because they can occur together.

c. P(female) = 800/2000 = .400 and P(female | have shopped) = 300/800 = .375. Since these probabilities are not equal, the events "female" and "have shopped" are dependent.

4.55
a. i. P(in favor) = 695/2000 = .3475

ii. P(against) = 1085/2000 = .5425

iii. P(in favor | female) = 300/1100 = .2727

iv. P(male | no opinion) = 100/220 = .4545

b. The events "male" and "in favor" are not mutually exclusive because they can occur together. The events "in favor" and "against" are mutually exclusive because they cannot occur together.

c. P(female) = 1100/2000 = .5500 and P(female | no opinion) = 120/220 = .5455. Since these two probabilities are not equal, the events "female" and "no opinion" are dependent.

4.57 a. i. P(more than 1 hour late) = 172/1700 = .1012

ii. P(less than 30 minutes late) = 822/1700 = .4835

iii. P(Airline A's flight | 30 minutes to 1 hour late) = 390/706 = .5524

iv. P(more than 1 hour late | Airline B's flight) = 80/789 = .1014

b. The events "Airline A" and "more than 1 hour late" are not mutually exclusive because they can occur together. The events "less than 30 minutes late" and "more than 1 hour late" are mutually exclusive because they cannot occur together.

c. P(Airline B) = 789/1700 = .4641 and P(Airline B | 30 minutes to 1 hour late) = 316/706 = .4476. Since these two probabilities are not equal, the events "Airline B" and "30 minutes to 1 hour late" are not independent.

4.59 P(pediatrician) = 25/160 = .1563 and P(pediatrician | female) = 20/75 = .2667. Since these two probabilities are not equal, the events "female" and "pediatrician" are dependent. The events are not mutually exclusive because they can occur together.

4.61 P(business major) = 11/30 = .3667 and P(business major | female) = 9/16 = .5625. Since these two probabilities are not equal, the events "female" and "business major" are dependent. The events are not mutually exclusive because they can occur together.

4.63 Event A will occur if either a 1-spot or a 2-spot is obtained on the die. Thus, $P(A)$ = 2/6 = .3333. The complementary event of A is that a 3-spot, or a 4-spot, or a 5-spot, or a 6-spot is obtained on the die. Hence, $P(\overline{A}) = 1 - .3333 = .6667$.

4.65 The complementary event is that the college student attended no major league baseball games last year. The probability of this complementary event is 1 – .12 = .88

Section 4.8

4.67 The **joint probability** of two events is the probability of the intersection of the two events. For example, suppose a die is rolled once. Let A = a number greater than 4 occurs = {5, 6} and B = an odd number occurs = {1, 3, 5}. Then $P(A \text{ and } B) = P(\{5\}) = 1/6$ is the joint probability of A and B.

4.69 The joint probability of two mutually exclusive events is zero. For example, consider one roll of a die. Let A = a number less than 4 occurs = {1, 2, 3} and B = a number greater than 3 occurs =

{4, 5, 6}. Then, A and B are mutually exclusive events, since they have no outcomes in common. The event (A and B) is impossible, and so $P(A \text{ and } B) = 0$.

4.71 a. $P(A \text{ and } B) = P(B \text{ and } A) = P(B)P(A|B) = (.59)(.77) = .4543$

b. $P(A \text{ and } B) = P(A)P(B|A) = (.28)(.35) = .0980$

4.73 a. $P(A \text{ and } B) = P(A)P(B) = (.20)(.76) = .1520$

b. $P(A \text{ and } B) = P(A)P(B) = (.57)(.32) = .1824$

4.75 a. $P(A \text{ and } B \text{ and } C) = P(A)P(B)P(C) = (.49)(.67)(.75) = .2462$

b. $P(A \text{ and } B \text{ and } C) = P(A)P(B)P(C) = (.71)(.34)(.45) = .1086$

4.77 $P(A|B) = P(A \text{ and } B) / P(B) = .45/.65 = .6923$

4.79 $P(A) = P(A \text{ and } B) / P(B|A) = .58/.80 = .725$

4.81 Let M = male, F = female, G = graduated, and N = did not graduate.

a. i. $P(F \text{ and } G) = P(F)P(G|F) = \left(\frac{165}{346}\right)\left(\frac{133}{165}\right) = .3844$

ii. $P(M \text{ and } N) = P(M)P(N|M) = \left(\frac{181}{346}\right)\left(\frac{55}{181}\right) = .1590$

b. $P(G \text{ and } N) = 0$ since G and N are mutually exclusive events.

4.83 Let M = male, F = female, Y = has shopped at least once on the internet, and N = has never shopped on the internet.

a. i. $P(N \text{ and } M) = P(N)P(M|N) = \left(\frac{1200}{2000}\right)\left(\frac{700}{1200}\right) = .350$

ii. $P(Y \text{ and } F) = P(Y)P(F|Y) = \left(\frac{800}{2000}\right)\left(\frac{300}{800}\right) = .150$

b.

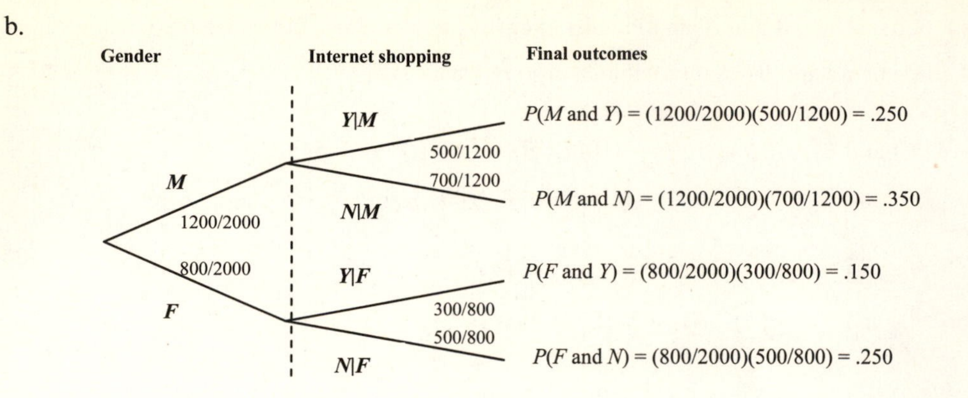

4.85 a. i. P(better off and high school)

$$= P(\text{better off})\, P(\text{high school} \mid \text{better off}) = \left(\frac{1010}{2000}\right)\left(\frac{450}{1010}\right) = .225$$

ii. P(more than high school and worse off)

$$= P(\text{more than high school})\, P(\text{worse off} \mid \text{more than high school}) = \left(\frac{600}{2000}\right)\left(\frac{70}{600}\right) = .035$$

b. P(worse off and better off) = 0 because "worse off" and "better off" are mutually exclusive.

4.87 Let A = first selected student favors abolishing the Electoral College, B = first selected student favors keeping the Electoral College, C = second selected student favors abolishing the Electoral College, and D = second selected student favors keeping the Electoral College.

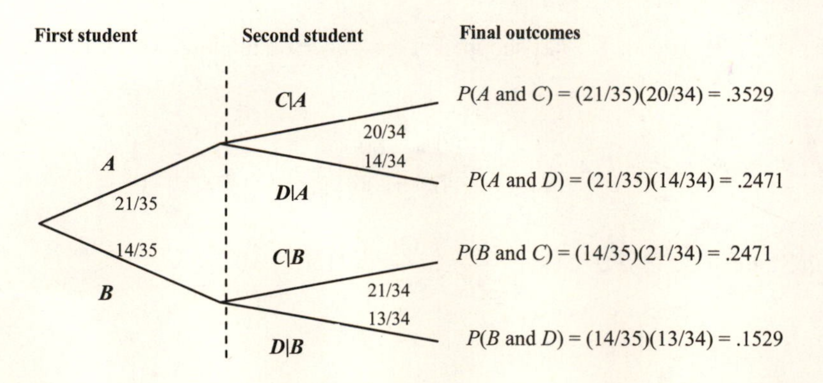

Then, P(both student favors abolishing the Electoral College) = $P(A \text{ and } C) = P(A)P(C|A) = .3529$.

4.89 Let C = first selected person has a type A personality, D = first selected person has a type B personality, E = second selected person has a type A personality, and F = second selected person has a type B personality.

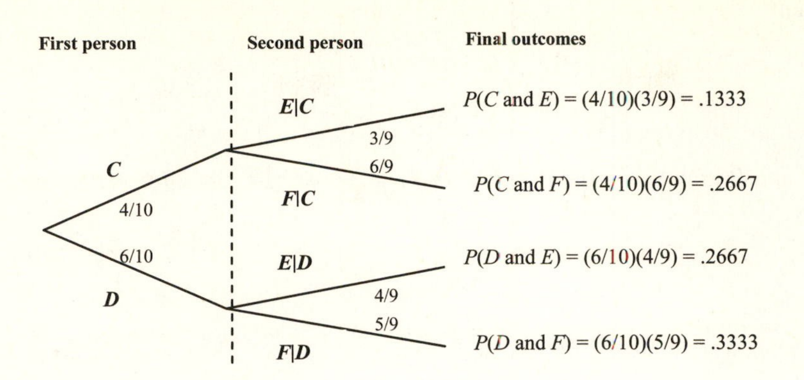

Then P(the first person has a type A personality and the second has a type B personality) = $P(C$ and $F) = P(C)P(F|C) = .2667$.

4.91 Let A = first student selected has student loans to pay off, B = first student selected does not have student loans to pay off, C = second student selected has student loans to pay off, and D = second student selected does not have student loans to pay off. Because students are independent, P(neither student selected has loans to pay off) = $P(B \text{ and } D) = P(B)P(D) = (.4)(.4) = .16$

4.93 Let A = first item is returned, B = first item is not returned, C = second item is returned, and D = second item is not returned.

a. $P(A \text{ and } C) = P(A)P(C) = (.05)(.05) = .0025$

b. $P(B \text{ and } D) = P(B)P(D) = (.95)(.95) = .9025$

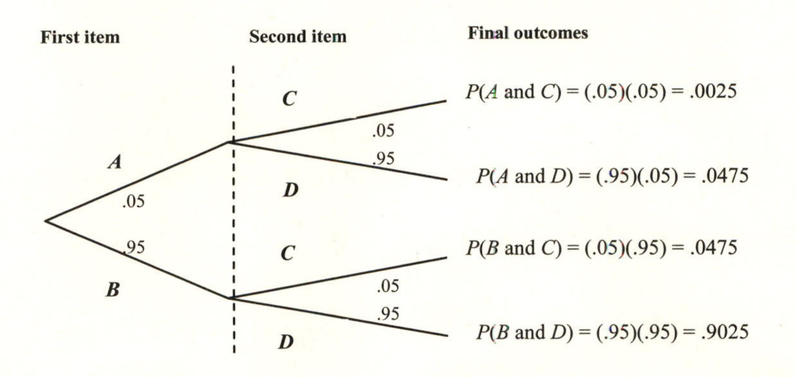

4.95 Let D_1 = first farmer selected is in debt, D_2 = second farmer selected is in debt, and D_3 = third farmer selected is in debt. Then,

$P(D_1 \text{ and } D_2 \text{ and } D_3) = P(D_1)P(D_2)\,P(D_3) = (.80)(.80)(.80) = .5120$.

4.97 Let F = employee selected is a female and M = employee selected is married. Since $P(F) = .36$ and $P(F \text{ and } M) = .19$, $P(M \mid F) = P(F \text{ and } M) / P(F) = .19/.36 = .5278$.

4.99 Let A = adult in small town lives alone and Y = adult in small town has at least one pet. Since $P(A) = .20$ and $P(A \text{ and } Y) = .08$, $P(Y|A) = P(A \text{ and } Y) / P(A) = .08/.20 = .400$

Section 4.9

4.101 When two events are mutually exclusive, their joint probability is zero and is dropped from the formula. So, if A and B are mutually nonexclusive events, then $P(A \text{ or } B) = P(A) + P(B) - P(A \text{ and } B)$. However, if A and B are mutually exclusive events, then $P(A \text{ or } B) = P(A) + P(B) - 0 = P(A) + P(B)$.

4.103 The formula $P(A \text{ or } B) = P(A) + P(B)$ is used when A and B are mutually exclusive events. For example, consider Table 4.11 in the text. Let A = an odd number on both rolls and B = the sum is an odd number. Since A and B are mutually exclusive,

$$P(A \text{ or } B) = P(A) + P(B) = \frac{9}{36} + \frac{18}{36} = \frac{27}{36} = .75 .$$

4.105 a. $P(A \text{ or } B) = P(A) + P(B) - P(A \text{ and } B) = .18 + .49 - .11 = .56$

b. $P(A \text{ or } B) = P(A) + P(B) - P(A \text{ and } B) = .73 + .71 - .68 = .76$

4.107 a. $P(A \text{ or } B) = P(A) + P(B) = .25 + .27 = .52$

b. $P(A \text{ or } B) = P(A) + P(B) = .58 + .09 = .67$

4.109 Let M = basketball player selected is a male, F = basketball player selected is a female, G = player selected has graduated, and N = player selected has not graduated.

a. $P(F \text{ or } N) = P(F) + P(N) - P(F \text{ and } N) = \frac{165}{346} + \frac{87}{346} - \frac{32}{346} = .6358$

b. $P(G \text{ or } M) = P(G) + P(M) - P(G \text{ and } M) = \frac{259}{346} + \frac{181}{346} - \frac{126}{346} = .9075$

4.111 Let M = male, F = female, Y = this adult has shopped on the internet, and N = this adult has never shopped on the internet.

a. $P(N \text{ or } F) = P(N) + P(F) - P(N \text{ and } F) = \frac{1200}{2000} + \frac{800}{2000} - \frac{500}{2000} = .750$

b. $P(M \text{ or } Y) = P(M) + P(Y) - P(M \text{ and } Y) = \frac{1200}{2000} + \frac{800}{2000} - \frac{500}{2000} = .750$

c. Since Y and N are mutually exclusive events, $P(Y \text{ or } N) = P(Y) + P(N) = \frac{800}{2000} + \frac{1200}{2000} = 1.0$.

In fact, these two events are complementary.

4.113 Let B = better off, S = same, W = worse off, L = less than high school, H = high school, and M = more than high school.

a. $P(B \text{ or } H) = P(B) + P(H) - P(B \text{ and } H) = \frac{1010}{2000} + \frac{1000}{2000} - \frac{450}{2000} = .780$

b. $P(M \text{ or } W) = P(M) + P(W) - P(M \text{ and } W) = \frac{600}{2000} + \frac{570}{2000} - \frac{70}{2000} = .550$

c. Since B and W are mutually exclusive events, $P(B \text{ or } W) = P(B) + P(W) = \frac{1010}{2000} + \frac{570}{2000} = .790$.

4.115 Let W = family selected owns a washing machine and V = family selected owns a DVD player. Then, $P(W \text{ or } V) = P(W) + P(V) - P(W \text{ and } V) = .68 + .81 - .58 = .91$

4.117 Let F = teacher selected is a female and S = teacher selected holds a second job. Then, $P(F \text{ or } S) = P(F) + P(S) - P(F \text{ and } S) = .68 + .38 - .29 = .77$

4.119 Let A = person washes his/her hands more than 10 times a day, B = person washes his/her 7 to 10 times a day, and C = person washes his/her hands 5 to 6 times a day. Since A, B and C are mutually exclusive, $P(A \text{ or } B \text{ or } C) = P(A) + P(B) + P(C) = \frac{540}{1500} + \frac{360}{1500} + \frac{345}{1500} = .83$.

This probability is not equal to 1.0 because some people wash their hands between 0 and 4 times per day.

4.121 Let A = voter is against the discount store and I = voter is indifferent about the discount store. Since these events are mutually exclusive, $P(A \text{ or } I) = P(A) + P(I) = .63 + .17 = .80$.

This probability is not equal to 1.0 because some voters favor letting a major discount store move into their neighborhood.

4.123 Let A = first open-heart operation is successful, B = first open-heart operation is not successful, C = second open-heart operation is successful, and D = second open-heart operation is not successful.

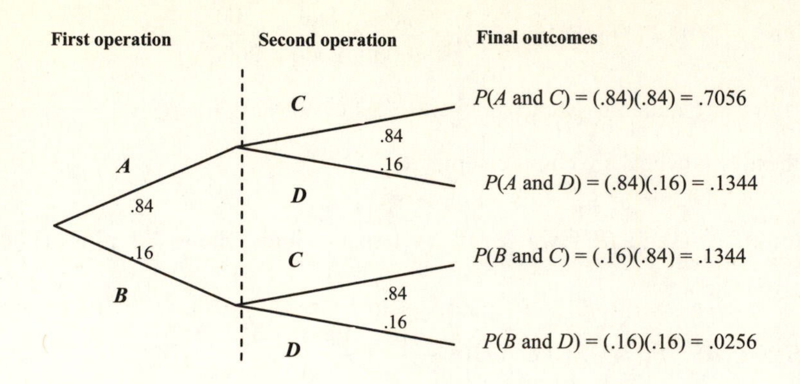

P(at least one open-heart operation is successful)

$= P(A \text{ and } C) + P(A \text{ and } D) + P(B \text{ and } C) = .7056 + .1344 + .1344 = .9744$

Supplementary Exercises

4.125 a. P(student selected is a junior) = 9/35 = .2571

b. P(student selected is a freshman) = 5/35 = .1429

4.127 Let M = adult selected is a male, F = adult selected is a female, A = adult selected prefers watching sports, and B = adult selected prefers watching opera.

a. i. $P(B) = 109/250 = .4360$

ii. $P(M) = 120/250 = .4800$

iii. $P(A|F) = 45/130 = .3462$

iv. $P(M|A) = 96/141 = .6809$

v. $P(F \text{ and } B) = P(F)P(B|F) = \left(\frac{130}{250}\right)\left(\frac{85}{130}\right) = .3400$

vi. $P(A \text{ or } M) = P(A) + P(M) - P(A \text{ and } M) = \frac{141}{250} + \frac{120}{250} - \frac{96}{250} = .6600$

b. $P(F) = 130/250 = .5200$ and $P(F|A) = 45/141 = .3191$. Since these two probabilities are not equal, the events "female" and "prefers watching sports" are dependent. The events "female" and "prefers watching sports" are not mutually exclusive because they can occur together.

4.129 Let A = student selected is an athlete, B = student selected is a nonathlete, F = student selected favors paying college athletes, and N = student selected is against paying college athletes.

a. i. $P(F) = 300/400 = .750$

ii. $P(F|B) = 210/300 = .700$

iii. $P(A \text{ and } F) = P(A)P(F|A) = \left(\frac{100}{400}\right)\left(\frac{90}{100}\right) = .225$

iv. $P(B \text{ or } N) = P(B) + P(N) - P(B \text{ and } N) = \frac{300}{400} + \frac{100}{400} - \frac{90}{400} = .775$

b. $P(A) = 100/400 = .250$ and $P(A|F) = 90/300 = .300$

Since these two probabilities are not equal, the events "athlete" and "should be paid" are dependent. The events "athlete" and "should be paid" are not mutually exclusive because they can occur together.

4.131 Let A = first employee selected has confidence in senior management, B = first employee selected does not have confidence in senior management, C = second employee selected has confidence in senior management, and D = second employee selected does not have confidence in senior management.

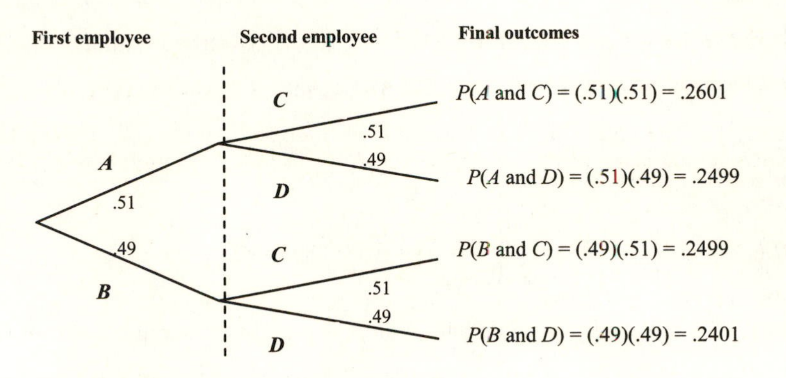

a. $P(A \text{ and } C) = .2601$

b. P(at most one has confidence in his or her senior management)
$= P(A \text{ and } D) + P(B \text{ and } C) + P(B \text{ and } D) = .2499 + .2499 + .2401 = .7399$

4.133 Let J_1 = first student selected is a junior and S_2 = second student selected is a sophomore.

$P(J_1 \text{ and } S_2) = P(J_1)P(S_2|J_1) = \left(\frac{9}{35}\right)\left(\frac{8}{34}\right) = .0605$

4.135 P(both machines are not working properly)
$= P$(first machine is not working properly)P(second machine is not working properly)
$= (.08)(.06) = .0048$

4.137 a. There are 26 possibilities for each letter and 10 possibilities for each digit. Hence, there are $26^3 \cdot 10^3 = 17{,}576{,}000$ possible different license places.

b. There are 2 possibilities for the second letter, 26 possibilities for the third letter, and 10 possibilities for each of the two missing numbers. There are $2 \times 26 \times 10 \times 10 = 5200$ license plates which fit the description.

4.139 Note: This exercise requires the use of combinations, which students may have studied in algebra. Combinations are covered in Section 5.5.2 in the text.

a. Let A = the player's first five numbers match the numbers on the five white balls drawn by the lottery organization, B = the player's powerball number matches the powerball number drawn by the lottery organization, and D = the player's powerball number does not match the powerball number drawn by the lottery organization. There are ${}_{53}C_5$ ways for the lottery organization to draw five different white balls from a set of 53 balls. Thus the sample space for this phase of the drawing consists of ${}_{53}C_5 = 2{,}869{,}685$ equally likely outcomes and $P(A) = 1/({}_{53}C_5)$. Since the sample space for the drawing of the powerball number consists of 42 equally likely outcomes, $P(B) = 1/42 = .0238$. Because the powerball number is drawn independently of the five white balls, A and B are independent events. Therefore,

P(player wins jackpot) = $P(A$ and $B)$

$$= P(A)P(B) = \left(\frac{1}{{}_{53}C_5}\right)\left(\frac{1}{42}\right) = \frac{1}{120{,}526{,}770} = .0000000083$$

b. To win the \$100,000 prize, events A and D must occur. In the sample space of 42 equally likely outcomes for the drawing of the powerball numbers, there are 41 outcomes which do not match the powerball number, and thus result in event D. Hence, $P(D) = 41/42$. Therefore,

P(player wins the \$100,000 prize) = $P(A$ and $D)$

$$P(A)\,P(D) = \left(\frac{1}{{}_{53}C_5}\right)\left(\frac{41}{42}\right) = \frac{41}{120{,}526{,}770} = .00000034$$

4.141 a. P(sixth marble is red) = 10/20 = .5000

b. P(sixth marble is red) = 5/15 = .3333

c. The probability of obtaining a head on the sixth toss is .5, since each toss is independent of the previous outcomes. Tossing a coin is mathematically equivalent to the situation in part a. Each drawing in part a is independent of previous drawings and the probability of drawing a red marble is .5 each time.

4.143 a. The thief has three attempts to guess the correct PIN. Since there are 100 possible numbers in the beginning, the probability that he finds the number on the first attempt is 1/100. Assuming that the first guess is wrong, there are 99 numbers left, etc. Hence,

P(thief succeeds) = 1 – P(thief fails) = 1 – P(thief guessed incorrectly on all three attempts)

$$= 1-\left(\frac{99}{100}\cdot\frac{98}{99}\cdot\frac{97}{98}\right)=1-\left(\frac{97}{100}\right)=\frac{3}{100}=.030$$

b. Since the first two digits of the four-digit PIN must be 3 and 5, respectively, and the third digit must be 1 or 7, the possible PINs are 3510 to 3519 and 3570 to 3579, a total of 20 possible PINs. Hence, P(thief succeeds) = 1 – P(thief fails) = $1-\left(\frac{19}{20}\cdot\frac{18}{19}\cdot\frac{17}{18}\right)=1-\left(\frac{17}{20}\right)=\frac{3}{20}=.150$

4.145 Let J_1 = your friend selects the first jar, J_2 = your friend selects the second jar, R = your friend selects a red marble, and G = your friend selects a green marble.

a. $P(R) = P(J_1)\,P(R|J_1) + P(J_2)\,P(R|J_2) = (1/2)(5/10) + (1/2)(5/10) = .5000$

b. $P(R) = P(J_1)\,P(R|J_1) + P(J_2)\,P(R|J_2) = (1/2)(2/4) + (1/2)(8/16) = .5000$

c. Put one red marble in one jar, and the rest of the marbles in the other jar. Then,

$P(R) = P(J_1)P(R|J_1) + P(J_2)P(R|J_2) = (1/2)(1) + (1/2)(9/19) = .7368$

4.147 a. Let E = neither topping is anchovies, A = customer's first selection is not anchovies, and B = customer's second selection is not anchovies. For A to occur, the customer may choose any of 11 toppings from the 12 available. Thus $P(A) = \frac{11}{12}$. For B to occur, given that A has occurred, the customer may choose any of 10 toppings from the remaining 11 and $P(B|A) = \frac{10}{11}$. Therefore, $P(E) = P(A \text{ and } B) = P(A \text{ and } B) = P(A)P(B|A) = \left(\frac{11}{12}\right)\left(\frac{10}{11}\right) = .8333$

b. Let C = pepperoni is one of the toppings. Then $\overline{C}$ = neither topping is pepperoni. By a similar argument used to find $P(E)$ in part a, we obtain $P(\overline{C}) = \left(\frac{11}{12}\right)\left(\frac{10}{11}\right) = .8333$. Then,

$P(C) = 1 - P(\overline{C}) = 1 - .8333 = .1667$.

4.149 a. $P(\text{win}) = \frac{1}{10\times10\times10\times10} = .0001$

b. i. Since all four digits are unique and order does not matter, $P(\text{win}) = \frac{1}{{}_{10}C_4} = .0048$.

ii. There are ${}_{10}C_3$ ways to choose 3 distinct numbers, and for each of those sets of 3 numbers there are 3 digits that can be repeated. Then, $P(\text{win}) = \dfrac{1}{{}_{10}C_3 \times 3} = .0028$.

iii. There are ${}_{10}C_2$ ways to choose 2 distinct numbers and both of the numbers must be repeated. Then, $P(\text{win}) = \dfrac{1}{{}_{10}C_2} = .0222$.

iv. There are ${}_{10}C_2$ ways to choose 2 distinct numbers, and for each of those sets of 2 numbers there are 2 digits that can be repeated. Then, $P(\text{win}) = \dfrac{1}{{}_{10}C_2 \times 2} = .0111$.

4.151 Let W_1 = the first machine on the first line works, W_2 = the second machine on the first line works, W_3 = the first machine on the second line works, W_4 = the second machine on the second line works, D_1 = the first machine on the first line does not work, D_2 = the second machine on the first line does not work, D_3 = the first machine on the second line does not work, and D_4 = the second machine on the second line does not work.

a. $P(W_1W_2W_3W_4) = P(W_1)P(W_2)P(W_3)P(W_4) = (.98)(.96)(.98)(.96) = .8851$

b. P(at least one machine in each production line is not working properly)
$= P(W_1D_2W_3D_4) + P(W_1D_2D_3W_4) + P(D_1W_2W_3D_4) + P(D_1W_2D_3W_4) + P(W_1D_2D_3D_4) + P(D_1W_2D_3D_4) + P(D_1D_2W_3D_4) + P(D_1D_2D_3W_4) + P(D_1D_2D_3D_4) = (.98)(.02)(.96)(.04) + (.98)(.02)(.04)(.96) + (.02)(.98)(.96)(.04) + (.02)(.98)(.04)(.96) + (.98)(.02)(.04)(.04) + (.02)(.98)(.04)(.04) + (.02)(.02)(.96)(.04) + (.02)(.02)(.04)(.96) + (.02)(.02)(.04)(.04) = .0031$

Self-Review Test

1. a **2.** b **3.** c **4.** a **5.** a **6.** b

7. c **8.** b **9.** b **10.** c **11.** b

12. Total outcomes = 4 × 3 × 5 × 2 = 120

13. a. P(job offer selected is from the insurance company) = 1/3 = .3333

b. P(job offer selected is not from the accounting firm) = 2/3 = .6667

14. a. P(out of state) = 125/200 = .6250 and P(out of state | female) = 70/110 = .6364. Since these two probabilities are not equal, the two events are dependent. Events "female" and "out of state" are not mutually exclusive because they can occur together.

b. i. In 200 students, there are 90 males. Hence, P(a male is selected) = 90/200 = .4500.

ii. There are a total of 110 female students and 70 of them are out of state students. Hence, P(out of state | female) = 70/110 = .6364.

15. P(out of state or female) = P(out of state) + P(female) – P(out of state and female)
= P(out of state) + P(female) – P(out of state) P(female |out of state)=

$$\frac{125}{200}+\frac{110}{200}-\left(\frac{110}{200}\right)\left(\frac{70}{110}\right)=.825$$

16. Let S_1 = first student selected is from out of state and S_2 = second student selected is from out of state.

Then, $P(S_1 \text{ and } S_2) = P(S_1)P(S_2|S_1) = \left(\frac{125}{200}\right)\left(\frac{124}{199}\right) = .3894$

17. Let F_1 = first adult selected has experienced a migraine headache, N_1 = first adult selected has never experienced a migraine headache, F_2 = second adult selected has experienced a migraine headache, and N_2 = second adult selected has never experienced a migraine headache. Note that the two adults are independent. From the given information: $P(F_1) = .35$ and $P(F_2) = .35$. Hence, $P(N_1) = 1 - .35 = .65$ and $P(N_2) = 1 - .35 = .65$. Then, $P(N_1 \text{ and } N_2) = P(N_1)P(N_2) = (.65)(.65) = .4225$.

18. $P(A)$ = 8/20 = .400. The complementary event of A is that the selected marble is not red, that is, the selected marble is either green or blue. Hence, $P(\bar{A}) = 1 - .400 = .600$.

19. Let M = male, F = female, W = works at least 10 hours, and N = does not work 10 hours.

a. $P(M \text{ and } W) = P(M)P(W) = (.45)(.62) = .279$

b. $P(F \text{ or } W) = P(F) + P(W) - P(F \text{ and } W) = .55 + .62 - (.55)(.62) = .829$

20. a. i. $P(Y) = (77 + 104)/506 = .3577$

ii. $P(Y|W) = 104/(104 + 119 + 34) = .4047$

iii. $P(W \text{ and } N) = P(W)P(N/W) = \left(\frac{257}{506}\right)\left(\frac{119}{257}\right) = .2352$

iv. $P(N \text{ or } M) = P(N) + P(M) - P(N \text{ and } M) = \frac{66}{506}+\frac{249}{506}-\frac{32}{506} = .5593$

b. $P(W) = 257/506 = .5079$ and $P(W|Y) = 104 / 181 = .5746$. Since these two probabilities are not equal, the events "woman" and "yes" are dependent. Events "woman" and "yes" are not mutually exclusive because they can occur together.

Chapter Five

Section 5.1

5.1 A variable whose value is determined by the outcome of a random experiment is called a **random variable**. A random variable whose values that assumes countable values is called a **discrete random variable**. The number of cars owned by randomly selected individuals is an example of a discrete random variable. A random variable that can assume any value contained in one or more intervals is called a **continuous random variable**. An example of a continuous random variable is the amount of time taken by a randomly selected student to complete a statistics exam.

5.3
a. discrete — b. continuous
c. continuous — d. discrete
e. discrete — f. continuous

5.5 The number of cars x that stop at the Texaco station is a discrete random variable because the values of x are countable: 0, 1, 2, 3, 4, 5 and 6.

Section 5.2

5.7
1. The probability assigned to each value of a random variable x lies in the range 0 to 1, that is, $0 \le P(x) \le 1$ for each x.
2. The sum of the probabilities assigned to all possible values of x is equal to 1, that is, $\sum P(x) = 1$.

5.9
a. This table does not satisfy the first condition of a probability distribution because the probability of $x = 5$ is negative. Hence, it does not represent a valid probability distribution of x.
b. This table represents a valid probability distribution of x because it satisfies both conditions required for a valid probability distribution.

c. This table does not represent a valid probability distribution of x because the sum of the probabilities of all outcomes listed in the table is not 1, which violates the second condition of a probability distribution.

5.11 a. $P(x = 1) = .17$

b. $P(x \leq 1) = P(x = 0) + P(x = 1) = .03 + .17 + = .20$

c. $P(x \geq 3) = P(x = 3) + P(x = 4) + P(x = 5) = .31 + .15 + .12 = .58$

d. $P(0 \leq x \leq 2) = P(x = 0) + P(x = 1) + P(x = 2) = .03 + .17 + .22 = .42$

e. $P(x < 3) = P(x = 0) + P(x = 1) + P(x = 2) = .03 + .17 + .22 = .42$

f. $P(x > 3) = P(x = 4) + P(x = 5) = .15 + .12 = .27$

g. $P(2 \leq x \leq 4) = P(x = 2) + P(x = 3) + P(x = 4) = .22 + .31 + .15 = .68$

5.13 a.

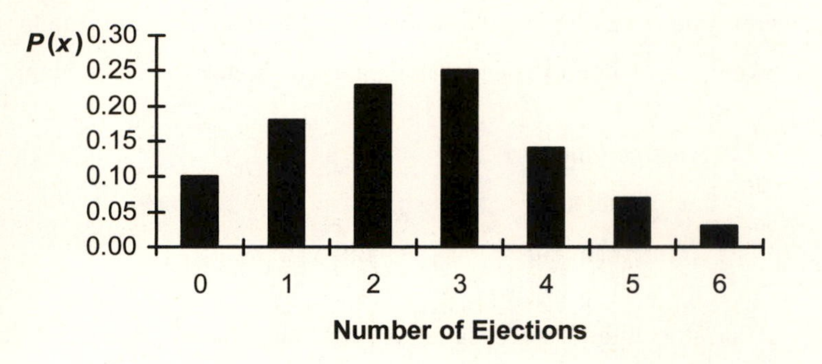

b. i. $P(\text{exactly } 3) = P(x = 3) = .25$

ii. $P(\text{at least } 4) = P(x \geq 4) = P(x = 4) + P(x = 5) + P(x = 6) = .14 + .07 + .03 = .24$

iii. $P(\text{less than } 3) = P(x < 3) = P(x = 0) + P(x = 1) + P(x = 2) = .10 + .18 + .23 = .51$

iv. $P(2 \text{ to } 5) = P(2 \leq x \leq 5) = P(x = 2) + P(x = 3) + P(x = 4) + P(x = 5) = .23 + .25 + .14 + .07 = .69$

5.15 a.

x	$P(x)$
1	8/80 = .10
2	20/80 = .25
3	24/80 = .30
4	16/80 = .20
5	12/80 = .15

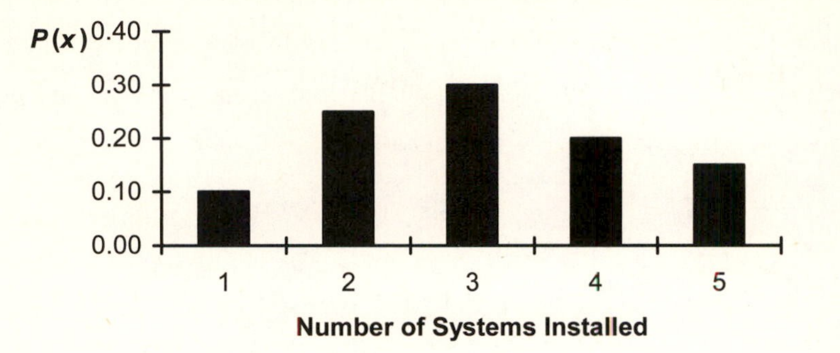

b. The probabilities listed in the table of part a are approximate because they are obtained from a sample of 80 days.

c. i. $P(x = 3) = .30$

ii. $P(x \geq 3) = P(x = 3) + P(x = 4) + P(x = 5) = .30 + .20 + .15 = .65$

iii. $P(2 \leq x \leq 4) = P(x = 2) + P(x = 3) + P(x = 4) = .25 + .30 + .20 = .75$

iv. $P(x < 4) = P(x = 1) + P(x = 2) + P(x = 3) = .10 + .25 + .30 = .65$

5.17 Let D = adult selected uses prescription drugs regularly and N = adult selected does not use prescription drugs regularly.

Then $P(D) = .45$ and $P(N) = 1 - .45 = .55$. Note that the first and second events are independent.

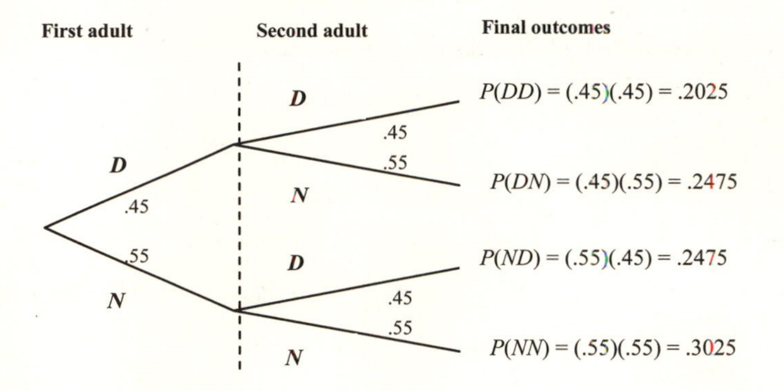

$P(x = 0) = P(NN) = .3025$, $P(x = 1) = P(DN) + P(ND) = .2475 + .2475 = .4950$,
$P(x = 2) = P(DD) = .2025$

x	$P(x)$
0	.3025
1	.4950
2	.2025

5.19 Let M = adult is more likely to see movies in a theater if tickets and concessions cost less and N = adult is not more likely to see movies in a theater if tickets and concessions cost less.

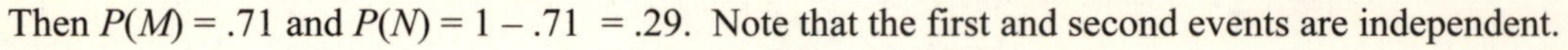

Then $P(M) = .71$ and $P(N) = 1 - .71 = .29$. Note that the first and second events are independent.

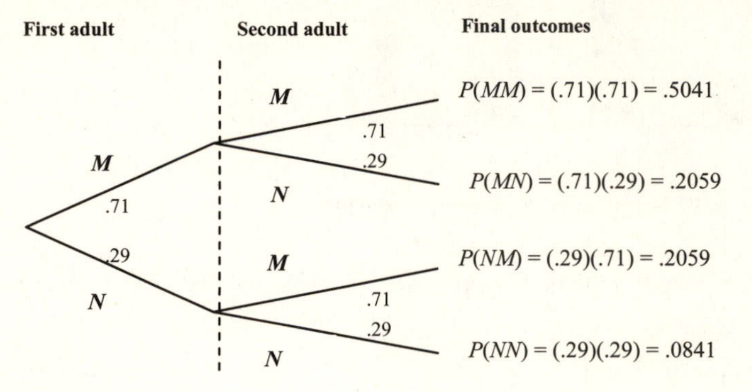

$P(x = 0) = P(NN) = .0841$, $P(x = 1) = P(MN) + P(NM) = .2059 + .2059 = .4118$,
$P(x = 2) = P(MM) = .5041$

x	$P(x)$
0	.0841
1	.4118
2	.5041

5.21 Let A = first athlete selected used drugs, B = first athlete selected did not use drugs, C = second athlete selected used drugs, and D = second athlete selected did not use drugs. Note that the first and second events are independent.

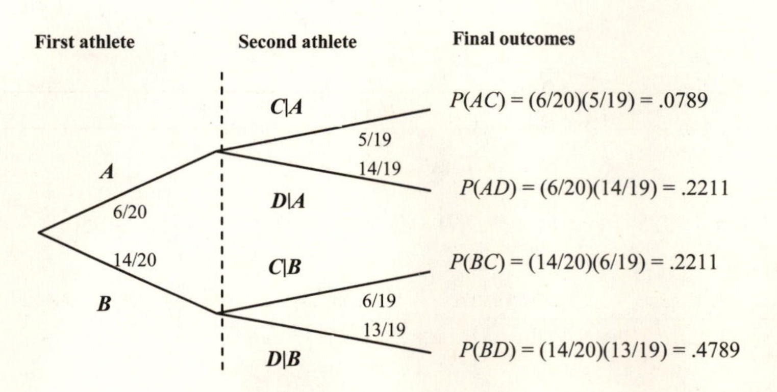

$P(x = 0) = P(BD) = .4789$, $P(x = 1) = P(AD) + P(BC) = .2211 + .2211 = .4422$,
$P(x = 2) = P(AC) = .0789$

x	$P(x)$
0	.4789
1	.4422
2	.0789

Sections 5.3 - 5.4

5.23 a.

x	$P(x)$	$xP(x)$	x^2	$x^2P(x)$
0	.16	.00	0	.00
1	.27	.27	1	.27
2	.39	.78	4	1.56
3	.18	.54	9	1.62
		$\Sigma xP(x) = 1.59$		$\Sigma x^2P(x) = 3.45$

$\mu = \Sigma xP(x) = 1.590$

$\sigma = \sqrt{\sum x^2 P(x) - \mu^2)} = \sqrt{3.45 - (1.59)^2} = .960$

b.

x	$P(x)$	$xP(x)$	x^2	$x^2P(x)$
6	.40	2.40	36	14.40
7	.26	1.82	49	12.74
8	.21	1.68	64	13.44
9	.13	1.17	81	10.53
		$\Sigma xP(x) = 7.07$		$\Sigma x^2P(x) = 51.11$

$\mu = \Sigma xP(x) = 7.070$

$\sigma = \sqrt{\Sigma x^2 P(x) - \mu^2} = \sqrt{51.11 - (7.07)^2} = 1.061$

5.25

x	$P(x)$	$xP(x)$	x^2	$x^2P(x)$
0	.73	.00	0	.00
1	.16	.16	1	.16
2	.06	.12	4	.24
3	.04	.12	9	.36
4	.01	.04	16	.16
		$\Sigma xP(x) = .44$		$\Sigma x^2P(x) = .92$

$\mu = \Sigma xP(x) = .440$ errors

$\sigma = \sqrt{\Sigma x^2 P(x) - \mu^2} = \sqrt{.92 - (.44)^2} = .852$ errors

5.27 Let x be the number of camcorders sold on a given day at an electronics store.

x	$P(x)$	$xP(x)$	x^2	$x^2P(x)$
0	.05	.00	0	.00
1	.12	.12	1	.12
2	.19	.38	4	.76
3	.30	.90	9	2.70
4	.20	.80	16	3.20
5	.10	.50	25	2.50
6	.04	.24	36	1.44
		$\Sigma xP(x) = 2.94$		$\Sigma x^2P(x) = 10.72$

$\mu = \Sigma xP(x) = 2.94$ camcorders

$\sigma = \sqrt{\Sigma x^2 P(x) - \mu^2} = \sqrt{10.72 - (2.94)^2} = 1.441$ camcorders

On average, 2.94 camcorders are sold per day at this store.

5.29

x	$P(x)$	$xP(x)$	x^2	$x^2P(x)$
0	.25	.00	0	.00
1	.50	.50	1	.50
2	.25	.50	4	1.00
		$\Sigma xP(x) = 1.00$		$\Sigma x^2P(x) = 1.50$

$\mu = \Sigma xP(x) = 1.00$ head

$\sigma = \sqrt{\Sigma x^2 P(x) - \mu^2} = \sqrt{1.50 - (1.00)^2} = .707$ head

On average, we will obtain 1 head in every two tosses of the coin.

5.31 Let x be the number of TV sets owned by a family.

x	$P(x)$	$xP(x)$	x^2	$x^2P(x)$
0	.048	.000	0	.000
1	.388	.388	1	.388
2	.292	.584	4	1.168
3	.164	.492	9	1.476
4	.108	.432	16	1.728
		$\Sigma xP(x) = 1.896$		$\Sigma x^2P(x) = 4.760$

$\mu = \Sigma xP(x) = 1.896$ TV sets

$\sigma = \sqrt{\Sigma x^2 P(x) - \mu^2} = \sqrt{4.760 - (1.896)^2} = 1.079$ TV sets

There is an average of 1.896 TV sets per family, with a standard deviation of 1.079 sets.

5.33

x	$P(x)$	$xP(x)$	x^2	$x^2P(x)$
0	.9025	.0000	0	.0000
1	.0950	.0950	1	.0950
2	.0025	.0050	4	.0100
		$\Sigma xP(x) = .100$		$\Sigma x^2P(x) = .105$

$\mu = \Sigma xP(x) = .10$ car

$\sigma = \sqrt{\Sigma x^2 P(x) - \mu^2} = \sqrt{.105 - (.100)^2} = .308$ car

5.35

x	$P(x)$	$xP(x)$	x^2	$x^2P(x)$
0	.10	.00	0	.00
2	.45	.90	4	1.80
5	.30	1.50	25	7.50
10	.15	1.50	100	15.00
		$\Sigma xP(x) = 3.9$		$\Sigma x^2P(x) = 24.30$

$\mu = \Sigma xP(x) = \$3.9$ million

$\sigma = \sqrt{\Sigma x^2 P(x) - \mu^2} = \sqrt{24.30 - (3.9)^2} = \$3.015$ million

The contractor is expected to make \$3.9 million profit with a standard deviation of \$3.015 million.

5.37

x	$P(x)$	$xP(x)$	x^2	$x^2P(x)$
0	.5455	.0000	0	.0000
1	.4090	.4090	1	.4090
2	.0455	.0910	4	.1820
		$\Sigma xP(x) = .500$		$\Sigma x^2P(x) = .591$

$\mu = \Sigma xP(x) = .500$ person

$\sigma = \sqrt{\Sigma x^2 P(x) - \mu^2} = \sqrt{.591 - (.500)^2} = .584$ person

Section 5.5

5.39 $3! = 3 \cdot 2 \cdot 1 = 6$

$(9 - 3)! = 6! = 6 \cdot 5 \cdot 4 \cdot 3 \cdot 2 \cdot 1 = 720$

$9! = 9 \cdot 8 \cdot 7 \cdot 6 \cdot 5 \cdot 4 \cdot 3 \cdot 2 \cdot 1 = 362{,}880$

$(14 - 12)! = 2! = 2 \cdot 1 = 2$

$${}_5C_3 = \frac{5!}{3!(5-3)!} = \frac{5!}{3!\,2!} = \frac{120}{(6)(2)} = 10 \qquad {}_7C_4 = \frac{7!}{4!(7-4)!} = \frac{7!}{4!\,3!} = \frac{5040}{(24)(6)} = 35$$

$${}_9C_3 = \frac{9!}{3!(9-3)!} = \frac{9!}{3!\,6!} = \frac{362{,}880}{(6)(720)} = 84 \qquad {}_4C_0 = \frac{4!}{0!(4-0)!} = \frac{4!}{0!\,4!} = \frac{24}{(1)(24)} = 1$$

$${}_3C_3 = \frac{3!}{3!(3-3)!} = \frac{3!}{3!\,0!} = \frac{6}{(6)(1)} = 1 \qquad {}_6P_2 = \frac{6!}{(6-2)!} = \frac{6!}{4!} = \frac{720}{24} = 30$$

$${}_8P_4 = \frac{8!}{(8-4)!} = \frac{8!}{4!} = \frac{40{,}320}{24} = 1680$$

5.41 $${}_9C_2 = \frac{9!}{2!(9-2)!} = \frac{9!}{2!\,7!} = \frac{362{,}880}{(2)(5040)} = 36\,;\ {}_9P_2 = \frac{9!}{(9-2)!} = \frac{9!}{7!} = \frac{362{,}880}{5040} = 72$$

5.43 $${}_{12}C_3 = \frac{12!}{3!(12-3)!} = \frac{12!}{3!\,9!} = 220\,;\ {}_{12}P_3 = \frac{12!}{(12-3)!} = \frac{12!}{9!} = 1320$$

5.45 $${}_{20}C_6 = \frac{20!}{6!(20-6)!} = \frac{20!}{6!\,14!} = 38{,}760\,;\ {}_{20}P_6 = \frac{20!}{(20-6)!} = \frac{20!}{14!} = 27{,}907{,}200$$

5.47 $${}_{20}C_9 = \frac{20!}{9!(20-9)!} = \frac{20!}{9!\,11!} = 167{,}960$$

Section 5.6

5.49

a. An experiment that satisfies the following four conditions is called a **binomial experiment**:

   i. There are n identical trials. In other words, the given experiment is repeated n times. All these repetitions are performed under identical conditions.

   ii. Each trial has two and only two outcomes. These outcomes are usually called a *success* and a *failure*.

   iii. The probability of success is denoted by p and that of failure by q, and $p + q = 1$. The probability of p and q remain constant for each trial.

   iv. The trials are independent. In other words, the outcome of one trial does not affect the outcome of another trial.

b. Each repetition of a binomial experiment is called a **trial**.

c. A **binomial random variable** x represents the number of successes in n independent trials of a binomial experiment.

5.51

a. This is not a binomial experiment because there are more than two outcomes for each repetition.

b. This is an example of a binomial experiment because it satisfies all four conditions of a binomial experiment:

  i. There are many identical rolls of the die.

  ii. Each trial has two outcomes: an even number and an odd number.

  iii. The probability of obtaining an even number is 1/2 and that of an odd number is 1/2. These probabilities add up to 1, and they remain constant for all trials.

  iv. All rolls of the die are independent.

c. This is an example of a binomial experiment because it satisfies all four conditions of a binomial experiment:

  i There are a few identical trials (selection of voters).

  ii. Each trial has two outcomes: a voter favors the proposition and a voter does not favor the proposition.

  iii. The probability of the two outcomes are .54 and .46, respectively. These probabilities add up to 1. These two probabilities remain the same for all selections.

  iv. All voters are independent.

5.53 a. $n = 8, x = 5, n - x = 8 - 5 = 3, p = .70$, and $q = 1 - p = 1 - .70 = .30$

$P(x = 5) = {}_nC_x p^x q^{n-x} = {}_8C_5 (.70)^5(.30)^3 = (56)(.16807)(.027) = .2541$

b. $n = 4, x = 3, n - x = 4 - 3 = 1, p = .40$, and $q = 1 - p = 1 - .40 = .60$

$P(x = 3) = {}_nC_x p^x q^{n-x} = {}_4C_3 (.40)^3(.60)^1 = (4)(.064)(.60) = .1536$

c. $n = 6, x = 2, n - x = 6 - 2 = 4, p = .30$, and $q = 1 - p = 1 - .30 = .70$

$P(x = 2) = {}_nC_x p^x q^{n-x} = {}_6C_2 (.30)^2(.70)^4 = (15)(.09)(.2401) = .3241$

5.55 a.

x	$P(x)$
0	.0824
1	.2471
2	.3177
3	.2269
4	.0972
5	.0250
6	.0036
7	.0002

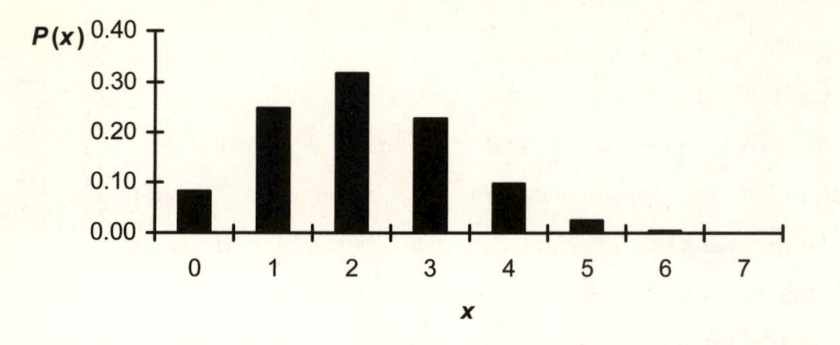

b. $\mu = np = (7)(.30) = 2.100$

$\sigma = \sqrt{npq} = \sqrt{(7)(.30)(.70)} = 1.212$

5.57 Answers will vary depending on the values of n and p selected.

Let $n = 5$. The probability distributions for $p = .30$ (skewed right), $p = .50$ (symmetric), and $p = .80$ (skewed left) are displayed in the tables below followed by the graphs.

$p = .30$

x	$P(x)$
0	.1681
1	.3602
2	.3087
3	.1323
4	.0283
5	.0024

$p = 0.50$

x	$P(x)$
0	.0312
1	.1562
2	.3125
3	.3125
4	.1562
5	.0312

$p = 0.80$

x	$P(x)$
0	.0003
1	.0064
2	.0512
3	.2048
4	.4096
5	.3277

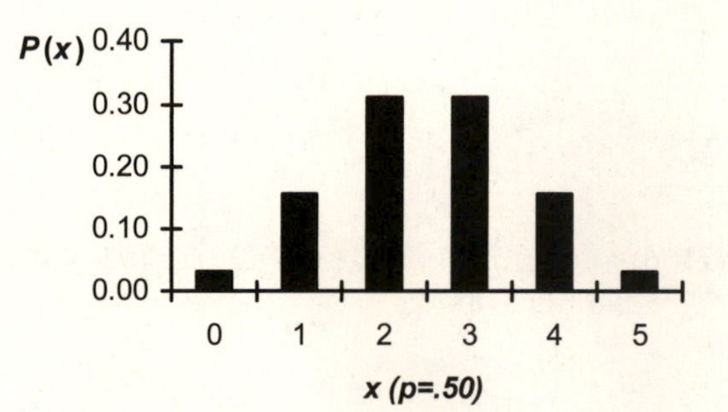

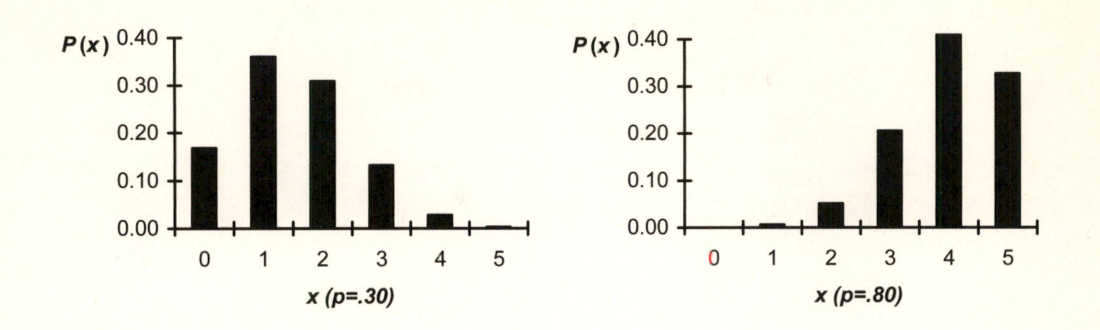

5.59 a. The random variable x can assume any of the values 0, 1, 2, 3, 4, 5, 6, 7, 8, 9, or 10.

b. $n = 10$ and $p = .63$

$P(x = 7) = {}_nC_x p^x q^{n-x} = {}_{10}C_7 (.63)^7 (.37)^3 = (120)(.03938981)(.050653) = .2394$

5.61 $n = 16$ and $p = .60$

Let x denote the number of U.S. companies in a random sample of 16 who paid no federal taxes from 1996 to 2000.

a. $P(\text{at most } 7) = P(x \le 7) = P(x = 0) + P(x = 1) + P(x = 2) + P(x = 3) + P(x = 4) + P(x = 5)$
$+ P(x = 6) + P(x = 7) = .0000 + .0000 + .0001 + .0008 + .0040 + .0142 + .0392 + .0840 = .1423$

b. $P(\text{at least } 10) = P(x \ge 10) = P(x = 10) + P(x = 11) + P(x = 12) + P(x = 13) + P(x = 14) +$
$P(x = 15) + P(x = 16) = .1983 + .1623 + .1014 + .0468 + .0150 + .0030 + .0003 = .5271$

c. $P(8 \text{ to } 11) = P(8 \le x \le 11) = P(x = 8) + P(x = 9) + P(x = 10) + P(x = 11)$
$= .1417 + .1889 + .1983 + .1623 = .6912$

5.63 $n = 12$ and $p = .68$

Let x denote the number of children in a random sample of 12 who live in homes with two married parents.

a. $P(\text{exactly } 6) = P(x = 6) = {}_nC_x p^x q^{n-x} = {}_{12}C_6 (.68)^6 (.32)^6 = (924)(.09886748)(.00107374) =$
$.0981$

b. $P(\text{none}) = P(x = 0) = {}_nC_x p^x q^{n-x} = {}_{12}C_0 (.68)^0 (.32)^{12} = (1)(1)(.00000115) = .0000$

c. $P(\text{exactly } 9) = P(x = 9) = {}_nC_x p^x q^{n-x} = {}_{12}C_9 (.68)^9 (.32)^3 = (220)(.0310871)(.032768) = .2241$

5.65 $n = 8$ and $p = .85$

a. $P(\text{exactly } 8) = P(x = 8) = {}_nC_x p^x q^{n-x} = {}_8C_8 (.85)^8 (.15)^0 = (1)(.27249053)(1) = .2725$

b. $P(\text{exactly } 5) = P(x = 5) = {}_nC_x p^x q^{n-x} = {}_8C_5 (.85)^5 (.15)^3 = (56)(.4437053)(.003375) = .0839$

5.67 a. $n = 7$ and $p = .80$

x	$P(x)$
0	.0000
1	.0004
2	.0043
3	.0287
4	.1147
5	.2753
6	.3670
7	.2097

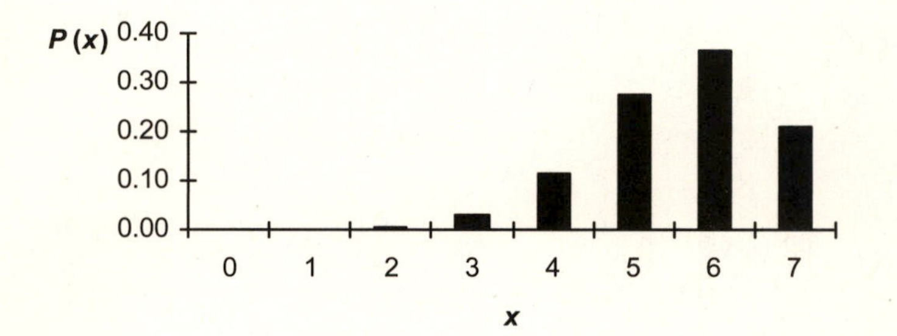

$\mu = np = (7)(.80) = 5.6$ customers

$\sigma = \sqrt{npq} = \sqrt{(7)(.80)(.20)} = 1.058$ customers

b. $P(\text{exactly } 4) = P(x = 4) = .1147$

5.69 a. $n = 8$ and $p = .70$

x	$P(x)$
0	.0001
1	.0012
2	.0100
3	.0467
4	.1361
5	.2541
6	.2965
7	.1977
8	.0576

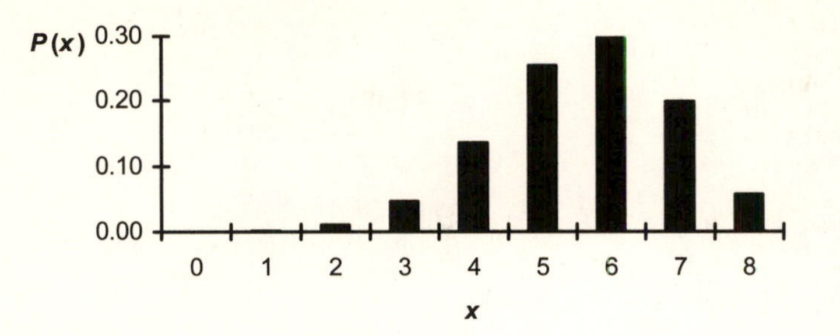

$\mu = np = (8)(.70) = 5.600$ customers

$\sigma = \sqrt{npq} = \sqrt{(8)(.70)(.30)} = 1.296$ customers

b. $P(\text{exactly } 3) = P(x = 3) = .0467$

Section 5.7

5.71 $N = 8, r = 3, N - r = 5$, and $n = 4$

a. $P(x=2) = \dfrac{{}_rC_x \; {}_{N-r}C_{n-x}}{{}_NC_n} = \dfrac{{}_3C_2 \; {}_5C_2}{{}_8C_4} = \dfrac{(3)(10)}{70} = .4286$

b. $P(x=0) = \dfrac{{}_rC_x \; {}_{N-r}C_{n-x}}{{}_NC_n} = \dfrac{{}_3C_0 \; {}_5C_4}{{}_8C_4} = \dfrac{(1)(5)}{70} = .0714$

c. $P(x \le 1) = P(x=0) + P(x=1) = .0714 + \dfrac{{}_3C_1 \; {}_5C_3}{{}_8C_4} = .0714 + \dfrac{(3)(10)}{70} = .0714 + .4286 = .5000$

5.73 $N = 11, r = 4, N - r = 7$, and $n = 4$

a. $P(x=2) = \dfrac{{}_rC_x \; {}_{N-r}C_{n-x}}{{}_NC_n} = \dfrac{{}_4C_2 \; {}_7C_2}{{}_{11}C_4} = \dfrac{(6)(21)}{330} = .3818$

b. $P(x=4) = \dfrac{{}_rC_x \; {}_{N-r}C_{n-x}}{{}_NC_n} = \dfrac{{}_4C_4 \; {}_7C_0}{{}_{11}C_4} = \dfrac{(1)(1)}{330} = .0030$

c. $P(\text{at most } 1) = P(x \le 1) = P(x = 0) + P(x = 1)$

$= \dfrac{{}_4C_0 \; {}_7C_4}{{}_{11}C_4} + \dfrac{{}_4C_1 \; {}_7C_3}{{}_{11}C_4} = \dfrac{(1)(35)}{330} + \dfrac{(4)(35)}{330} = .1061 + .4242 = .5303$

5.75 $N = 15, r = 9, N - r = 6$, and $n = 3$

Let x be the number of corporations that incurred losses in a random sample of 3 corporations, and r be the number of corporations in 15 that incurred losses.

a. $P(\text{exactly } 2) = P(x=2) = \frac{{}_rC_x \; {}_{N-r}C_{n-x}}{{}_NC_n} = \frac{{}_9C_2 \; {}_6C_1}{{}_{15}C_3} = \frac{(36)(6)}{455} = .4747$

b. $P(\text{none}) = P(x=0) = \frac{{}_rC_x \; {}_{N-r}C_{n-x}}{{}_NC_n} = \frac{{}_9C_0 \; {}_6C_3}{{}_{15}C_3} = \frac{(1)(20)}{455} = .0440$

c. $P(\text{at most } 1) = P(x \le 1) = P(x=0) + P(x=1) =$

$= .0440 + \frac{{}_9C_1 \; {}_6C_2}{{}_{15}C_3} = .0440 + \frac{(9)(15)}{455} = .0440 + .2967 = .3407$

5.77 $N = 11$, $r = 4$, and $N - r = 7$, and $n = 3$

Let x be the number of extremely violent games in a random sample of 3, and r be the number of extremely violent games in 11.

a. $P(\text{exactly } 2) = P(x=2) = \frac{{}_rC_x \; {}_{N-r}C_{n-x}}{{}_NC_n} = \frac{{}_4C_2 \; {}_7C_1}{{}_{11}C_3} = \frac{(6)(7)}{165} = .2545$

b. $P(\text{more than one}) = P(x > 1) = P(x=2) + P(x=3)$

$= .2545 + \frac{{}_4C_3 \; {}_7C_0}{{}_{11}C_3} = .2545 + \frac{(4)(1)}{165} = .2545 + .0242 = .2787$

c. $P(\text{none}) = P(x=0) = \frac{{}_rC_x \; {}_{N-r}C_{n-x}}{{}_NC_n} = \frac{{}_4C_0 \; {}_7C_3}{{}_{11}C_3} = \frac{(1)(35)}{165} = .2121$

Section 5.8

5.79 The following three conditions must be satisfied to apply the Poisson probability distribution:

1) x is a discrete random variable.
2) The occurrences are random.
3) The occurrences are independent.

5.81 a. $P(x \le 1) = P(x=0) + P(x=1) = \frac{(5)^0 e^{-5}}{0!} + \frac{(5)^1 e^{-5}}{1!} = \frac{(1)(.00673795)}{1} + \frac{(5)(.00673795)}{1}$

$= .0067 + .0337 = .0404$

Note that the value of e^{-5} is obtained from Table II of Appendix C of the text.

b. $P(x=2) = \frac{\lambda^x e^{-\lambda}}{x!} = \frac{(2.5)^2 e^{-2.5}}{2!} = \frac{(6.25)(.08208500)}{2} = .2565$

5.83 a.

x	$P(x)$
0	.2725
1	.3543
2	.2303
3	.0998
4	.0324
5	.0084
6	.0018
7	.0003
8	.0001

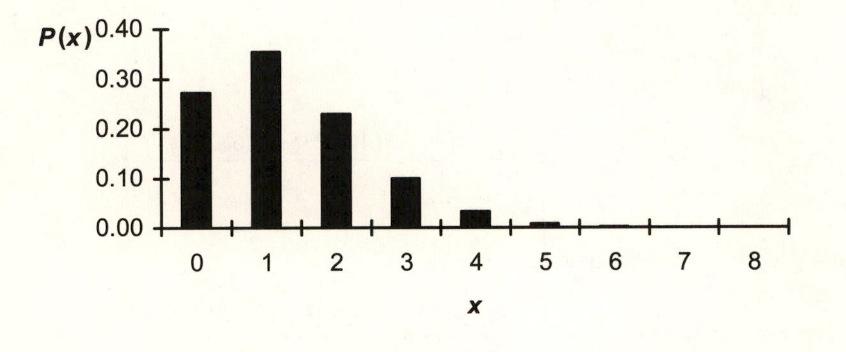

$\mu = \lambda = 1.3$, $\sigma^2 = \lambda = 1.3$, and $\sigma = \sqrt{\lambda} = \sqrt{1.3} = 1.140$

b.

x	$P(x)$
0	.1225
1	.2572
2	.2700
3	.1890
4	.0992
5	.0417
6	.0146
7	.0044
8	.0011
9	.0003
10	.0001

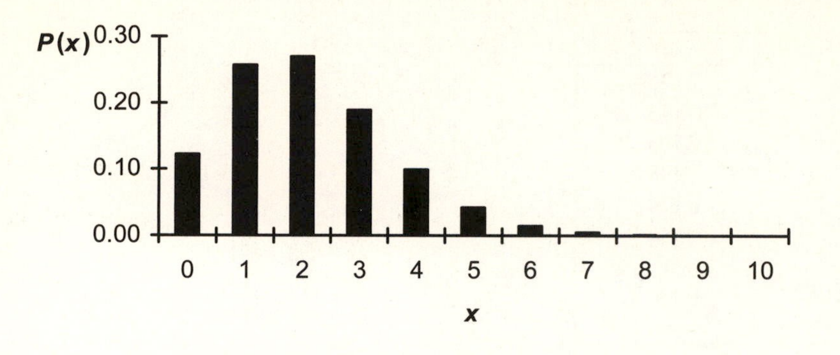

$\mu = \lambda = 2.1, \sigma^2 = \lambda = 2.1, \text{ and } \sigma = \sqrt{\lambda} = \sqrt{2.1} = 1.449$

5.85 $\lambda = 1.7$ pieces of junk mail per day

$$P(\text{exactly three}) = P(x=3) = \frac{\lambda^x e^{-\lambda}}{x!} = \frac{(1.7)^3 e^{-1.7}}{3!} = \frac{(4.913)(.18268352)}{6} = .1496$$

5.87 $\lambda = 5.4$ shoplifting incidents per day

$$P(\text{exactly three}) = P(x=3) = \frac{\lambda^x e^{-\lambda}}{x!} = \frac{(5.4)^3 e^{-5.4}}{3!} = \frac{(157.464)(.00451658)}{6} = .1185$$

5.89 $\lambda = 3.7$ reports of lost students' ID cards per week

a. $P(\text{at most } 1) = P(x \le 1) = P(x=0) + P(x=1)$

$$= \frac{(3.7)^0 e^{-3.7}}{0!} + \frac{(3.7)^1 e^{-3.7}}{1!} = \frac{(1)(.02472353)}{1} + \frac{(3.7)(.02472353)}{1} = .0247 + .0915 = .1162$$

b. i. $P(1 \text{ to } 4) = P(1 \le x \le 4) = P(x=1) + P(x=2) + P(x=3) + P(x=4)$

$= .0915 + .1692 + .2087 + .1931 = .6625$

ii. $P(\text{at least } 6) = P(x \ge 6)$

$= P(x=6) + P(x=7) + P(x=8) + P(x=9) + P(x=10) + P(x=11) + P(x=12) + P(x=13)$

$= .0881 + .0466 + .0215 + .0089 + .0033 + .0011 + .0003 + .0001 = .1699$

iii. $P(\text{at most } 3) = P(x \le 3) = P(x=0) + P(x=1) + P(x=2) + P(x=3)$

$= .0247 + .0915 + .1692 + .2087 = .4941$

5.91 Let x be the number of defects in a given 500-yard piece of fabric. $\lambda = .5$ defect per 500 yards

a. $P(\text{exactly } 1) = P(x=1) = \dfrac{\lambda^x e^{-\lambda}}{x!} = \dfrac{(.5)^1 e^{-.5}}{1!} = \dfrac{(.5)(.60653066)}{1} = .3033$

b. i. $P(2 \text{ to } 4) = P(2 \le x \le 4) = P(x=2) + P(x=3) + P(x=4) = .0758 + .0126 + .0016 = .0900$

ii. $P(\text{more than 3}) = P(x > 3) = P(x = 4) + P(x = 5) + P(x = 6) + P(x = 7)$

$= .0016 + .0002 + .0000 + .0000 = .0018$

iii. $P(\text{less than 2}) = P(x < 2) = P(x = 0) + P(x = 1) = .6065 + .3033 = .9098$

5.93 Let x be the number of customers that come to this savings and loan during a given hour. Since the average number of customers per half hour is 4.8, $\lambda = (2)(4.8) = 9.6$ customers per hour.

a. $P(\text{exactly 2}) = P(x = 2) = \dfrac{\lambda^x e^{-\lambda}}{x!} = \dfrac{(9.6)^2 e^{-9.6}}{2!} = \dfrac{(92.16)(.00006773)}{2} = .0031$

b. i. $P(\text{2 or fewer}) = P(x \le 2) = P(x = 0) + P(x = 1) + P(x = 2) = .0001 + .0007 + .0031 = .0039$

ii. $P(\text{10 or more}) = P(x \ge 10) = P(x = 10) + P(x = 11) + P(x = 12) + \ldots + P(x = 24)$

$= .1241 + .1083 + .0866 + .0640 + .0439 + .0281 + .0168 + .0095 + .0051 + .0026 + .0012$

$+ .0006 + .0002 + .0001 + .0000 = .4911$

5.95 Let x be the number of policies sold by this salesperson on a given day. $\lambda = 1.4$ policies per day

a. $P(\text{none}) = P(x = 0) = \dfrac{\lambda^x e^{-\lambda}}{x!} = \dfrac{(1.4)^0 e^{-1.4}}{0!} = \dfrac{(1)(.24659696)}{1} = .2466$

b.

x	$P(x)$
0	.2466
1	.3452
2	.2417
3	.1128
4	.0395
5	.0111
6	.0026
7	.0005
8	.0001

c. $\mu = \lambda = 1.4$, $\sigma^2 = \lambda = 1.4$, and $\sigma = \sqrt{\lambda} = \sqrt{1.4} = 1.183$

5.97 Let x denote the number of households in a random sample of 50 who own answering machines. $\lambda = 20$ households in 50

a. $P(\text{exactly 25}) = P(x = 25) = \dfrac{\lambda^x e^{-\lambda}}{x!} = \dfrac{(20)^{25} e^{-20}}{25!} = .0446$

b. i. $P(\text{at most 12}) = P(x \le 12) = P(x = 0) + P(x = 1) + P(x = 2) + P(x = 3) + P(x = 4) + P(x = 5)$

$+ P(x = 6) + P(x = 7) + P(x = 8) + P(x = 9) + P(x = 10) + P(x = 11)$

$+ P(x = 12) = .0000 + .0000 + .0000 + .0000 + .0000 + .0001 + .0002 + .0005 + .0013$

$+ .0029 + .0058 + .0106 + .0176 = .0390$

ii. $P(13 \text{ to } 17) = P(13 \le x \le 17) = P(x = 13) + P(x = 14) + P(x = 15) + P(x = 16) + P(x = 17)$

$= .0271 + .0387 + .0516 + .0646 + .0760 = .2580$

iii. $P(\text{at least } 30) = P(x \ge 30) = P(x = 30) + P(x = 31) + P(x = 32) + P(x = 33) + P(x = 34)$

$+ P(x = 35) + P(x = 36) + P(x = 37) + P(x = 38) + P(x = 39)$

$= .0083 + .0054 + .0034 + .0020 + .0012 + .0007 + .0004 + .0002 + .0001 + .0001 = .0218$

Supplementary Exercises

5.99

x	$P(x)$	$xP(x)$	x^2	$x^2P(x)$
2	.05	.10	4	.20
3	.22	.66	9	1.98
4	.40	1.60	16	6.40
5	.23	1.15	25	5.75
6	.10	.60	36	3.60
		$\Sigma xP(x) = 4.11$		$\Sigma x^2P(x) = 17.93$

$\mu = \Sigma xP(x) = 4.11$ cars

$\sigma = \sqrt{\Sigma x^2 P(x) - \mu^2} = \sqrt{17.93 - (4.11)^2} = 1.019$ cars

This mechanic repairs, on average, 4.11 cars per day.

5.101 a.

x	$P(x)$	$xP(x)$	x^2	$x^2P(x)$
-1.2	.17	-.204	1.44	.2448
-.7	.21	-.147	.49	.1029
.9	.37	.333	.81	.2997
2.3	.25	.575	5.29	1.3225
		$\Sigma xP(x) = .557$		$\Sigma x^2P(x) = 1.9699$

b. $\mu = \Sigma xP(x) = \$.557 \text{ million} = \$557{,}000$

$\sigma = \sqrt{\Sigma x^2 P(x) - \mu^2} = \sqrt{1.9699 - (557)^2} = \$1.288274 \text{ million} = \$1{,}288{,}274$

The company has an expected profit of $557,000 for next year.

5.103 Let x denote the number of machines that are broken down at a given time. Assuming machines are independent, x is a binomial random variable with $n = 8$ and $p = .04$.

a. $P(\text{exactly } 8) = P(x = 8) = {}_nC_x p^x q^{n-x} = {}_8C_8 (.04)^8 (.96)^0 = (1)(.000000000007)(1) \approx .0000$

b. $P(\text{exactly } 2) = P(x = 2) = {}_nC_x p^x q^{n-x} = {}_8C_2 (.04)^2 (.96)^6 = (28)(.0016)(.78275779) = .0351$

c. $P(\text{none}) = P(x = 0) = {}_nC_x p^x q^{n-x} = {}_8C_0 (.04)^0 (.96)^8 = (1)(1)(.72138958) = .7214$

5.105 Let x denote the number of defective motors in a random sample of 20. Then x is a binomial random variable with $n = 20$ and $p = .05$.

a. $P(\text{shipment accepted}) = P(x \le 2) = P(x = 0) + P(x = 1) + P(x = 2)$

$= .3585 + .3774 + .1887 = .9246$

b. $P(\text{shipment rejected}) = 1 - P(\text{shipment accepted}) = 1 - .9246 = .0754$

5.107 Let x denote the number of households who own homes in the random sample of 4 households. Then x is a hypergeometric random variable with $N = 15$, $r = 9$, $N - r = 6$ and $n = 4$.

a. $P(\text{exactly 3}) = P(x = 3) = \dfrac{{}_rC_x \; {}_{N-r}C_{n-x}}{{}_NC_n} = \dfrac{{}_9C_3 \; {}_6C_1}{{}_{15}C_4} = \dfrac{(84)(6)}{1365} = .3692$

b. $P(\text{at most 1}) = P(x \le 1) = P(x = 0) + P(x = 1)$

$$= \frac{{}_9C_0 \; {}_6C_4}{{}_{15}C_4} + \frac{{}_9C_1 \; {}_6C_3}{{}_{15}C_4} = \frac{(1)(15)}{1365} + \frac{(9)(20)}{1365} = .0110 + .1319 = .1429$$

c. $P(\text{exactly 4}) = P(x = 4) = \dfrac{{}_rC_x \; {}_{N-r}C_{n-x}}{{}_NC_n} = \dfrac{{}_9C_4 \; {}_6C_0}{{}_{15}C_4} = \dfrac{(126)(1)}{1365} = .0923$

5.109 Let x denote the number of defective parts in a random sample of 4. Then x is a hypergeometric random variable with $N = 16$, $r = 3$, $N - r = 13$, and $n = 4$.

a. $P(\text{shipment accepted}) = P(x \le 1) = P(x = 0) + P(x = 1)$

$$= \frac{{}_3C_0 \; {}_{13}C_4}{{}_{16}C_4} + \frac{{}_3C_1 \; {}_{13}C_3}{{}_{16}C_4} = \frac{(1)(715)}{1820} + \frac{(3)(286)}{1820} = .3929 + .4714 = .8643$$

b. $P(\text{shipment not accepted}) = 1 - P(\text{shipment accepted}) = 1 - .8643 = .1357$

5.111 $\lambda = 7$ cases per day

a. $P(\text{exactly 4}) = P(x = 4) = \dfrac{\lambda^x e^{-\lambda}}{x!} = \dfrac{(7)^4 e^{-7}}{4!} = \dfrac{(2401)(.00091188)}{24} = .0912$

b. i. $P(\text{at least 7}) = P(x \ge 7) = P(x = 7) + P(x = 8) + P(x = 9) + P(x = 10) + P(x = 11)$

$+ P(x = 12) + P(x = 13) + P(x = 14) + P(x = 15) + P(x = 16) + P(x = 17) + P(x = 18)$

$= .1490 + .1304 + .1014 + .0710 + .0452 + .0263 + .0142 + .0071 + .0033 + .0014 + .0006$

$+ .0002 + .0001 = .5502$

ii. $P(\text{at most 3}) = P(x \le 3) = P(x = 0) + P(x = 1) + P(x = 2) + P(x = 3)$

$= .0009 + .0064 + .0223 + .0521 = .0817$

iii. $P(\text{2 to 5}) = P(2 \le x \le 5) = P(x = 2) + P(x = 3) + P(x = 4) + P(x = 5)$

$= .0223 + .0521 + .0912 + .1277 = .2933$

5.113 $\lambda = 1.4$ airplanes per hour

a. $P(\text{none}) = P(x=0) = \frac{\lambda^x e^{-\lambda}}{x!} = \frac{(1.4)^0 e^{-1.4}}{0!} = \frac{(1)(.24659696)}{1} = .2466$

b.

x	$P(x)$
0	.2466
1	.3452
2	.2417
3	.1128
4	.0395
5	.0111
6	.0026
7	.0005
8	.0001

5.115 Let x be a random variable that denotes the gain you have from this game. There are 36 different outcomes for two dice: (1, 1), (1, 2), (1, 3), (1, 4), (1, 5), (1, 6), (2, 1), (2, 2),…, (6, 6).

$P(sum = 2) = P(sum = 12) = \frac{1}{36}$ $\qquad$ $P(sum = 3) = P(sum = 11) = \frac{2}{36}$

$P(sum = 4) = P(sum = 10) = \frac{3}{36}$ $\qquad$ $P(sum = 9) = \frac{4}{36}$

$P(x = 20) = P(\text{you win}) = P(sum = 2) + P(sum = 3) + P(sum = 4) + P(sum = 9) + P(sum = 10) +$

$P(sum = 11) + P(sum = 12) = \frac{1}{36} + \frac{2}{36} + \frac{3}{36} + \frac{4}{36} + \frac{3}{36} + \frac{2}{36} + \frac{1}{36} = \frac{16}{36}$

$P(x = -20) = P(\text{you lose}) = 1 - P(\text{you win}) = 1 - \frac{16}{36} = \frac{20}{36}$

x	$P(x)$	$xP(x)$
20	$\frac{16}{36} = .4444$	8.89
–20	$\frac{20}{36} = .5556$	–11.11
		$\Sigma xP(x) = -2.22$

The value of $\sum xP(x) = -2.22$ indicates that your expected "gain" is –$2.22, so you should not accept this offer. This game is not fair to you since you are expected to lose $2.22.

5.117 a. Team A needs to win four games in a row, each with probability .5, so

P(team A wins the series in four games) = $.5^4 = .0625$

b. In order to win in five games, Team A needs to win 3 of the first four games as well as the fifth game, so P(team A wins the series in five games) = ${}_4C_3(.5)^3(.5)(.5) = .125$.

c. If seven games are required for a team to win the series, then each team needs to win three of the first six games, so P(seven games are required to win the series) = ${}_6C_3(.5)^3(.5)^3 = .3125$.

5.119 a. Let x denote the number of drug deals on this street on a given night. Note that x is discrete. This text has covered two discrete distributions, the binomial and the Poisson. The binomial distribution does not apply here, since there is no fixed number of "trials". However, the Poisson distribution might be appropriate since we have an estimated average number occurrences over a particular interval.

b. To use the Poisson distribution we would have to assume that the drug deals occur randomly and independently.

c. The mean number of drug deals per night is three; if the residents tape for two nights, then $\lambda = (2)(3) = 6$.

P(film at least 5 drug deals) = $P(x \geq 5) = 1 - P(x < 5)$

$= 1 - [P(x = 0) + P(x = 1) + P(x = 2) + P(x = 3) + P(x = 4)]$

$= 1 - (.0025 + .0149 + .0446 + .0892 + .1339) = .7149$

d. Part c. shows that two nights of taping are insufficient, since $P(x \geq 5) = .7149 < .90$. Try taping for three nights; then $\lambda = (3)(3) = 9$.

$P(x \geq 5) = 1 - (.0001 + .0011 + .0050 + .0150 + .0337) = .9451$.

This exceeds the required probability of .90, so the camera should be rented for three nights.

5.121 Let x be the number of sales per day, and let λ be the mean number of cheesecakes sold per day. Here λ=5. We want to find k such that $P(x > k) < .1$. Using the Poisson probability distribution we find that $P(x > 7) = 1 - P(x \leq 7) = 1 - .867 = .133$ and $P(x > 8) = 1 - P(x \leq 8) = 1 - .932 = .068$. So, if the baker wants the probability of losing a sale to be less than .1, he needs to make 8 cheesecakes.

5.123 a. There are ${}_7C_4 = 35$ ways to choose four questions from the set of seven.

b. The teacher must choose both questions that the student did not study, and any two of the remaining five questions. Thus, there are ${}_2C_2\,{}_5C_2 = (1)\,(10) = 10$ ways to choose four questions that include the two that the student did not study.

c. From the answers to parts a and b,

P(the four questions on the test include both questions the student did not study) = 10/35 = .2857.

5.125 Let x_1 = the number of contacts on the first day and x_2 = the number of contacts on the second day. The following table, which may be constructed with the help of a tree diagram, lists the various combinations of contacts during the two days and their probabilities. Note that the probability of each combination is obtained by multiplying the probabilities of the two events included in that combination since the events are independent.

x_1, x_2	Probability	y
(1, 1)	.0144	2
(1, 2)	.0300	3
(1, 3)	.0672	4
(1, 4)	.0084	5
(2, 1)	.0300	3
(2, 2)	.0625	4
(2, 3)	.1400	5
(2. 4)	.0175	6
(3, 1)	.0672	4
(3, 2)	.1400	5
(3, 3)	.3136	6
(3, 4)	.0392	7
(4, 1)	.0084	5
(4, 2)	.0175	6
(4, 3)	.0392	7
(4, 4)	.0049	8

y	$P(y)$
2	.0144
3	.0600
4	.1969
5	.2968
6	.3486
7	.0784
8	.0049

5.127 There are a total of 27 outcomes for the game which can be determined utilizing a tree diagram. Three outcomes are favorable to Player A, 18 outcomes are favorable to Player B, and 6 outcomes are favorable to Player C. Player B's expected winnings are $\Sigma xP(x) = (0)(9/27) + (1)(18/27) = .67$ or 67¢. Player C's expected winnings are also 67¢: $\Sigma xP(x) = (0)(21/27) + (3)(6/27) = .67$. Since Player A has a probability of winning of 3/27, this player should be paid \$6 for winning so that $\Sigma xP(x) = (0)(24/27) + (6)(3/27) = .67$ or 67¢.

5.129 a. .0211285 b. .047539 c. .4225690

Self-Review Test

1. See solution to Exercise 5.1.

2. The probability distribution table. **3.** a **4.** b **5.** a

6. See solution to Exercise 5.49.

7. b **8.** a **9.** b **10.** a **11.** c **13.** a

12. A hypergeometric probability distribution is used to find probabilities for the number of successes in a fixed number of trials, when the trials are not independent (such as sampling without replacement from a finite population.) Example 5-23 is an example of a hypergeometric probability distribution.

14. See solution to Exercise 5.79.

15.

x	$P(x)$	$xP(x)$	x^2	$x^2P(x)$
0	.15	.00	0	.00
1	.24	.24	1	.24
2	.29	.58	4	1.16
3	.14	.42	9	1.26
4	.10	.40	16	1.60
5	.08	.40	25	2.00
		$\sum xP(x) = 2.04$		$\sum x^2P(x) = 6.26$

$\mu = \sum xP(x) = 2.04$ homes

$\sigma = \sqrt{\sum x^2P(x) - \mu^2} = \sqrt{6.26 - (2.04)^2} = 1.449$ homes

The four real estate agents sell an average of 2.04 homes per week.

16. $n = 12$ and $p = .60$

a. i. $P(\text{exactly } 8) = P(x = 8) = {}_nC_x p^x q^{n-x} = {}_{12}C_8(..60)^8(.40)^4 = (495)(.01679616)(.0256) = .2128$

ii. $P(\text{at least } 6) = P(x \geq 6) = P(x = 6) + P(x = 7) + P(x = 8) + P(x = 9) + P(x = 10) + P(x = 11) + P(x = 12) = .1766 + .2270 + .2128 + .1419 + .0639 + .0174 + .0022 = .8418$

iii. $P(\text{less than } 4) = P(x < 4) = P(x = 0) + P(x = 1) + P(x = 2) + P(x = 3) = .0000 + .0003 + .0025 + .0125 = .0153$

b.

x	$P(x)$
0	.0000
1	.0003
2	.0025
3	.0125
4	.0420
5	.1009
6	.1766
7	.2270
8	.2128
9	.1419
10	.0639
11	.0174
12	.0022

$\mu = np = 12(.60) = 7.2$ adults

$\sigma = \sqrt{npq} = \sqrt{12(.60)(.40)} = 1.697$ adults

17. Let x denote the number of females in a sample of 4 volunteers from the 12 nominees. Then x is a hypergeometric random variable with: $N = 12$, $r = 8$, $N - r = 4$ and $n = 4$.

a. $P(\text{exactly } 3) = P(x = 3) = \dfrac{{}_rC_x\ {}_{N-r}C_{n-x}}{{}_NC_n} = \dfrac{{}_8C_3\ {}_4C_1}{{}_{12}C_4} = \dfrac{(56)(4)}{495} = .4525$

b. $P(\text{exactly } 1) = P(x = 1) = \dfrac{{}_rC_x\ {}_{N-r}C_{n-x}}{{}_NC_n} = \dfrac{{}_8C_1\ {}_4C_3}{{}_{12}C_4} = \dfrac{(8)(4)}{495} = .0646$

c. $P(\text{at most one}) = P(x \le 1) = P(x = 0) + P(x = 1)$

$= \dfrac{{}_8C_0\ {}_4C_4}{{}_{12}C_4} + .0646 = \dfrac{(1)(1)}{495} + .0646 = .0020 + .0646 = .0666$

18. $\lambda = 10$ red light runners are caught per day.

Let x = number of drivers caught during rush hour on a given weekday.

a. i. $P(x = 14) = \dfrac{\lambda^x e^{-\lambda}}{x!} = \dfrac{(10)^{14} e^{-10}}{14!} = \dfrac{(100000000000000)(.0000453999)}{87{,}178{,}291{,}200} = .0521$

ii. $P(\text{at most } 7) = P(x \le 7) = P(x = 0) + P(x = 1) + P(x = 2) + P(x = 3) + P(x = 4) + P(x = 5)$
$+ P(x = 6) + P(x = 7) = .0000 + .0005 + .0023 + .0076 + .0189 + .0378 + .0631 + .090$
$= .2203$

iii. $P(13 \text{ to } 18) = P(13 \le x \le 18) = P(x = 13) + P(x = 14) + P(x = 15) + P(x = 16) + P(x = 17)$
$+ P(x = 18) = .0729 + .0521 + .0347 + .0217 + .0128 + .0071 = .2013$

b.

x	$P(x)$
0	.0000
1	.0005
2	.0023
3	.0076
4	.0189
5	.0378
6	.0631
7	.0901
8	.1126
9	.1251
10	.1251
11	.1137
12	.0948
13	.0729
14	.0521
15	.0347
16	.0217
17	.0128
18	.0071
19	.0037
20	.0019
21	.0009
22	.0004
23	.0002
24	.0001

19. See solution to Exercise 5.57.

Chapter Six

Sections 6.1 - 6.3

6.1 The probability distribution of a discrete random variable assigns probabilities to points while that of a continuous random variable assigns probabilities to intervals.

6.3 Since $P(a) = 0$ and $P(b) = 0$ for a continuous random variable, $P(a \leq x \leq b) = P(a < x < b)$.

6.5 The **standard normal distribution** is a special case of the normal distribution. For the standard normal distribution, the value of the mean is equal to zero and the value of the standard deviation is 1. The units of the standard normal distribution curve are denoted by z and are called the *z values* or *z scores*. The *z values* on the right side of the mean (which is zero) are positive and those on the left side are negative. A specific value of z gives the distance between the mean and the point represented by z in terms of the standard deviation.

6.7 As its standard deviation decreases, the width of a normal distribution curve decreases and its height increases.

6.9 For a standard normal distribution, z gives the distance between the mean and the point represented by z in terms of the standard deviation. The *z values* on the right side of the mean are positive and those on the left side are negative.

6.11 Area between $\mu - 1.5\sigma$ and $\mu + 1.5\sigma$ is the area between $z = -1.5$ and $z = 1.5$. Then,
$P(-1.5 < z < 1.5) = P(z < 1.5) - P(z < -1.5) = .9332 - .0668 = .8664$

6.13 Area within 2.5 standard deviations of the mean is:
$P(-2.5 < z < 2.5) = P(z < 2.5) - P(z < -2.5) = .9938 - .0062 = .9876$

6.15
a. $P(0 < z < 1.95) = P(z < 1.95) - P(z < 0) = .9744 - .5000 = .4744$
b. $P(-1.85 < z < 0) = P(z < 0) - P(z < -1.85) = .5000 - .0322 = .4678$

c. $P(1.15 < z < 2.37) = P(z < 2.37) - P(z < 1.15) = .9911 - .8749 = .1162$

d. $P(-2.88 \le z \le -1.53) = P(z \le -1.53) - P(z \le -2.88) = .0630 - .0020 = .0610$

e. $P(-1.67 \le z \le 2.44) = P(z \le 2.44) - P(z \le -1.67) = .9927 - .0475 = .9452$

6.17 a. $P(z > 1.56) = 1 - P(z \le 1.56) = 1 - .9406 = .0594$

b. $P(z < -1.97) = .0244$

c. $P(z > -2.05) = 1 - P(z \le -2.05) = 1 - .0202 = .9798$

d. $P(z < 1.86) = .9686$

6.21 a. $P(-1.83 \le z \le 2.57) = P(z \le 2.57) - P(z \le -1.83) = .9949 - .0336 = .9613$

b. $P(0 \le z \le 2.02) = P(z \le 2.02) - P(z \le 0) = .9783 - .5000 = .4783$

c. $P(-1.99 \le z \le 0) = P(z \le 0) - P(z \le -1.99) = .5000 - .0233 = .4767$

d. $P(z \ge 1.48) = 1 - P(z \le 1.48) = 1 - .9306 = .0694$

6.23 a. $P(z < -2.14) = .0162$

b. $P(.67 \le z \le 2.49) = P(z \le 2.49) - P(z \le .67) = .9936 - .7486 = .2450$

c. $P(-2.07 \le z \le -.93) = P(z \le -.93) - P(z \le -2.07) = .1762 - .0192 = .1570$

d. $P(z < 1.78) = .9625$

6.25 a. $P(z > -.98) = 1 - P(z \le -.98) = 1 - .1635 = .8365$

b. $P(-2.47 \le z \le 1.19) = P(z \le 1.19) - P(z \le -2.47) = .8830 - .0068 = .8762$

c. $P(0 \le z \le 4.25) = P(z \le 4.25) - P(z \le 0) = 1 - .5 = .5$ approximately

d. $P(-5.36 \le z \le 0) = P(z \le 0) - P(z \le -5.36) = .5 - 0 = .5$ approximately

e. $P(z > 6.07) = 1 - P(z \le 6.07) = 1 - 1 = 0$ approximately

f. $P(z < -5.27) = 0$ approximately

Section 6.4

6.27 $\mu = 30$ and $\sigma = 5$

a. $z = (x - \mu)/\sigma = (39 - 30)/5 = 1.80$

b. $z = (x - \mu)/\sigma = (17 - 30)/5 = -2.60$

c. $z = (x - \mu)/\sigma = (22 - 30)/5 = -1.60$

d. $z = (x - \mu)/\sigma = (42 - 30)/5 = 2.40$

6.29 $\mu = 20$ and $\sigma = 4$

a. For $x = 20$: $z = (x - \mu)/\sigma = (20 - 20)/4 = 0$

For $x = 27$: $z = (x - \mu)/\sigma = (27 - 20)/4 = 1.75$

$P(20 < x < 27) = P(0 < z < 1.75) = P(z < 1.75) - P(z < 0) = .9599 - .5000 = .4599$

b. For $x = 23$: $z = (x - \mu)/\sigma = (23 - 20)/4 = .75$

For $x = 25$: $z = (x - \mu)/\sigma = (25 - 20)/4 = 1.25$

$P(23 \le x \le 25) = P(.75 \le z \le 1.25) = P(z \le 1.25) - P(z \le .75) = .8944 - .7734 = .1210$

c. For $x = 9.5$: $z = (x - \mu)/\sigma = (9.5 - 20)/4 = -2.63$

For $x = 17$: $z = (x - \mu)/\sigma = (17 - 20)/4 = -.75$

$P(9.5 < x < 17) = P(-2.63 < z < -.75) = P(z < -.75) - P(z < -2.63) = .2266 - .0043 = .2223$

6.31 $\mu = 55$ and $\sigma = 7$

a. For $x = 58$: $z = (x - \mu)/\sigma = (58 - 55)/7 = .43$

$P(x > 58) = P(z > .43) = 1 - P(z \le .43) = 1 - .6664 = .3336$

b. For $x = 43$: $z = (x - \mu)/\sigma = (43 - 55)/7 = -1.71$

$P(x > 43) = P(z > -1.71) = 1 - P(z \le -1.71) = 1 - .0436 = .9564$

c. For $x = 67$: $z = (x - \mu)/\sigma = (67 - 55)/7 = 1.71$

$P(x < 67) = P(z < 1.71) = .9564$

d. For $x = 24$: $z = (x - \mu)/\sigma = (24 - 55)/7 = -4.43$

$P(x < 24) = P(z < -4.43) = 0$ approximately

6.33 $\mu = 25$ and $\sigma = 6$

a. For $x = 29$: $z = (x - \mu)/\sigma = (29 - 25)/6 = .67$

For $x = 36$: $z = (x - \mu)/\sigma = (36 - 25)/6 = 1.83$

$P(29 < x < 36) = P(.67 < z < 1.83) = P(z < 1.83) - P(z < .67) = .9664 - .7486 = .2178$

b. For $x = 22$: $z = (x - \mu)/\sigma = (22 - 25)/6 = -.50$

For $x = 33$: $z = (x - \mu)/\sigma = (33 - 25)/6 = 1.33$

$P(22 < x < 33) = P(-.50 < z < 1.33) = P(z < 1.33) - P(z < -.50) = .9082 - .3085 = .5997$

6.35 $\mu = 80$ and $\sigma = 12$

a. For $x = 69$: $z = (x - \mu)/\sigma = (69 - 80)/12 = -.92$

$P(x > 69) = P(z > -.92) = 1 - P(z \le -.92) = 1 - .1788 = .8212$

b. For $x = 74$: $z = (x - \mu)/\sigma = (74 - 80)/12 = -.50$

$P(x < 74) = P(z < -.50) = .3085$

c. For $x = 101$: $z = (x - \mu)/\sigma = (101 - 80)/12 = 1.75$

$P(x > 101) = P(z > 1.75) = 1 - P(z \le 1.75) = 1 - .9599 = .0401$

d. For $x = 88$: $z = (x - \mu)/\sigma = (88 - 80)/12 = .67$

$P(x < 88) = P(z < .67) = .7486$

Section 6.5

6.37 $\mu = 190$ minutes and $\sigma = 21$ minutes

a. For $x = 150$: $z = (x - \mu)/\sigma = (150 - 190)/21 = -1.90$

$P(x < 150) = P(z < -1.90) = .0287$

b. For $x = 205$: $z = (x - \mu)/\sigma = (205 - 190)/21 = .71$

For $x = 245$: $z = (x - \mu)/\sigma = (245 - 190)/21 = 2.62$

$P(205 \le x \le 245) = P(.71 \le z \le 2.62) = P(z \le 2.62) - P(z \le .71) = .9956 - .7611 = .2345$

6.39 $\mu = \$810$ and $\sigma = \$155$

a. For $x = 1000$: $z = (x - \mu)/\sigma = (1000 - 810)/155 = 1.23$

$P(x > 1000) = P(z > 1.23) = 1 - P(z \le 1.23) = 1 - .8907 = .1093$

b. For $x = 620$: $z = (x - \mu)/\sigma = (620 - 810)/155 = -1.23$

For $x = 940$: $z = (x - \mu)/\sigma = (940 - 810)/155 = .84$

$P(620 < x < 940) = P(-1.23 < z < .84) = P(z < .84) - P(z < -1.23) = .7995 - .1093 = .6902$

6.41 $\mu = 46$ miles per hour and $\sigma = 4$ miles per hour

a. For $x = 40$: $z = (x - \mu)/\sigma = (40 - 46)/4 = -1.50$

$P(x > 40) = P(z > -1.50) = 1 - P(z \le -1.50) = 1 - .0668 = .9332$ or 93.32%

b. For $x = 50$: $z = (x - \mu)/\sigma = (50 - 46)/4 = 1.00$

For $x = 55$: $z = (x - \mu)/\sigma = (55 - 46)/4 = 2.25$

$P(50 < x < 55) = P(1.00 < z < 2.25) = P(z < 2.25) - P(z < 1.00) = .9878 - .8413 = .1465$ or 14.65%

6.43 $\mu = 48$ minutes and $\sigma = 11$ minutes

a. For $x = 68$: $z = (x - \mu)/\sigma = (68 - 48)/11 = 1.82$

$P(x > 68) = P(z > 1.82) = 1 - P(z \le 1.82) = 1 - .9656 = .0344$

b. For $x = 30$: $z = (x - \mu)/\sigma = (30 - 48)/11 = -1.64$

For $x = 73$: $z = (x - \mu)/\sigma = (73 - 48)/11 = 2.27$

$P(30 < x < 73) = P(-1.64 < z < 2.27) = P(z < 2.27) - P(z < -1.64) = .9884 - .0505 = .9379$ or 93.79%

6.45 $\mu = 1650$ kwh and $\sigma = 320$ kwh

a. For $x = 1850$: $z = (x - \mu)/\sigma = (1850 - 1650)/320 = .63$

$P(x < 1850) = P(z < .63) = .7357$

b. For $x = 900$: $z = (x - \mu)/\sigma = (900 - 1650)/320 = -2.34$

For $x = 1340$: $z = (x - \mu)/\sigma = (1340 - 1650)/320 = -.97$

$P(900 \le z \le 1340) = P(-2.34 \le z \le -.97) = P(z \le -.97) - P(z \le -2.34) = .1660 - .0096 =$ $.1564$ or 15.64%

6.47 $\mu = \$19{,}800$ and $\sigma = \$300$

a. For $x = 19{,}445$: $z = (x - \mu)/\sigma = (19{,}445 - 19{,}800)/300 = -1.18$

$P(x < 19{,}445) = P(z < -1.18) = .1190$ or 11.90%

b. For $x = 20{,}300$: $z = (x - \mu)/\sigma = (20{,}300 - 19{,}800)/300 = 1.67$

$P(x > 20{,}300) = P(z > 1.67) = 1 - P(z \le 1.67) = 1 - .9525 = .0475$ or 4.75%

6.49 $\mu = 231$ minutes and $\sigma = 45$ minutes

a. For $x = 290$: $z = (x - \mu)/\sigma = (290 - 231)/45 = 1.31$

$P(x > 290) = P(z > 1.31) = 1 - P(z \le 1.31) = 1 - .9049 = .0951$ or 9.51%

b. For $x = 150$: $z = (x - \mu)/\sigma = (150 - 231)/45 = -1.80$

$P(x < 150) = P(z < -1.80) = .0359$ or 3.59%

c. For $x = 180$: $z = (x - \mu)/\sigma = (180 - 231)/45 = -1.13$

For $x = 320$: $z = (x - \mu)/\sigma = (320 - 231)/45 = 1.98$

$P(180 \le x \le 320) = P(-1.13 \le z \le 1.98) = P(z \le 1.98) - P(z \le -1.13) = .9761 - .1292 = .8469$ or 84.69%

d. For $x = 270$: $z = (x - \mu)/\sigma = (270 - 231)/45 = .87$

For $x = 350$: $z = (x - \mu)/\sigma = (350 - 231)/45 = 2.64$

$P(270 \le x \le 350) = P(.87 \le z \le 2.64) = P(z \le 2.64) - P(z \le .87) = .9959 - .8078 = .1881$ or 18.81%

6.51 $\mu = 3.0$ inches and $\sigma = .009$ inch

For $x = 2.98$: $z = (x - \mu)/\sigma = (2.98 - 3.0)/.009 = -2.22$

For $x = 3.02$: $z = (x - \mu)/\sigma = (3.02 - 3.0)/.009 = 2.22$

$P(x < 2.98) + P(x > 3.02) = 1 - [P(2.98 \le x \le 3.02)] = 1 - [P(-2.22 \le z \le 2.22)]$

$= 1 - [P(z \le 2.22) - P(z \le -2.22)] = 1 - [.9868 - .0132] = 1 - .9736 = .0264$ or 2.64%

Section 6.6

6.53 a. $z = 2.00$ b. $z = -2.02$ approximately

c. $z = -.37$ approximately d. $z = 1.02$ approximately

6.55 a. $z = 1.65$ approximately b. $z = -1.96$

c. $z = -2.33$ approximately d. $z = 2.58$ approximately

6.57 $\mu = 200$ and $\sigma = 25$

a. $z = .34, x = \mu + z\sigma = 200 + (.34)(25) = 208.50$

b. $z = 1.65, x = \mu + z\sigma = 200 + (1.65)(25) = 241.25$

c. $z = -.86, x = \mu + z\sigma = 200 + (-.86)(25) = 178.50$

d. $z = -2.17, x = \mu + z\sigma = 200 + (-2.17)(25) = 145.75$

e. $z = -1.67, x = \mu + z\sigma = 200 + (-1.67)(25) = 158.25$

f. $z = 2.05, x = \mu + z\sigma = 200 + (2.05)(25) = 251.25$

6.59 $\mu = 15$ minutes and $\sigma = 2.4$ minutes

Let x denote the time to service a randomly chosen car. We are to find x so that the area in the right tail of the normal distribution curve is .05. Thus, $z = 1.65$ and $x = \mu + z\sigma = 15 + (1.65)(2.4) = 18.96$ minutes. The maximum guaranteed waiting time should be approximately 19 minutes.

6.61 $\mu = 1650$ kwh and $\sigma = 320$ kwh

Let x denote the amount of electric consumption during the winter by a randomly selected customer. We are to find x such that the area to the right of x in the normal distribution curve is .90. Thus, $z = 1.28$ and $x = \mu + z\sigma = 1650 + (1.28)(320) = 2059.6$ kwh. Bill Johnson's monthly electric consumption is approximately 2060 kwh.

6.63 $\sigma = \$9.50$ and $P(x \geq 90) = .20$

The area to the left of $x = 90$ is $1 - .20 = .80$ and $z = .84$ approximately. Then, from $x = \mu + z\sigma$ obtain $\mu = x - z\sigma = 90 - (.84)(9.50) = \$82.02$. The mean price of all college textbooks is approximately \$82.

Section 6.7

6.65 The normal distribution may be used as an approximation to a binomial distribution when both $np > 5$ and $nq > 5$.

6.67 a. From Table I of Appendix C, for $n = 25$ and $p = .40$,

$P(8 \leq x \leq 13) = P(x = 8) + P(x = 9) + P(x = 10) + P(x = 11) + P(x = 12) + P(x = 13)$

$= .1200 + .1511 + .1612 + .1465 + .1140 + .0760 = .7688$

b. $\mu = np = 25(.40) = 10$ and $\sigma = \sqrt{npq} = \sqrt{25(.40)(.60)} = 2.44948974$

For $x = 7.5$: $z = (7.5 - 10)/2.44948974 = -1.02$

For $x = 13.5$: $z = (13.5 - 10)/2.44948974 = 1.43$

$P(7.5 \le x \le 13.5) = P(-1.02 \le z \le 1.43) = P(z \le 1.43) - P(z \le -1.02) = .9236 - .1539 = .7697$

The difference between this approximation and the exact probability is $.7697 - .7688 = .0009$

6.69 a. $\mu = np = 120(.60) = 72$ and $\sigma = \sqrt{npq} = \sqrt{120(.60)(.40)} = 5.36656315$

b. For $x = 72.5$: $z = (72.5 - 72)/5.36656315 = .09$

$P(x \le 72.5) = P(z \le .09) = .5359$

c. For $x = 66.5$: $z = (66.5 - 72)/5.36656315 = -1.02$

For $x = 73.5$: $z = (73.5 - 72)/5.36656315 = .28$

$P(66.5 \le x \le 73.5) = P(-1.02 \le z \le .28) = P(z \le .28) - P(z \le -1.02) = .6103 - .1539 = .4564$

6.71 a. $\mu = np = 70(.30) = 21$ and $\sigma = \sqrt{npq} = \sqrt{70(.30)(.70)} = 3.38405790$

For $x = 17.5$: $z = (17.5 - 21)/3.38405790 = -.91$

For $x = 18.5$: $z = (18.5 - 21)/3.38405790 = -.65$

$P(17.5 \le x \le 18.5) = P(-.91 \le z \le -.65) = P(z \le -.65) - P(z \le -.91) = .2578 - .1814 = .0764$

b. $\mu = np = 200(.70) = 140$ and $\sigma = \sqrt{npq} = \sqrt{200(.70)(.30)} = 6.48074070$

For $x = 132.5$: $z = (132.5 - 140)/6.48074070 = -1.16$

For $x = 145.5$: $z = (145.5 - 140)/6.48074070 = .85$

$P(132.5 \le x \le 145.5) = P(-1.16 \le z \le .85) = P(z \le .85) - P(z \le -1.16) = .8023 - .1230$
$= .6793$

c. $\mu = np = 85(.40) = 34$ and $\sigma = \sqrt{npq} = \sqrt{85(.40)(.60)} = 4.51663592$

For $x = 29.5$: $z = (29.5 - 34)/4.51663592 = -1.00$

$P(x \ge 29.5) = P(z \ge -1.00) = 1 - P(z \le -1.00) = 1 - .1587 = .8413$

d. $\mu = np = 150(.38) = 57$ and $\sigma = \sqrt{npq} = \sqrt{150(.38)(.62)} = 5.94474558$

For $x = 62.5$: $z = (62.5 - 57)/5.94474558 = .93$

$P(x \le 62.5) = P(z \le .93) = .8238$

6.73 $\mu = np = 300(.34) = 102$ and $\sigma = \sqrt{npq} = \sqrt{300(.34)(.66)} = 8.20487660$

For $x = 83.5$: $z = (83.5 - 102)/8.20487660 = -2.25$

For $x = 94.5$: $z = (94.5 - 102)/8.20487660 = -.91$

$P(83.5 \le x \le 94.5) = P(-2.25 \le z \le -.91) = P(z \le -.91) - P(z \le -2.25) = .1814 - .0122 = .1692$

6.75 $\mu = np = 400(.43) = 172$ and $\sigma = \sqrt{npq} = \sqrt{400(.43)(.57)} = 9.90151504$

a. For $x = 174.5$: $z = (174.5 - 172)/9.90151504 = .25$

For $x = 175.5$: $z = (175.5 - 172)/9.90151504 = .35$

$P(174.5 \le x \le 175.5) = P(.25 \le z \le .35) = P(z \le .35) - P(z \le .25) = .6368 - .5987 = .0381$

b. For $x = 184.5$: $z = (184.5 - 172)/9.90151504 = 1.26$

$P(x \ge 184.5) = P(z \ge 1.26) = 1 - P(z \le 1.26) = 1 - .8961 = .1039$

c. For $x = 159.5$: $z = (159.5 - 172)/9.90151504 = -1.26$

For $x = 180.5$: $z = (180.5 - 172)/9.90151504 = .86$

$P(159.5 \le x \le 180.5) = P(-1.26 \le z \le .86) = P(z \le .86) - P(z \le -1.26) = .8051 - .1038$

$= .7013$

6.77 $\mu = np = 100(.80) = 80$ and $\sigma = \sqrt{npq} = \sqrt{100(.80)(.20)} = 4$

a. For $x = 74.5$: $z = (74.5 - 80)/4 = -1.38$

For $x = 75.5$: $z = (75.5 - 80)/4 = -1.13$

$P(74.5 \le x \le 75.5) = P(-1.38 \le z \le -1.13) = P(z \le -1.13) - P(z \le -1.38) = .1292 - .0838$

$= .0454$

b. For $x = 73.5$: $z = (73.5 - 80)/4 = -1.63$

$P(x \le 73.5) = P(z \le -1.63) = .0516$

c. For $x = 73.5$: $z = (73.5 - 80)/4 = -1.63$

For $x = 85.5$: $z = (85.5 - 80)/4 = 1.38$

$P(73.5 \le x \le 85.5) = P(-1.63 \le z \le 1.38) = P(z \le 1.38) - P(z \le -1.63) = .9162 - .0516$

$= .8646$

6.79 $\mu = np = 100(.05) = 5$ and $\sigma = \sqrt{npq} = \sqrt{100(.05)(.95)} = 2.17944947$

a. For $x = 6.5$: $z = (6.5 - 5)/2.17944947 = .69$

P(shipment is accepted) $= P(x \le 6.5) = P(z \le .69) = .7549$

b. P(shipment is not accepted) $= 1 - P$(shipment is accepted) $= 1 - .7549 = .2451$

Supplementary Exercises

6.81 $\mu = \$87$ and $\sigma = \$22$

a. For $x = 114$: $z = (114 - 87)/22 = 1.23$

$P(x > 114) = P(z > 1.23) = 1 - P(z \le 1.23) = 1 - .8907 = .1093$

b. For $x = 40$: $z = (40 - 87)/22 = -2.14$

For $x = 60$: $z = (60 - 87)/22 = -1.23$

$P(40 < x < 60) = P(-2.14 < z < -1.23) = P(z \leq -1.23) - P(z \leq -2.14) = .1093 - .0162 = .0931$ or 9.31%

c. For $x = 70$: $z = (70 - 87)/22 = -.77$

For $x = 105$: $z = (105 - 87)/22 = .82$

$P(70 < x < 105) = P(-.77 < z < .82) = P(z \leq .82) - P(z \leq -.77) = .7939 - .2206 = .5733$ or 57.33%

d. For $x = 185$: $z = (185 - 87)/22 = 4.45$

$P(x > 185) = P(z > 4.45) = 1 - P(z \leq 4.45) = 1 - 1 = 0$ approximately

Although it is possible for a customer to write a check for more than $185, the probability of this is very close to zero.

6.83 $\mu = 50$ inches and $\sigma = .06$

For $x = 49.85$: $z = (49.85 - 50)/.06 = -2.50$

For $x = 50.15$: $z = (50.15 - 50)/.06 = 2.50$

$P(x < 49.85) + P(x > 50.15) = 1 - [P(49.85 \leq x \leq 50.15)] = 1 - [P(-2.50 \leq z \leq 2.50)]$

$= 1 - [P(z \leq 2.50) - P(z \leq -2.50)] = 1 - [.9938 - .0062] = 1 - .9876 = .0124$ or 1.24%

6.85 $\mu = 750$ hours and $\sigma = 50$ hours

a. The area in the right tail of the normal distribution curve is given to be .025, which gives $z = 1.96$. Then,

$x = \mu + z\sigma = 750 + (1.96)(50) = 848$ hours

b. Area to the left of x is .80, which gives $z = .84$ approximately. Then,

$x = \mu + z\sigma = 750 + (.84)(50) = 792$ hours

6.87 $\mu = np = 100(.80) = 80$ and $\sigma = \sqrt{npq} = \sqrt{100(.80)(.20)} = 4$

a. For $x = 84.5$: $z = (84.5 - 80)/4 = 1.13$

For $x = 85.5$: $z = (85.5 - 80)/4 = 1.38$

$P(84.5 \leq x \leq 85.5) = P(1.13 \leq z \leq 1.38) = P(z \leq 1.38) - P(z \leq 1.13) = .9162 - .8708 = .0454$

b. For $x = 74.5$: $z = (74.5 - 80)/4 = -1.38$

$P(x \leq 74.5) = P(z \leq -1.38) = .0838$

c. For $x = 74.5$: $z = (74.5 - 80)/4 = -1.38$

For $x = 87.5$: $z = (87.5 - 80)/4 = 1.88$

$P(74.5 \leq x \leq 87.5) = P(-1.38 \leq z \leq 1.88) = P(z \leq 1.88) - P(z \leq -1.38) = .9699 - .0838$

$= .8861$

d. For $x = 71.5$: $z = (71.5 - 80)/4 = -2.13$

For $x = 77.5$: $z = (77.5 - 80) / 4 = -.63$

$P(71.5 \le x \le 77.5) = P(-2.13 \le z \le -.63) = P(z \le -.63) - P(z \le -2.13) = .2643 - .0166$
$= .2477$

6.89 $\sigma = \$350$ and $P(x > 2500) = .15$

The area to the left of $x = 2500$ is $1 - .15 = .85$ and $z = 1.04$ approximately. Then, $\mu = x - z\sigma$ $= 2500 - (1.04)(350) = \$2136$. Thus, the mean month mortgage is approximately \$2136.

6.91 a. If $\$3500 = \$1000 + \frac{\$12}{\text{ft}} \cdot x$, where x is depth in feet, then $x = 208.33$. Hence, Company B charges more for depths of more than 208.33 ft.

$\mu = 250$ and $\sigma = 40$

For $x = 208.33$: $z = (208.33 - 250)/40 = -1.04$

$P(x > 208.33) = P(z > -1.04) = 1 - P(z \le -1.04) = 1 - .1492 = .8508$

The probability that Company B charges more than Company A to drill a well is .8508.

b. $\mu = 250$, so the mean amount charged by Company B is $\$1000 + \frac{\$12}{\text{ft}} \cdot 250 \text{ ft} = \$4000$.

6.93 $\mu = 45{,}000$ and $\sigma = 2000$

First, we find the probability that one tire lasts at least 46,000 miles.

For $x = 46{,}000$: $z = (46{,}000 - 45{,}000)/2000 = .50$

$P(x \ge 46{,}000) = P(z \ge .50) = 1 - P(z \le .50) = 1 - .6915 = .3085$

So, the probability of one tire lasting at least 46,000 miles is .3085. Then,

$P(\text{all four tires last more than 46,000 miles}) = (.3085)^4 = .0091$.

6.95 $\mu = 0$ and $\sigma = 2$ mph

a. Let x be the error of these estimates in mph.

For $x = 5$: $z = (5 - 0)/2 = 2.50$

$P(x \ge 5) = P(z \ge 2.50) = 1 - P(z \le 2.50) = 1 - .9938 = .0062$

b. We are given that the area to the left of x is .99, which gives $z = 2.33$ approximately. Then, $x = \mu + z\sigma = 0 + (2.33)(2) = 4.66$ mph ≈ 5 mph. So, the minimum estimate of speed at which a car should be cited for speeding is $60 + 5 = 65$ mph.

6.97 $\sigma = .07$ ounce and $P(x \geq 8) = .99$

The area to the left of $x = 8$ is $1 - .99 = .01$ and $z = -2.33$ approximately. Then, $\mu = x - z\sigma$ $= 8 - (-2.33)(.07) = 8.16$ ounces. Thus, the mean should be set at approximately 8.16 ounces.

6.99 Company A: $\mu = 8$ mm and $\sigma = .15$ mm

For $x = 7.8$: $z = (7.8 - 8)/.15 = -1.33$

For $x = 8.2$: $z = (8.2 - 8)/.15 = 1.33$

$P(7.8 \leq x \leq 8.2) = P(-1.33 \leq z \leq 1.33) = P(z \leq 1.33) - P(z \leq -1.33) = .9082 - .0918 = .8164$

Price per usable rod = $400/(.8164 \times 10{,}000) \approx$ \$0.048996

Company B: $\mu = 8$ mm and $\sigma = .12$ mm

For $x = 7.8$: $z = (7.8 - 8) / .12 = -1.67$

For $x = 8.2$: $z = (8.2 - 8) / .12 = 1.67$

$P(7.8 \leq x \leq 8.2) = P(-1.67 \leq z \leq 1.67) = P(z \leq 1.67) - P(z \leq -1.67) = .9525 - .0475 = .9050$

Price per usable rod = 460/(.9050 x 10,000) ≈ \$0.050829

Hence, the Alpha Corporation should choose Company A as a supplier.

6.101 a. Let x = number of ticket holders who show up.

Then, x is a binomial random variable with $n = 65$ and $p = 1 - .10 = .90$

$\mu = np = 65(.90) = 58.5$ and $\sigma = \sqrt{npq} = \sqrt{65(.90)(.10)} = 2.41867732$

Using a normal approximation with correction for continuity:

For $x = 60.5$: $z = (60.5 - 58.5)/2.41867732 = .83$

$P(x \leq 60.5) = P(z \leq .83) = .7967$

b. Let n = number of tickets sold

Since μ and σ both depend on n, it is not easy to solve for n directly. Instead, we will use trial and error.

In part a, $n = 65$ was too large, since $P(x \leq 60.5) = P(z \leq .83) = .7967 < .95$.

Try $n = 62$: $\mu = np = 62(.90) = 55.8$ and $\sigma = \sqrt{npq} = \sqrt{62(.90)(.10)} = 2.36220236$

For $x = 60.5$: $z = (60.5 - 55.8)/2.36220236 = 1.99$

$P(x \leq 60.5) = P(z \leq 1.99) = .9767 > .95$.

Thus, $n = 62$ satisfies the requirement.

To see if n may be increased, try $n = 63$.

$\mu = np = 63(.90) = 56.7$ and $\sigma = \sqrt{npq} = \sqrt{63(.90)(.10)} = 2.38117618$

For $x = 60.5$: $z = (60.5 - 56.7)/2.38117618 = 1.60$

$P(x \leq 60.5) = P(z \leq 1.60) = .9452 < .95$

Thus, $n = 63$ is too large.

Therefore, the largest number of tickets the company can sell and be at least 95% sure that the bus can hold all ticket holders who show up is 62.

6.103 $\mu = np = 15(.02) = .30$ and $\sigma = \sqrt{npq} = \sqrt{15(.02)(.98)} = .54221767$

Since $np < 5$, the normal approximation to the binomial is not appropriate. The Empirical Rule requires a bell-shaped distribution, and this distribution is skewed right. By the Empirical Rule, approximately 68% of the observations fall in the interval $\mu \pm \sigma$, approximately 95% fall in the interval $\mu \pm 2\sigma$, and about 99.7% fall in the interval $\mu \pm 3\sigma$. These intervals are −.24 to .84, −.78 to 1.38, and −1.33 to 1.93, respectively. Using the normal approximation with continuity correction,

For $x = -.74$: $z = (-.74 - .3)/.54221767 = -1.92$

For $x = 1.34$: $z = (1.34 - .3)/.54221767 = 1.92$

$P(-.74 \le x \le 1.34) = P(-1.92 \le z \le 1.92) = P(z \le 1.92) - P(z \le -1.92) = .9726 - .0274$

$= .9452 > .68$

For $x = -1.28$: $z = (-1.28 - .3)/.54221767 = -2.91$

For $x = 1.88$ $z = (1.88 - .3)/.54221767 = 2.91$

$P(-1.28 \le x \le 1.88) = P(-2.91 \le z \le 2.91) = P(z \le 2.91) - P(z \le -2.91) = .9982 - .0018 =$

$.9964 > .95$

For $x = -1.83$: $z = (-1.83 - .3)/.54221767 = -3.93$

For $x = 2.43$: $z = (2.43 - .3)/.54221767 = 3.93$

$P(-1.83 \le x \le 2.43) = P(-3.93 \le z \le 3.93) = P(z \le 3.93) - P(z \le -3.93) = 1 - 0 \approx 1 > .997$

6.105 $\mu = 8$ and $P(x > 8.2) = .03$. Then, $z = 1.88$ approximately. Now $x = \mu + z\sigma$, so

$\sigma = (x - \mu)/z = (8.2 - 8)/(1.88) = .106$.

Self-Review Test

1. a **2.** a **3.** d **4.** b **5.** a **6.** c **7.** b **8.** b

9.
a. $P(.85 \le z \le 2.33) = P(z \le 2.33) - P(z \le .85) = .9901 - .8023 = .1878$
b. $P(-2.97 \le z \le 1.49) = P(z \le 1.49) - P(z \le -2.97) = .9319 - .0015 = .9304$
c. $P(z \le -1.29) = .0985$
d. $P(z > -.74) = 1 - P(z \le -.74) = 1 - .2296 = .7704$

10. a. $z = -1.28$ approximately b. $z = .61$

c. $z = 1.65$ approximately d. $z = -1.07$ approximately

11. $\mu = 20.9$ minutes and $\sigma = 5$ minutes

a. For $x = 15$: $z = (15 - 20.9)/5 = -1.18$

For $x = 30$: $z = (30 - 20.9)/5 = 1.82$

$P(15 < x < 30) = P(-1.18 < z < 1.82) = P(z < 1.82) - P(z < -1.18) = .9656 - .1190 = .8466$

b. For $x = 13$: $z = (13 - 20.9)/5 = -1.58$

$P(x < 13) = P(z < -1.58) = .0571$

c. For $x = 26$: $z = (26 - 20.9)/5 = 1.02$

$P(x > 26) = P(z > 1.02) = 1 - P(z \leq 1.02) = 1 - .8461 = .1539$

d. For $x = 30$: $z = (30 - 20.9)/5 = 1.82$

For $x = 35$: $z = (35 - 20.9)/5 = 2.82$

$P(30 < x < 35) = P(1.82 < z < 2.82) = P(z < 2.82) - P(z < 1.82) = .9976 - .9656 = .0320$

12. $\mu = 20.9$ minutes and $\sigma = 5$ minutes

a. For .10 area in the left tail of the normal distribution curve, $z \approx -1.28$. Then,

$x = \mu + z\sigma = 20.9 + (-1.28)(5) = 14.5$ minutes

b. For .05 area in the right tail of the normal distribution curve, $z \approx 1.65$. Then,

$x = \mu + z\sigma = 20.9 + (1.65)(5) = 29.15$ minutes

13. $\mu = np = 500(.70) = 350$ and $\sigma = \sqrt{npq} = \sqrt{500(.70)(.30)} = 10.2469508$

a. i. For $x = 344.5$: $z = (344.5 - 350)/10.2469508 = -.54$

For $x = 345.5$: $z = (345.5 - 350)/10.2469508 = -.44$

$P(344.5 \leq x \leq 345.5) = P(-.54 \leq z \leq -.44) = P(z \leq -.44) - P(z \leq -.54) = .3300 - .2946$
$= .0354$

ii. For $x = 337.5$: $z = (337.5 - 350)/10.2469508 = -1.22$

For $x = 370.5$: $z = (370.5 - 350)/10.2469508 = 2.00$

$P(337.5 \leq x \leq 370.5) = P(-1.22 \leq z \leq 2.00) = P(z \leq 2.00) - P(z \leq -1.22) = .9772 - .1112$
$= .8660$

iii. For $x = 342.5$: $z = (342.5 - 350)/10.2469508 = -.73$

$P(x \leq 342.5) = P(z \leq -.73) = .2327$

iv. For $x = 334.5$: $z = (334.5 - 350)/10.2469508 = -1.51$

$P(x \geq 334.5) = P(z \geq -1.51) = 1 - P(z \leq -1.51) = 1 - .0655 = .9345$

v. For $x = 321.5$: $z = (321.5 - 350)/10.2469508 = -2.78$

For $x = 340.5$: $z = (340.5 - 350)/10.2469508 = -.93$

$P(321.5 \leq x \leq 340.5) = P(-2.78 \leq z \leq -.93) = P(z \leq -.93) - P(z \leq -2.78) = .1762 - .0027$
$= .1735$

b. P(at most 165 of the 500 have not gambled) = P(at least 335 have gambled) = $P(x \geq 335)$

For $x = 334.5$: $z = (334.5 - 350)/10.2469508 = -1.51$

$P(x \geq 334.5) = P(z \geq -1.51) = 1 - P(z \leq -1.51) = 1 - .0655 = .9345$

c. P(between 155 and 175 of the 500 have not gambled) =

P(between 325 and 345 have gambled) = $P(324.5 \leq x \leq 345.5)$

For $x = 324.5$: $z = (324.5 - 350)/10.2469508 = -2.49$

For $x = 345.5$: $z = (345.5 - 350)/10.2469508 = -.44$

$P(324.5 \leq x \leq 345.5) = P(-2.49 \leq z \leq -.44) = P(z \leq -.44) - P(z \leq -2.49) = .3300 - .0064$
$= .3236$

Chapter Seven

Sections 7.1 - 7.2

7.1 The probability distribution of the population data is called the **population distribution**. Table 7.2 on page 297 of the text provides an example of such a distribution. The probability distribution of a sample statistic is called its **sampling distribution**. Table 7.5 on page 299 of the text provides an example of the sampling distribution of the sample mean.

7.3 **Nonsampling errors** are errors that may occur in the collection, recording, and tabulation of data. Example 7–1 on page 301 of the text exhibits nonsampling error. Nonsampling errors can occur in both a sample survey and a census.

7.5

a. $\mu = (20 + 25 + 13 + 19 + 9 + 15 + 11 + 7 + 17 + 30)/10 = 166/10 = 16.60$

b. $\bar{x} = (20 + 25 + 13 + 9 + 15 + 11 + 7 + 17 + 30)/9 = 147/9 = 16.33$

Sampling error $= \bar{x} - \mu = 16.33 - 16.60 = -.27$

c. Rich's incorrect $\bar{x} = (20 + 25 + 13 + 9 + 15 + 11 + 17 + 17 + 30)/9 = 157/9 = 17.44$

$\bar{x} - \mu = 17.44 - 16.60 = .84$

Sampling error (from part b) = –.27

Nonsampling error = .84 – (–.27) = 1.11

d.

Sample	$\bar{x}$	$\bar{x} - \mu$
25, 13 19, 9,15,11,7,17,30	16.22	–.38
20, 13, 19, 9, 15, 11, 7, 17, 30	15.67	–.93
20, 25 19, 9, 15, 11, 7, 17, 30	17.00	.40
20, 25, 13, 9, 15, 11, 7, 17, 30	16.33	–.27
20, 25, 13, 19, 15, 11, 7, 17, 30	17.44	.84
20, 25, 13, 19, 9, 11, 7, 17, 30	16.78	.18
20, 25, 13, 19, 9, 15, 7, 17, 30	17.22	.62
20, 25, 13, 19, 9, 15, 11, 17, 30	17.67	1.07
20, 25, 13, 19, 9, 15, 11, 7, 30	16.56	–.04
20, 25, 13, 19, 9, 15, 11, 7, 17	15.11	–1.49

7.7 a.

x	$P(x)$
15	1/6=.167
21	1/6=.167
25	1/6=.167
28	1/6=.167
53	1/6=.167
55	1/6=.167

b.

Sample	$\bar{x}$
55, 53, 28, 25, 21	36.4
55, 53, 28, 25, 15	35.2
55, 53, 28, 21, 15	34.4
55, 53, 25, 21, 15	33.8
55, 28, 25, 21, 15	28.8
53, 28, 25, 21, 15	28.4

$\bar{x}$	$P(\bar{x})$
28.4	1/6=.167
28.8	1/6=.167
33.8	1/6=.167
34.4	1/6=.167
35.2	1/6=.167
36.4	1/6=.167

c. The mean for the population data is $\mu = (55 + 53 + 28 + 25 + 21 + 15)/6 = 197/6 = 32.83$
Suppose the random sample of five family members includes the observations: 55, 28, 25, 21, and 15. The mean for this sample is $\bar{x} = (55 + 28 + 25 + 21 + 15)/5 = 144/5 = 28.80$
Sampling error $= \bar{x} - \mu = 28.80 - 32.83 = -4.03$

Section 7.3

7.9 a. Mean of $\bar{x} = \mu_{\bar{x}} = \mu$

b. Standard deviation of $\bar{x} = \sigma_{\bar{x}} = \sigma/\sqrt{n}$ where σ = population standard deviation and n = sample size.

7.11 An estimator is **consistent** when its standard deviation decreases as the sample size is increased. The sample mean $\bar{x}$ is a consistent estimator of μ because its standard deviation decreases as the sample size increases. As n increases, $\sqrt{n}$ increases, and, consequently, the value of $\sigma_{\bar{x}} = \sigma/\sqrt{n}$ decreases.

7.13 $\mu = 60$ and $\sigma = 10$

a. $\mu_{\bar{x}} = \mu = 60$ and $\sigma_{\bar{x}} = \sigma/\sqrt{n} = 10/\sqrt{18} = 2.357$

b. $\mu_{\bar{x}} = \mu = 60$ and $\sigma_{\bar{x}} = \sigma/\sqrt{n} = 10/\sqrt{90} = 1.054$

7.15 a. $n/N = 300/5000 = .06 > .05$

$$\sigma_{\bar{x}} = \frac{\sigma}{\sqrt{n}}\sqrt{\frac{N-n}{N-1}} = \frac{25}{\sqrt{300}}\sqrt{\frac{5000-300}{5000-1}} = 1.400$$

b. Since $n/N = 100/5000 = .02 < .05$, $\sigma_{\bar{x}} = \sigma/\sqrt{n} = 25/\sqrt{100} = 2.500$

7.17 $\mu = 125$ and $\sigma = 36$

a. Since $\sigma_{\bar{x}} = \sigma/\sqrt{n}$, $n = (\sigma/\sigma_{\bar{x}})^2 = (36/3.6)^2 = 100$

b. $n = (\sigma/\sigma_{\bar{x}})^2 = (36/2.25)^2 = 256$

7.19 $\mu = 3$ hours, $\sigma = .8$ hour, and $n = 75$

$\mu_{\bar{x}} = \mu = 3$ hours and $\sigma_{\bar{x}} = \sigma/\sqrt{n} = .8/\sqrt{75} = .092$ hour

7.21 $\mu = \$320$, $\sigma = \$72$, and $n = 25$

$\mu_{\bar{x}} = \mu = \$320$ and $\sigma_{\bar{x}} = \sigma/\sqrt{n} = 72/\sqrt{25} = \$14.40$

7.23 $\sigma = \$605$ and $\sigma_{\bar{x}} = \$55$

$n = (\sigma/\sigma_{\bar{x}})^2 = (605/55)^2 = 121$ players

7.25 a.

$\bar{x}$	$P(\bar{x})$	$\bar{x}P(\bar{x})$	$\bar{x}^2$	$\bar{x}^2P(\bar{x})$
76.00	.20	15.200	5776.0000	1155.200
76.67	.10	7.667	5878.2889	587.829
79.33	.10	7.933	6293.2489	629.325
81.00	.10	8.100	6561.0000	656.100
81.67	.20	16.334	6669.9889	1333.998
84.33	.20	16.866	7111.5489	1422.310
85.00	.10	8.500	7225.0000	722.500
		$\sum\bar{x}P(\bar{x}) = 80.60$		$\sum\bar{x}^2P(\bar{x}) = 6507.262$

$\sum\bar{x}P(\bar{x}) = 80.60$ is the same value found in Exercise 7.6 for μ.

b. $\sigma_{\bar{x}} = \sqrt{\sum\bar{x}^2P(\bar{x}) - \mu_{\bar{x}}^2} = \sqrt{6507.262 - (80.60)^2} = 3.302$

c. $\sigma/\sqrt{n} = 8.09/\sqrt{3} = 4.67$ is not equal to $\sigma_{\bar{x}} = 3.30$ in this case because $n/N = 3/5 = .60 > .05$.

d. $\sigma_{\bar{x}} = \dfrac{\sigma}{\sqrt{n}}\sqrt{\dfrac{N-n}{N-1}} = \dfrac{8.09}{\sqrt{3}}\sqrt{\dfrac{5-3}{5-1}} = 3.302$

Section 7.4

7.27 The **central limit theorem** states that for a large sample, the sampling distribution of the sample mean is approximately normal, irrespective of the shape of the population distribution. Furthermore, $\mu_{\bar{x}} = \mu$ and $\sigma_{\bar{x}} = \sigma/\sqrt{n}$, where μ and σ are the population mean and standard deviation, respectively. A sample size of 30 or more is considered large enough to apply the central limit theorem to $\bar{x}$.

7.29
a. Slightly skewed to the right
b Approximately normal because $n > 30$ and the central limit theorem applies
c. Close to normal with a slight skew to the right

7.31 a. and b. In both cases the sampling distribution of $\bar{x}$ would be normal because the population distribution is normal.

7.33 $\mu = 46$ mph, $\sigma = 3$ mph, and $n = 20$

$\mu_{\bar{x}} = \mu = 46$ mph and $\sigma_{\bar{x}} = \sigma/\sqrt{n} = 3/\sqrt{20} = .671$ mph

The sampling distribution of $\bar{x}$ is normal because the population is normally distributed.

7.35 $\mu = 3.02$, $\sigma = .29$, $N = 5540$ and $n = 48$

$\mu_{\bar{x}} = \mu = 3.02$

Since $n/N = 48/5540 = .009 < .05$, $\sigma_{\bar{x}} = \sigma/\sqrt{n} = .29/\sqrt{48} = .042$.

The sampling distribution of $\bar{x}$ is approximately normal because the population is approximately normally distributed.

7.37 $\mu = \$96$, $\sigma = \$27$, and $n = 90$

$\mu_{\bar{x}} = \mu = \$96$ and $\sigma_{\bar{x}} = \sigma/\sqrt{n} = 27/\sqrt{90} = \$2.846$

The sampling distribution of $\bar{x}$ is approximately normal because the sample size is large ($n > 30$).

7.39 $\mu = 62$ minutes, $\sigma = 14$ minutes, and $n = 400$

$\mu_{\bar{x}} = \mu = 62$ minutes and $\sigma_{\bar{x}} = \sigma/\sqrt{n} = 14/\sqrt{400} = .7$ minute

The sampling distribution of $\bar{x}$ is approximately normal because the sample size is large ($n > 30$).

Section 7.5

7.41 $P(\mu - 1.50\sigma_{\bar{x}} \le \bar{x} \le \mu + 1.50\sigma_{\bar{x}}) = P(-1.50 \le z \le 1.50) = P(z \le 1.50) - P(z \le -1.50) = .9332 - .0668 = .8664$ or 86.64%.

7.43 $\mu = 66$, $\sigma = 7$, $N = 205{,}000$, and $n = 49$

$\mu_{\bar{x}} = \mu = 66$

Since $n/N = 49/205{,}000 = .0002 < .05$, $\sigma_{\bar{x}} = \sigma/\sqrt{n} = 7/\sqrt{49} = 1$

a. $z = (\bar{x} - \mu)/\sigma_{\bar{x}} = (68.44 - 66)/1 = 2.44$

b. $z = (\bar{x} - \mu)/\sigma_{\bar{x}} = (58.75 - 66)/1 = -7.25$

c. $z = (\bar{x} - \mu)/\sigma_{\bar{x}} = (62.35 - 66)/1 = -3.65$

d. $z = (\bar{x} - \mu)/\sigma_{\bar{x}} = (71.82 - 66)/1 = 5.82$

7.45 $\mu = 48$, $\sigma = 8$, and $n = 16$

$\mu_{\bar{x}} = \mu = 48$ and $\sigma_{\bar{x}} = \sigma/\sqrt{n} = 8/\sqrt{16} = 2$

a. For $\bar{x} = 49.6$: $z = (\bar{x} - \mu)/\sigma_{\bar{x}} = (49.6 - 48)/2 = .80$

For $\bar{x} = 52.2$: $z = (\bar{x} - \mu)/\sigma_{\bar{x}} = (52.2 - 48)/2 = 2.10$

$P(49.6 < \bar{x} < 52.2) = P(.80 < z < 2.10) = P(z < 2.10) - P(z < .80) = .9821 - .7881 = .1940$

b. For $\bar{x} = 45.7$: $z = (\bar{x} - \mu)/\sigma_{\bar{x}} = (45.7 - 48)/2 = -1.15$

$P(\bar{x} > 45.7) = P(z > -1.15) = 1 - P(z \le -1.15) = 1 - .1251 = .8749$

7.47 $\mu = 90$, $\sigma = 18$, and $n = 64$

$\mu_{\bar{x}} = \mu = 90$ and $\sigma_{\bar{x}} = \sigma/\sqrt{n} = 18/\sqrt{64} = 2.25$

a. For $\bar{x} = 82.3$: $z = (\bar{x} - \mu)/\sigma_{\bar{x}} = (82.3 - 90)/2.25 = -3.42$

$P(\bar{x} < 82.3) = P(z < -3.42) = .0003$

b. For $\bar{x} = 86.7$: $z = (\bar{x} - \mu)/\sigma_{\bar{x}} = (86.7 - 90)/2.25 = -1.47$

$P(\bar{x} > 86.7) = P(z > -1.47) = 1 - P(z \le -1.47) = 1 - .0708 = .9292$

7.49 $\mu = 3.02$, $\sigma = .29$, and $n = 20$

$\mu_{\bar{x}} = \mu = 3.02$ and $\sigma_{\bar{x}} = \sigma/\sqrt{n} = .29/\sqrt{20} = .06484597$

a. For $\bar{x} = 3.10$: $z = (\bar{x} - \mu)/\sigma_{\bar{x}} = (3.10 - 3.02)/.06484597 = 1.23$

$P(\bar{x} \geq 3.10) = P(z \geq 1.23) = 1 - P(z \leq 1.23) = 1 - .8907 = .1093$

b. For $\bar{x} = 2.90$: $z = (\bar{x} - \mu)/\sigma_{\bar{x}} = (2.90 - 3.02)/.06484597 = -1.85$

$P(\bar{x} \leq 2.90) = P(z \leq -1.85) = .0322$

a. For $\bar{x} = 2.95$: $z = (\bar{x} - \mu)/\sigma_{\bar{x}} = (2.95 - 3.02)/.06484597 = -1.08$

For $\bar{x} = 3.11$: $z = (\bar{x} - \mu)/\sigma_{\bar{x}} = (3.11 - 3.02)/.06484597 = 1.39$

$P(2.95 \leq \bar{x} \leq 3.11) = P(-1.08 \leq z \leq 1.39) = P(z \leq 1.39) - P(z \leq -1.08) = .9177 - .1401 =$ $.7776$

7.51 $\mu = \$16,842$, $\sigma = \$3600$, and $n = 225$

$\mu_{\bar{x}} = \mu = \$16,842$ and $\sigma_{\bar{x}} = \sigma/\sqrt{n} = 3600/\sqrt{225} = \$240$

a. For $\bar{x} = 16,450$: $z = (\bar{x} - \mu)/\sigma_{\bar{x}} = (16,450 - 16,842)/240 = -1.63$

For $\bar{x} = 17,120$: $z = (\bar{x} - \mu)/\sigma_{\bar{x}} = (17,120 - 16,842)/240 = 1.16$

$P(16,450 < \bar{x} < 17,120) = P(-1.63 < z < 1.16) = P(z < 1.16) - P(z < -1.63) = .8770 - .0516$ $= .8254$

b. $P(\bar{x}$ within \$300 of $\mu) = P(16,542 \leq \bar{x} \leq 17,142)$

For $\bar{x} = 16,542$: $z = (\bar{x} - \mu)/\sigma_{\bar{x}} = (16,542 - 16,842)/240 = -1.25$

For $\bar{x} = 17,142$: $z = (\bar{x} - \mu)/\sigma_{\bar{x}} = (17,142 - 16,842)/240 = 1.25$

$P(16,542 \leq \bar{x} \leq 17,142) = P(-1.25 \leq z \leq 1.25) = P(z \leq 1.25) - P(z \leq -1.25) = .8944 - .1056$ $= .7888$

c. $P(\bar{x}$ greater than μ by at least \$200$) = P(\bar{x} \geq 17,042)$

For $\bar{x} = 17,042$: $z = (\bar{x} - \mu)/\sigma_{\bar{x}} = (17,042 - 16,842)/240 = .83$

$P(\bar{x} \geq 17,042) = P(z \geq .83) = 1 - P(z \leq .83) = 1 - .7967 = .2033$

7.53 $\mu = \$1840$, $\sigma = \$453$, and $n = 36$

$\mu_{\bar{x}} = \mu = \$1840$ and $\sigma_{\bar{x}} = \sigma/\sqrt{n} = 453/\sqrt{36} = 75.5$

a. For $\bar{x} = 1750$: $z = (\bar{x} - \mu)/\sigma_{\bar{x}} = (1750 - 1840)/75.5 = -1.19$

For $\bar{x} = 1950$: $z = (\bar{x} - \mu)/\sigma_{\bar{x}} = (1950 - 1840)/75.5 = 1.46$

$P(1750 < \bar{x} < 1950) = P(-1.19 < z < 1.46) = P(z < 1.46) - P(z < -1.19) = .9278 - .1170$
$= .8108$

b. For $\bar{x} = 1700$: $z = (\bar{x} - \mu)/\sigma_{\bar{x}} = (1700 - 1840)/75.5 = -1.85$

$P(\bar{x} < 1700) = P(z < -1.85) = .0322$

7.55 $\mu = \$120$, $\sigma = \$25$, and $n = 75$

$\mu_{\bar{x}} = \mu = \$120$ and $\sigma_{\bar{x}} = \sigma/\sqrt{n} = 25/\sqrt{75} = \$2.88675135$

a. For $\bar{x} = 112$: $z = (\bar{x} - \mu)/\sigma_{\bar{x}} = (112 - 120)/2.88675135 = -2.77$

For $\bar{x} = 117$: $z = (\bar{x} - \mu)/\sigma_{\bar{x}} = (117 - 120)/2.88675135 = -1.04$

$P(112 < \bar{x} < 117) = P(-2.77 < z < -1.04) = P(z < -1.04) - P(z < -2.77) = .1492 - .0028 =$
$.1464$

b. $P(\bar{x}$ within \$6 of $\mu) = P(114 \le \bar{x} \le 126)$

For $\bar{x} = 114$: $z = (\bar{x} - \mu)/\sigma_{\bar{x}} = (114 - 120)/2.88675135 = -2.08$

For $\bar{x} = 126$: $z = (\bar{x} - \mu)/\sigma_{\bar{x}} = (126 - 120)/2.88675135 = 2.08$

$P(114 \le \bar{x} \le 126) = P(-2.08 \le z \le 2.08) = P(z \le 2.08) - P(z \le -2.08) = .9812 - .0188 =$
$.9624$

c. $P(\bar{x}$ greater than μ by at least \$5$) = P(\bar{x} \ge 125)$

For $\bar{x} = 125$: $z = (\bar{x} - \mu)/\sigma_{\bar{x}} = (125 - 120)/2.88675135 = 1.73$

$P(\bar{x} \ge 125) = P(z \ge 1.73) = 1 - P(z \le 1.73) = 1 - .9582 = .0418$

7.57 $\mu = 68$ inches, $\sigma = 4$ inches, and $n = 100$

$\mu_{\bar{x}} = \mu = 68$ inches and $\sigma_{\bar{x}} = \sigma/\sqrt{n} = 4/\sqrt{100} = .4$ inch

a. For $\bar{x} = 67.8$: $z = (\bar{x} - \mu)/\sigma_{\bar{x}} = (67.8 - 68)/.4 = -.50$

$P(\bar{x} < 67.8) = P(z < -.50) = .3085$

b. For $\bar{x} = 67.5$: $z = (\bar{x} - \mu)/\sigma_{\bar{x}} = (67.5 - 68)/.4 = -1.25$

For $\bar{x} = 68.7$: $z = (\bar{x} - \mu)/\sigma_{\bar{x}} = (68.7 - 68)/.4 = 1.75$

$P(67.5 < \bar{x} < 68.7) = P(-1.25 < z < 1.75) = P(z < 1.75) - P(z < -1.25) = .9599 - .1056$
$= .8543$

c. $P(\bar{x}$ within .6 inch of $\mu) = P(67.4 \le \bar{x} \le 68.6)$

For $\bar{x} = 67.4$: $z = (\bar{x} - \mu)/\sigma_{\bar{x}} = (67.4 - 68)/.4 = -1.50$

For $\bar{x} = 68.6$: $z = (\bar{x} - \mu)/\sigma_{\bar{x}} = (68.6 - 68)/.4 = 1.50$

$P(67.4 \le \bar{x} \le 68.6) = P(-1.50 \le z \le 1.50) = P(z \le 1.50) - P(z \le -1.50) = .9332 - .0668 =$ .8664

d. $P(\bar{x}$ is lower than μ by .5 inch or more) $= P(\bar{x} \le 67.5)$

For $\bar{x} = 67.5$: $z = (\bar{x} - \mu)/\sigma_{\bar{x}} = (67.5 - 68)/.4 = -1.25$

$P(\bar{x} \le 67.5) = P(z \le -1.25) = .1056$

7.59 $\mu = 3$ inches, $\sigma = .1$ inch, and $n = 25$

$\mu_{\bar{x}} = \mu = 3$ inches and $\sigma_{\bar{x}} = \sigma/\sqrt{n} = .1/\sqrt{25} = .02$ inch

For $\bar{x} = 2.95$: $z = (\bar{x} - \mu)/\sigma_{\bar{x}} = (2.95 - 3)/.02 = -2.50$

For $\bar{x} = 3.05$: $z = (\bar{x} - \mu)/\sigma_{\bar{x}} = (3.05 - 3)/.02 = 2.50$

$P(x < 2.95) + P(x > 3.05) = 1 - [P(2.95 \le x \le 3.05)] = 1 - [P(-2.50 \le z \le 2.50)]$
$= 1 - [P(z \le 2.50) - P(z \le -2.50)] = 1 - [.9938 - .0062] = 1 - .9876 = .0124$

Sections 7.6 - 7.7

7.61 $p = 600/5000 = .12$ and $\hat{p} = 18/120 = .15$

7.63 Number with characteristic in population = (9500)(.75) = 7125

Number with characteristic in sample = (400)(.78) = 312

7.65 Sampling error = $\hat{p} - p = .66 - .71 = -.05$

7.67 The estimator of p is the sample proportion $\hat{p}$.

The sample proportion $\hat{p}$ is an unbiased estimator of p, since the mean of $\hat{p}$ is equal to p.

7.69 $\sigma_{\hat{p}} = \sqrt{pq/n}$, hence $\sigma_{\hat{p}}$ decreases as n increases.

7.71 $p = .21$ and $q = 1 - p = 1 - .21 = .79$

a. $n = 400$, $\mu_{\hat{p}} = p = .21$, and $\sigma_{\hat{p}} = \sqrt{pq/n} = \sqrt{(.21)(.79)/400} = .020$

b. $n = 750$, $\mu_{\hat{p}} = p = .21$, and $\sigma_{\hat{p}} = \sqrt{pq/n} = \sqrt{(.21)(.79)/750} = .015$

7.73 $p = .47, q = 1 - p = 1 - .47 = .53$, and $N = 1400$

a. $n/N = 90/1400 = .064 > .05$

$$\sigma_{\hat{p}} = \sqrt{\frac{pq}{n}}\sqrt{\frac{N-n}{N-1}} = \sqrt{\frac{(.47)(.53)}{90}}\sqrt{\frac{1400-90}{1400-1}} = .051$$

b. Since $n/N = 50/1400 = .036 < .05$, $\sigma_{\hat{p}} = \sqrt{pq/n} = \sqrt{(.47)(.53)/50} = .071$

7.75 a. $np = (400)(.28) = 112$ and $nq = (400)(.72) = 288$

Since $np > 5$ and $nq > 5$, the central limit theorem applies.

b. $np = (80)(.05) = 4$; since $np < 5$, the central limit theorem does not apply.

c. $np = (60)(.12) = 7.2$ and $nq = (60)(.88) = 52.8$

Since $np > 5$ and $nq > 5$, the central limit theorem applies.

d. $np = (100)(.035) = 3.5$; since $np < 5$, the central limit theorem does not apply.

7.77 a. $p = 4/6 = .667$

b. ${}_6C_5 = 6$.

c & d. Let: G = good TV set and D = defective TV set

Let the six TV sets be denoted as: $1 = G$, $2 = G$, $3 = D$, $4 = D$, $5 = G$, and $6 = G$. The six possible samples, their sample proportions, and the sampling errors are given in the table below.

Sample	TV sets	$\hat{p}$	Sampling error
1, 2, 3, 4, 5	G, G, D, D, G	3/5=.60	.60 – .667 = –.067
1, 2, 3, 4, 6	G, G, D, D, G	3/5=.60	.60 – .667 = –.067
1, 2, 3, 5, 6	G, G, D, G, G	4/5=.80	.80 – .667 = .133
1, 2, 4, 5, 6	G, G, D, G, G	4/5=.80	.80 – .667 = .133
1, 3, 4, 5, 6	G, D, D, G, G	3/5=.60	.60 – .667 = –.067
2, 3, 4, 5, 6	G, D, D, G, G	3/5=.60	.60 – .667 = –.067

$\hat{p}$	f	Relative Frequency
.60	4	4/6=.667
.80	2	2/6=.333
	$\sum f = 6$	

$\hat{p}$	$P(\hat{p})$
.60	.667
.80	.333

7.79 $p = .19, q = 1 - p = 1 - .19 = .81$, and $n = 500$

$\mu_{\hat{p}} = p = .19$, and $\sigma_{\hat{p}} = \sqrt{pq/n} = \sqrt{(.19)(.81)/500} = .0175$

$np = (500)(.19) = 95$ and $nq = (500)(.81) = 405$

Since $np > 5$ and $nq > 5$, the sampling distribution of $\hat{p}$ is approximately normal.

7.81 $p = .17, q = 1 - p = 1 - .17 = .83$, and $n = 60$

$\mu_{\hat{p}} = p = .17$, and $\sigma_{\hat{p}} = \sqrt{pq/n} = \sqrt{(.17)(.83)/60} = .0485$

$np = (60)(.17) = 10.2$ and $nq = (60)(.83) = 49.8$

Since $np > 5$ and $nq > 5$, the sampling distribution of $\hat{p}$ is approximately normal.

Section 7.8

7.83 $P(p - 2.0\sigma_{\hat{p}} \leq \hat{p} \leq p + 2.0\sigma_{\hat{p}}) = P(-2.00 \leq z \leq 2.00) = P(z \leq 2.00) - P(z \leq -2.00) = .9772 - .0228$

$= .9544$ or 95.44%.

7.85 $p = .59, q = 1 - p = 1 - .59 = .41, N = 30{,}000$, and $n = 400$

$n/N = 100/30{,}000 = .033 < .05$, $np = (100)(.59) = 59 > 5$, $nq = (100)(41) = 41 > 5$

$\sigma_{\hat{p}} = \sqrt{pq/n} = \sqrt{(.59)(.41)/100} = .04918333$

a. $z = (\hat{p} - p)/\sigma_{\hat{p}} = (.56 - .59)/.04918333 = -.61$

b. $z = (\hat{p} - p)/\sigma_{\hat{p}} = (.68 - .59)/.04918333 = 1.83$

c. $z = (\hat{p} - p)/\sigma_{\hat{p}} = (.53 - .59)/.04918333 = -1.22$

d. $z = (\hat{p} - p)/\sigma_{\hat{p}} = (.65 - .59)/.04918333 = 1.22$

7.87 $p = .258, q = 1 - p = 1 - .258 = .742$, and $n = 225$

$\sigma_{\hat{p}} = \sqrt{pq/n} = \sqrt{(.258)(.742)/225} = .02916893$

a. For $\hat{p} = .24$: $z = (\hat{p} - p)/\sigma_{\hat{p}} = (.24 - .258)/.02916893 = -.62$

For $\hat{p} = .30$: $z = (\hat{p} - p)/\sigma_{\hat{p}} = (.30 - .258)/.02916893 = 1.44$

$P(.24 < \hat{p} < .30) = P(-.62 < z < 1.44) = P(z < 1.44) - P(z < -.62) = .9251 - .2676 = .6575$

b. For $\hat{p} = .27$: $z = (\hat{p} - p)/\sigma_{\hat{p}} = (.27 - .258)/.02916893 = .41$

$P(\hat{p} > .27) = P(z > .41) = 1 - P(z \leq .41) = 1 - .6591 = .3409$

7.89 $p = .44, q = 1 - p = 1 - .44 = .56$, and $n = 900$

$\sigma_{\hat{p}} = \sqrt{pq/n} = \sqrt{(.44)(.56)/900} = .01654623$

a. For $\hat{p} = .43$: $z = (\hat{p} - p)/\sigma_{\hat{p}} = (.43 - .44)/.01654623 = -.60$

For $\hat{p} = .46$: $z = (\hat{p} - p)/\sigma_{\hat{p}} = (.46 - .44)/.01654623 = 1.21$

$P(.43 < \hat{p} < .46) = P(-.60 < z < 1.21) = P(z < 1.21) - P(z < -.60) = .8869 - .2743 = .6126$

b. For $\hat{p} = .48$: $z = (\hat{p} - p)/\sigma_{\hat{p}} = (.48 - .44)/.01654623 = 2.42$

$P(\hat{p} > .48) = P(z > 2.42) = 1 - P(z \leq 2.42) = 1 - .9922 = .0078$

7.91 $p = .06$, $q = 1 - p = 1 - .06 = .94$, and $n = 100$

$\sigma_{\hat{p}} = \sqrt{pq/n} = \sqrt{(.06)(.94)/100} = .02374868$

For $\hat{p} = .08$: $z = (\hat{p} - p)/\sigma_{\hat{p}} = (.08 - .06)/.02374868 = .84$

$P(\hat{p} \geq .08) = P(z \geq .84) = 1 - P(z \leq .84) = 1 - .7995 = .2005$

Supplementary Exercises

7.93 $\mu = 750$ hours, $\sigma = 55$ hours, and $n = 25$

$\mu_{\bar{x}} = \mu = 750$ hours and $\sigma_{\bar{x}} = \sigma/\sqrt{n} = 55/\sqrt{25} = 11$ hours

The sampling distribution of $\bar{x}$ is normal because the population is normally distributed.

7.95 $\mu = 750$ hours, $\sigma = 55$ hours, and $n = 25$

$\mu_{\bar{x}} = \mu = 750$ hours and $\sigma_{\bar{x}} = \sigma/\sqrt{n} = 55/\sqrt{25} = 11$ hours

a. For $\bar{x} = 735$: $z = (\bar{x} - \mu)/\sigma_{\bar{x}} = (735 - 750)/11 = -1.36$

$P(\bar{x} > 735) = P(z > -1.36) = 1 - P(z \leq -1.36) = 1 - .0869 = .9131$

b. For $\bar{x} = 725$: $z = (\bar{x} - \mu)/\sigma_{\bar{x}} = (725 - 750)/11 = -2.27$

For $\bar{x} = 740$: $z = (\bar{x} - \mu)/\sigma_{\bar{x}} = (740 - 750)/11 = -.91$

$P(725 < \bar{x} < 740) = P(-2.27 < z < -.91) = P(z < -.91) - P(z < -2.27) = .1814 - .0116$
$= .1698$

c. $P(\bar{x}$ within 15 hours of $\mu) = P(735 \leq \bar{x} \leq 765)$

For $\bar{x} = 735$: $z = (\bar{x} - \mu)/\sigma_{\bar{x}} = (735 - 750)/11 = -1.36$

For $\bar{x} = 765$: $z = (\bar{x} - \mu)/\sigma_{\bar{x}} = (765 - 750)/11 = 1.36$

$P(735 \leq \bar{x} \leq 765) = P(-1.36 \leq z \leq 1.36) = P(z \leq 1.36) - P(z \leq -1.36) = .9131 - .0869 = .8262$

d. $P(\bar{x}$ is less than μ by 20 hours or more$) = P(\bar{x} < 730)$

For $\bar{x} = 730$: $z = (\bar{x} - \mu)/\sigma_{\bar{x}} = (730 - 750)/11 = -1.82$

$P(\bar{x} < 730) = P(z < -1.82) = .0344$

7.97 $\mu = 102$ minutes, $\sigma = 26$ minutes, and $n = 200$

$\mu_{\bar{x}} = \mu = 102$ minutes and $\sigma_{\bar{x}} = \sigma/\sqrt{n} = 26/\sqrt{200} = 1.83847763$ minutes

a. For $\bar{x} = 99$: $z = (\bar{x} - \mu)/\sigma_{\bar{x}} = (99 - 102)/1.83847763 = -1.63$

$P(\bar{x} > 99) = P(z > -1.63) = 1 - P(z \leq -1.63) = 1 - .0516 = .9484$

b. For $\bar{x} = 103$: $z = (\bar{x} - \mu)/\sigma_{\bar{x}} = (103 - 102)/1.83847763 = .54$

For $\bar{x} = 106$: $z = (\bar{x} - \mu)/\sigma_{\bar{x}} = (106 - 102)/1.83847763 = 2.18$

$P(103 < \bar{x} < 106) = P(.54 < z < 2.18) = P(z < 2.18) - P(z < .54) = .9854 - .7054 = .2800$

c. $P(\bar{x}$ within 3.5 minute of $\mu) = P(98.5 \leq \bar{x} \leq 105.5)$

For $\bar{x} = 98.5$: $z = (\bar{x} - \mu)/\sigma_{\bar{x}} = (98.5 - 102)/1.83847763 = -1.90$

For $\bar{x} = 105.5$: $z = (\bar{x} - \mu)/\sigma_{\bar{x}} = (105.5 - 102)/1.83847763 = 1.90$

$P(98.5 \leq \bar{x} \leq 105.5) = P(-1.90 \leq z \leq 1.90) = P(z \leq 1.90) - P(z \leq -1.90) = .9713 - .0287$

$= .9426$

d. $P(\bar{x}$ is less than μ by 3 minutes or more$) = P(\bar{x} < 99)$

For $\bar{x} = 99$: $z = (\bar{x} - \mu)/\sigma_{\bar{x}} = (99 - 102)/1.83847763 = -1.63$

$P(\bar{x} < 99) = P(z < -1.63) = .0516$

7.99 $p = .88$, $q = 1 - p = 1 - .88 = .12$, and $n = 80$

$\mu_{\hat{p}} = p = .88$, and $\sigma_{\hat{p}} = \sqrt{pq/n} = \sqrt{(.88)(.12)/80} = .03633180$

$np = (80)(.88) = 70.4 > 5$, $nq = (80)(.12) = 9.6 > 5$

Since np and nq are both greater than 5, the sampling distribution of $\hat{p}$ is approximately normal.

7.101 $p = .70$, $q = 1 - p = 1 - .70 = .30$, and $n = 400$

$\mu_{\hat{p}} = p = .70$, and $\sigma_{\hat{p}} = \sqrt{pq/n} = \sqrt{(.70)(.30)/400} = .02291288$

a. i. For $\hat{p} = .65$: $z = (\hat{p} - p)/\sigma_{\hat{p}} = (.65 - .70)/.02291288 = -2.18$

$P(\hat{p} < .65) = P(z < -2.18) = .0146$

ii. For $\hat{p} = .73$: $z = (\hat{p} - p)/\sigma_{\hat{p}} = (.73 - .70)/.02291288 = 1.31$

For $\hat{p} = .76$: $z = (\hat{p} - p)/\sigma_{\hat{p}} = (.76 - .70)/.02291288 = 2.62$

$P(.73 < \hat{p} < .76) = P(1.31 < z < 2.62) = P(z < 2.62) - P(z < 1.31) = .9956 - .9049 = .0907$

b. $P(\hat{p}$ within .06 of $p) = P(.64 \le \hat{p} \le .76)$

For $\hat{p} = .64$: $z = (\hat{p} - p)/\sigma_{\hat{p}} = (.64 - .70)/.02291288 = -2.62$

For $\hat{p} = .76$: $z = (\hat{p} - p)/\sigma_{\hat{p}} = (.76 - .70)/.02291288 = 2.62$

$P(.64 \le \hat{p} \le .76) = P(-2.62 \le z \le 2.62) = P(z \le 2.62) - P(z \le -2.62) = .9956 - .0044 = .9912$

c. $P(\hat{p}$ greater than p by .05 or more$) = P(\hat{p} \ge .75)$

For $\hat{p} = .75$: $z = (\hat{p} - p)/\sigma_{\hat{p}} = (.75 - .70)/.02291288 = 2.18$

$P(\hat{p} \ge .75) = P(z \ge 2.18) = 1 - P(z \le 2.18) = 1 - .9854 = .0146$

7.103 $\sigma = \$105{,}000$, and $n = 32$

$\sigma_{\bar{x}} = \sigma/\sqrt{n} = 105{,}000/\sqrt{32} = \$18{,}561.55$

The required probability is: $P(\mu - 10{,}000 \le \bar{x} \le \mu + 10{,}000)$

For $\bar{x} = \mu - 10{,}000$: $z = (\bar{x} - \mu)/\sigma_{\bar{x}} = (\mu - 10{,}000 - \mu)/18{,}561.55 = -.54$

For $\bar{x} = \mu + 10{,}000$: $z = (\bar{x} - \mu)/\sigma_{\bar{x}} = (\mu + 10{,}000 - \mu)/18{,}561.55 = .54$

$P(\mu - 10{,}000 \le \bar{x} \le \mu + 10{,}000) = P(-.54 \le z \le .54) = P(z \le .54) - P(z \le -.54) = .7054 - .2946$
$= .4108$

7.105 $\mu = c$ and $\sigma = .8$ ppm

We want $P(\mu - .5 \le \bar{x} \le \mu + .5) = .95$. The corresponding z value is 1.96; then $1.96\,\sigma_{\bar{x}} = .5$ and

$\sigma_{\bar{x}} = .255$. Since $\sigma_{\bar{x}} = \sigma/\sqrt{n}$, $n = (\sigma/\sigma_{\bar{x}})^2 = (.8/.255)^2 = 9.84$. Thus, 10 measurements are necessary.

7.107 a. $p = .53$, $q = 1 - p = 1 - .53 = .47$, and $n = 200$

$\sigma_{\hat{p}} = \sqrt{pq/n} = \sqrt{(.53)(.47)/200} = .03529164$

For $\hat{p} = .50$: $z = (\hat{p} - p)/\sigma_{\hat{p}} = (.50 - .53)/.03529164 = -.85$

$P(\hat{p} > .50) = P(z > -.85) = 1 - P(z \le -.85) = 1 - .1977 = .8023$

b. For .95 or higher, $z = -1.65$. Now,

$z = (\hat{p} - p)/\sigma_{\hat{p}}$, so $\sigma_{\hat{p}} = \dfrac{\hat{p} - p}{z} = \dfrac{.5 - .53}{-1.65} = .01818182$. Then, since

$\sigma_{\hat{p}} = \sqrt{pq/n}$, $n = pq/(\sigma_{\hat{p}})^2 = .53(.47)/(.01818182)^2 = 753.53$. The politician should take a sample of at least 754 voters.

7.109 μ = 160 pounds, σ = 25 pounds, and n = 35

$\mu_{\bar{x}} = \mu = 160$ pounds and $\sigma_{\bar{x}} = \sigma/\sqrt{n} = 25/\sqrt{35} = 4.22577127$ minutes

Since $n > 30$, $\bar{x}$ is approximately normally distributed.

P (sum of 35 weights exceeds 6000 pounds) = P (mean weight exceeds 6000/35) = $P(\bar{x} > 171.43)$

For $\bar{x} = 171.43$: $z = (\bar{x} - \mu)/\sigma_{\bar{x}} = (171.43 - 160)/4.22577127 = 2.70$

$P(\bar{x} > 171.43) = P(z > 2.70) = 1 - P(z \leq 2.70) = 1 - .9965 = .0035$

7.111

Sample	Scores	Sample Median
ABC	70, 78, 80	78
ABD	70, 78, 80	78
ABE	70, 78, 95	78
ACD	70, 80, 80	80
ACE	70, 80, 95	80
ADE	70, 80, 95	80
BCD	78, 80, 80	80
BCE	78, 80, 95	80
BDE	78, 80, 95	80
CDE	80, 80, 95	80

Mean of the sample medians: (78 + 78 + 78 + 80 + 80 + 80 + 80 + 80 + 80 + 80)/10 = 79.4. This is not equal to the population mean of 80.6. We could change the 78 to 80 and change the 95 to 90. Then, each of the sample medians would be equal to 80, and therefore have a mean of 80. The population mean would become μ = (70 + 80 + 80 + 80 + 80 + 80 + 80 + 80 + 80 + 90)/10 = 80.

Self – Review Test

1. b **2.** b **3.** a **4.** a **5.** b **6.** b

7. c **8.** a **9.** a **10.** a **11.** c **12.** a

13. According to the central limit theorem, for a large sample size, the sampling distribution of the sample mean is approximately normal irrespective of the shape of the population distribution. The mean and standard deviation of the sampling distribution of the sample mean are $\mu_{\bar{x}} = \mu$ and $\sigma_{\bar{x}} = \sigma/\sqrt{n}$, respectively. The sample size is usually considered to be large if $n \geq 30$. From the same theorem, the sampling distribution of $\hat{p}$ is approximately normal for sufficiently large samples. In the case of proportion, the sample is sufficiently large if $np > 5$ and $nq > 5$.

14. $\mu = 145$ pounds and $\sigma = 18$ pounds

a. $\mu_{\bar{x}} = \mu = 145$ pounds and $\sigma_{\bar{x}} = \sigma/\sqrt{n} = 18/\sqrt{25} = 3.60$ pounds

b. $\mu_{\bar{x}} = \mu = 145$ pounds and $\sigma_{\bar{x}} = \sigma/\sqrt{n} = 18/\sqrt{100} = 1.80$ pounds

In both cases the sampling distribution of $\bar{x}$ is approximately normal because the population has an approximate normal distribution.

15. $\mu = 11$ minutes and $\sigma = 2.7$ minutes

a. $\mu_{\bar{x}} = \mu = 11$ minutes and $\sigma_{\bar{x}} = \sigma/\sqrt{n} = 2.7/\sqrt{25} = .54$ minute

Since the population has an unknown distribution and $n < 30$, we can draw no conclusion about the shape of the sampling distribution of $\bar{x}$.

b. $\mu_{\bar{x}} = \mu = 11$ minutes and $\sigma_{\bar{x}} = \sigma/\sqrt{n} = 2.7/\sqrt{75} = .312$ minute

Since $n > 30$, the sampling distribution of $\bar{x}$ is approximately normal.

16. $\mu = 42$ seconds, $\sigma = 10$ seconds, and $n = 50$

$\mu_{\bar{x}} = \mu = 42$ seconds and $\sigma_{\bar{x}} = \sigma/\sqrt{n} = 10/\sqrt{50} = 1.41421356$ seconds

a. For $\bar{x} = 38$: $z = (\bar{x} - \mu)/\sigma_{\bar{x}} = (38 - 42)/1.41421356 = -2.83$

For $\bar{x} = 41$: $z = (\bar{x} - \mu)/\sigma_{\bar{x}} = (41 - 42)/1.41421356 = -.71$

$P(38 < \bar{x} < 41) = P(-2.83 < z < -.71) = P(z < -.71) - P(z < -2.83) = .2389 - .0023 = .2366$

b. $P(\bar{x}$ within 2 seconds of $\mu) = P(40 \le \bar{x} \le 44)$

For $\bar{x} = 40$: $z = (\bar{x} - \mu)/\sigma_{\bar{x}} = (40 - 42)/1.41421356 = -1.41$

For $\bar{x} = 44$: $z = (\bar{x} - \mu)/\sigma_{\bar{x}} = (44 - 42)/1.41421356 = 1.41$

$P(40 \le \bar{x} \le 44) = P(-1.41 \le z \le 1.41) = P(z \le 1.41) - P(z \le -1.41) = .9207 - .0793 = .8414$

c. $P(\bar{x}$ greater than μ by 1 second or more$) = P(\bar{x} \ge 43)$

For $\bar{x} = 43$: $z = (\bar{x} - \mu)/\sigma_{\bar{x}} = (43 - 42)/1.41421356 = .71$

$P(\bar{x} \ge 43) = P(z \ge .71) = 1 - P(z \le .71) = 1 - .7611 = .2389$

d. For $\bar{x} = 39$: $z = (\bar{x} - \mu)/\sigma_{\bar{x}} = (39 - 42)/1.41421356 = -2.12$

For $\bar{x} = 44$: $z = (\bar{x} - \mu)/\sigma_{\bar{x}} = (44 - 42)/1.41421356 = 1.41$

$P(39 < \bar{x} < 44) = P(-2.12 < z < 1.41) = P(z < 1.41) - P(z < -2.12) = .9207 - .0170 = .9037$

e. For $\bar{x} = 43$: $z = (\bar{x} - \mu)/\sigma_{\bar{x}} = (43 - 42)/1.41421356 = .71$

$P(\bar{x} < 43) = P(z < .71) = .7611$

17. $\mu = 16$ ounces, $\sigma = .18$ ounce, and $n = 16$

$\mu_{\bar{x}} = \mu = 16$ ounces and $\sigma_{\bar{x}} = \sigma/\sqrt{n} = .18/\sqrt{16} = .045$ ounce

a. i. For $\bar{x} = 15.90$: $z = (\bar{x} - \mu)/\sigma_{\bar{x}} = (15.90 - 16)/.045 = -2.22$

For $\bar{x} = 15.95$: $z = (\bar{x} - \mu)/\sigma_{\bar{x}} = (15.95 - 16)/.045 = -1.11$

$P(15.90 < \bar{x} < 15.95) = P(-2.22 < z < -1.11) = P(z < -1.11) - P(z < -2.22) = .1335 - .0132$

$= .1203$

ii. For $\bar{x} = 15.95$: $z = (\bar{x} - \mu)/\sigma_{\bar{x}} = (15.95 - 16)/.045 = -1.11$

$P(\bar{x} < 15.95) = P(z < -1.11) = .1335$

iii. For $\bar{x} = 15.97$: $z = (\bar{x} - \mu)/\sigma_{\bar{x}} = (15.97 - 16)/.045 = -.67$

$P(\bar{x} > 15.97) = P(z > -.67) = 1 - P(z \le -.67) = 1 - .2514 = .7486$

b. $P(\bar{x}$ within .10 ounce of $\mu) = P(15.90 \le \bar{x} \le 16.10)$

For $\bar{x} = 15.90$: $z = (\bar{x} - \mu)/\sigma_{\bar{x}} = (15.90 - 16)/.045 = -2.22$

For $\bar{x} = 16.10$: $z = (\bar{x} - \mu)/\sigma_{\bar{x}} = (16.10 - 16)/.045 = 2.22$

$P(15.90 \le \bar{x} \le 16.10) = P(-2.22 \le z \le 2.22) = P(z \le 2.22) - P(z \le -2.22) = .9868 - .0132$

$= .9736$

c. $P(\bar{x}$ is less than μ by .135 ounce or more$) = P(\bar{x} < 15.865)$

For $\bar{x} = 15.865$: $z = (\bar{x} - \mu)/\sigma_{\bar{x}} = (15.865 - 16)/.045 = -3.00$

$P(\bar{x} < 15.865) = P(z < -3.00) = .0013$

18. $p = .49$, and $q = 1 - p = 1 - .49 = .51$

a. $n = 40$, $\mu_{\hat{p}} = p = .49$, and $\sigma_{\hat{p}} = \sqrt{pq/n} = \sqrt{(.49)(.51)/40} = .079$

$np = (40)(.49) = 19.6$ and $nq = (40)(.51) = 20.4$

Since $np > 5$ and $nq > 5$, the sampling distribution of $\hat{p}$ is approximately normal.

b. $n = 100$, $\mu_{\hat{p}} = p = .49$, and $\sigma_{\hat{p}} = \sqrt{pq/n} = \sqrt{(.49)(.51)/100} = .050$

$np = (100)(.49) = 49$ and $nq = (100)(.51) = 51$

Since $np > 5$ and $nq > 5$, the sampling distribution of $\hat{p}$ is approximately normal.

c. $n = 500$, $\mu_{\hat{p}} = p = .49$, and $\sigma_{\hat{p}} = \sqrt{pq/n} = \sqrt{(.49)(.51)/500} = .022$

$np = (500)(.49) = 245$ and $nq = (500)(.51) = 255$

Since $np > 5$ and $nq > 5$, the sampling distribution of $\hat{p}$ is approximately normal.

19. $p = .63, q = 1 - p = 1 - .63 = .37$, and $n = 300$

a. $\mu_{\hat{p}} = p = .63$, and $\sigma_{\hat{p}} = \sqrt{pq/n} = \sqrt{(.63)(.37)/300} = .02787472$

i. For $\hat{p} = .66$: $z = (\hat{p} - p)/\sigma_{\hat{p}} = (.66 - .63)/.02787472 = 1.08$

$P(\hat{p} > .66) = P(z > 1.08) = 1 - P(z \le 1.08) = 1 - .8599 = .1401$

ii. For $\hat{p} = .60$: $z = (\hat{p} - p)/\sigma_{\hat{p}} = (.60 - .63)/.02787472 = -1.08$

For $\hat{p} = .67$: $z = (\hat{p} - p)/\sigma_{\hat{p}} = (.67 - .63)/.02787472 = 1.43$

$P(.60 < \hat{p} < .67) = P(-1.08 < z < 1.43) = P(z < 1.43) - P(z < -1.08) = .9236 - .1401 = .7835$

iii. For $\hat{p} = .65$: $z = (\hat{p} - p)/\sigma_{\hat{p}} = (.65 - .63)/.02787472 = .72$

$P(\hat{p} < .65) = P(z < .72) = .7642$

iv. For $\hat{p} = .57$: $z = (\hat{p} - p)/\sigma_{\hat{p}} = (.57 - .63)/.02787472 = -2.15$

For $\hat{p} = .61$: $z = (\hat{p} - p)/\sigma_{\hat{p}} = (.61 - .63)/.02787472 = -.72$

$P(.57 < \hat{p} < .61) = P(-2.15 < z < -.72) = P(z < -.72) - P(z < -2.15) = .2358 - .0158 = .2200$

b. $P(\hat{p}$ within .035 of $p) = P(.595 \le \hat{p} \le .665)$

For $\hat{p} = .595$: $z = (\hat{p} - p)/\sigma_{\hat{p}} = (.595 - .63)/.02787472 = -1.26$

For $\hat{p} = .665$: $z = (\hat{p} - p)/\sigma_{\hat{p}} = (.665 - .63)/.02787472 = 1.26$

$P(.595 \le \hat{p} \le .665) = P(-1.26 \le z \le 1.26) = P(z \le 1.26) - P(z \le -1.26) = .8962 - .1038 = .7924$

c. $P(\hat{p}$ is less than p by .025 or more$) = P(\hat{p} \le .605)$

For $\hat{p} = .605$: $z = (\hat{p} - p)/\sigma_{\hat{p}} = (.605 - .63)/.02787472 = -.90$

$P(\hat{p} \le .605) = P(z \le -.90) = .1841$

d. $P(\hat{p}$ greater than p by .03 or more$) = P(\hat{p} \ge .66)$

For $\hat{p} = .66$: $z = (\hat{p} - p)/\sigma_{\hat{p}} = (.66 - .63)/.02787472 = 1.08$

$P(\hat{p} \ge .66) = P(z \ge 1.08) = 1 - P(z \le 1.08) = 1 - .8599 = .1401$

Chapter Eight

Sections 8.1 – 8.3

8.1 An **estimator** is a sample statistic used to estimate a population parameter. The value(s) assigned to a population parameter based on the value of a sample statistic is called an **estimate**.

8.3 The sample mean $\bar{x}$ is the **point estimator of the population mean μ.** The **margin of error** is $E = z\sigma_{\bar{x}}$.

8.5 The width of a confidence interval depends on $E = z\sigma_{\bar{x}} = z\left(\sigma/\sqrt{n}\right)$. When n increases, $z\left(\sigma/\sqrt{n}\right)$ decreases and the width of the confidence interval decreases. From Example 8–1 in the text, n = 25, $\bar{x}$ = \$90.50, and σ = \$7.50. Then, $\sigma_{\bar{x}} = \sigma/\sqrt{n} = 7.50/\sqrt{25} = \$1.50$. The 90% confidence interval for μ is $\bar{x} \pm z\sigma_{\bar{x}} = 90.50 \pm 1.65(1.50) = 90.50 \pm 2.48 = \$88.02$ to $\$92.98$. The width of this interval is \$92.98 – \$88.02 = \$4.96. If n = 100, but all other values remain the same, $\sigma_{\bar{x}} = \sigma/\sqrt{n} = 7.50/\sqrt{100} = \$.75$, and the 90% confidence interval for μ is $\bar{x} \pm z\sigma_{\bar{x}} = 90.50 \pm 1.65(.75) = 90.50 \pm 1.24 = \$89.26$ to $\$91.74$. The width of this interval is \$91.74 – \$89.26 = \$2.48. Thus, the 90% confidence interval for μ is narrower when n = 100 than when n = 25.

8.7 A **confidence interval** is an interval constructed around a point estimate. A **confidence level** indicates how confident we are that the confidence interval contains the population parameter.

8.9 If we take all possible samples of a given size and construct a 99% confidence interval for μ from each sample, we can expect about 99% of these confidence intervals will contain μ and 1% will not.

8.11 $n = 20$, $\bar{x} = 24.5$, $\sigma = 3.1$, and $\sigma_{\bar{x}} = \sigma/\sqrt{n} = 3.1/\sqrt{20} = .69318107$

a. $\bar{x} = 24.5$

b. The 99% confidence interval for μ is

$\bar{x} \pm z\sigma_{\bar{x}} = 24.5 \pm 2.58(.69318107) = 24.5 \pm 1.79 = 22.71$ to 26.29

c. $E = z\sigma_{\bar{x}} = 2.58(.69318107) = 1.79$

8.13 $n = 36$, $\bar{x} = 74.8$, $\sigma = 15.3$, and $\sigma_{\bar{x}} = \sigma/\sqrt{n} = 15.3/\sqrt{36} = 2.55$

a. The 90% confidence interval for μ is $\bar{x} \pm z\sigma_{\bar{x}} = 74.8 \pm 1.65(2.55) = 74.8 \pm 4.21 = 70.59$ to 79.01

b. The 95% confidence interval for μ is

$\bar{x} \pm z\sigma_{\bar{x}} = 74.8 \pm 1.96(2.55) = 74.8 \pm 5.00 = 69.80$ to 79.80

c. The 99% confidence interval for μ is

$\bar{x} \pm z\sigma_{\bar{x}} = 74.8 \pm 2.58(2.55) = 74.8 \pm 6.58 = 68.22$ to 81.38

d. Yes, the width of the confidence intervals increases as the confidence level increases. This occurs because as the confidence level increases, the value of z increases.

8.15 $\bar{x} = 81.90$ and $\sigma = 6.30$

a. $n = 16$, so $\sigma_{\bar{x}} = \sigma/\sqrt{n} = 6.30/\sqrt{16} = 1.575$

The 99% confidence interval for μ is

$\bar{x} \pm z\sigma_{\bar{x}} = 81.90 \pm 2.58(1.575) = 81.90 \pm 4.06 = 77.84$ to 85.96

b. $n = 20$, so $\sigma_{\bar{x}} = \sigma/\sqrt{n} = 6.30/\sqrt{20} = 1.40872283$

The 99% confidence interval for μ is

$\bar{x} \pm z\sigma_{\bar{x}} = 81.90 \pm 2.58(1.40872283) = 81.90 \pm 3.63 = 78.27$ to 85.53

c. $n = 25$, so $\sigma_{\bar{x}} = \sigma/\sqrt{n} = 6.30/\sqrt{25} = 1.26$

The 99% confidence interval for μ is

$\bar{x} \pm z\sigma_{\bar{x}} = 81.90 \pm 2.58(1.26) = 81.90 \pm 3.25 = 78.65$ to 85.15

d. Yes, the width of the confidence intervals decreases as the sample size increases. This occurs because the standard deviation of the sample mean decreases as the sample size increases.

8.17 $n = 35$, $\bar{x} = (\Sigma x)/n = 1342/35 = 38.34$, $\sigma = 2.65$, and $\sigma_{\bar{x}} = \sigma/\sqrt{n} = 2.65/\sqrt{35} = .44793176$

a. $\bar{x} = 38.34$

b. The 98% confidence interval for μ is

$\bar{x} \pm z\sigma_{\bar{x}} = 38.34 \pm 2.33(.44793176) = 38.34 \pm 1.04 = 37.30$ to 39.38

c. $E = z\sigma_{\bar{x}} = 2.33(.44793176) = 1.04$

8.19 a. $E = 2.50$, $\sigma = 12.5$, and $z = 2.58$ for 99% confidence level

$$n = \frac{z^2\sigma^2}{E^2} = \frac{(2.58)^2(12.5)^2}{(2.50)^2} = 166.41 \approx 167$$

b. $E = 3.20$, $\sigma = 12.5$, and $z = 2.05$ for 96% confidence level

$$n = \frac{z^2\sigma^2}{E^2} = \frac{(2.05)^2(12.5)^2}{(3.20)^2} = 64.13 \approx 65$$

8.21 a. $E = 2.3$, $\sigma = 15.40$, and $z = 2.58$ for 99% confidence level

$$n = \frac{z^2\sigma^2}{E^2} = \frac{(2.58)^2(15.40)^2}{(2.3)^2} = 298.42 \approx 299$$

b. $E = 4.1$, $\sigma = 23.45$, and $z = 1.96$ for 95% confidence level

$$n = \frac{z^2\sigma^2}{E^2} = \frac{(1.96)^2(23.45)^2}{(4.1)^2} = 125.67 \approx 126$$

c. $E = 25.9$, $\sigma = 122.25$, and $z = 1.65$ for 90% confidence level

$$n = \frac{z^2\sigma^2}{E^2} = \frac{(1.65)^2(122.25)^2}{(25.9)^2} = 60.65 \approx 61$$

8.23 $n = 1500$, $\bar{x} = \$269{,}720$, $\sigma = \$68{,}650$, and $\sigma_{\bar{x}} = \sigma/\sqrt{n} = 68{,}650/\sqrt{1500} = \$1772.535378$

The 99% confidence interval for μ is

$\bar{x} \pm z\sigma_{\bar{x}} = 269{,}720 \pm 2.58(1772.535378) = 269{,}720 \pm 4573.14 = \$265{,}146.86$ to $\$274{,}293.14$

8.25 $n = 25$, $\bar{x} = 70$ hours, $\sigma = 16$ hours, and $\sigma_{\bar{x}} = \sigma/\sqrt{n} = 16/\sqrt{25} = 3.2$ hours

a. The 95% confidence interval for μ is $\bar{x} \pm z\sigma_{\bar{x}} = 70 \pm 1.96(3.2) = 70 \pm 6.27 = 63.73$ to 76.27 hours

b. The sample mean of 70 hours is an estimate of μ based on a random sample. Because of sampling error, this estimate might differ from the true mean, μ, so we make an interval estimate to allow for this uncertainty and sampling error.

8.27 $n = 32$, $\bar{x} = 31.94$ ounces, $\sigma = .15$ ounce, and $\sigma_{\bar{x}} = \sigma/\sqrt{n} = .15/\sqrt{25} = .03$ ounce

The 99% confidence interval for μ is

$\bar{x} \pm z\sigma_{\bar{x}} = 31.94 \pm 2.58(.03) = 31.94 \pm .08 = 31.86$ to 32.02 ounces

Since the upper limit, 32.02, is less than 32.15, and the lower limit, 31.86, is greater than 31.85, the machine does not need an adjustment.

8.29 a. $n = 120$, $\bar{x} = \$1575$, $\sigma = \$215$, and $\sigma_{\bar{x}} = \sigma/\sqrt{n} = 215/\sqrt{120} = \$19.62672498$

The 97% confidence interval for μ is

$\bar{x} \pm z\sigma_{\bar{x}} = 1575 \pm 2.17(19.62672498) = 1575 \pm 42.59 = \$1532.41$ to $\$1617.59$

b. The width of the confidence interval obtained in part a may be reduced by:

1. Lowering the confidence level
2. Increasing the sample size

The second alternative is better because lowering the confidence level results in a less reliable estimate for μ.

8.31 $E = .04$ ounce, $\sigma = .20$ ounce, and $z = 2.58$ for 99% confidence level

$$n = \frac{z^2\sigma^2}{E^2} = \frac{(2.58)^2(.20)^2}{(.04)^2} = 166.41 \approx 167$$

8.33 $E = .75$ hour, $\sigma = 2.5$ hours, and $z = 2.33$ for 98% confidence level

$$n = \frac{z^2\sigma^2}{E^2} = \frac{(2.33)^2(2.5)^2}{(.75)^2} = 60.32 \approx 61$$

8.35 To estimate μ, the mean age of cars at a 95% confidence level:

1. Take a random sample of 30 or more U.S. car owners.
2. Determine age of each car.
3. Find $\bar{x}$, the mean of the sample.
4. Calculate $\sigma_{\bar{x}} = \sigma/\sqrt{n}$.
5. Calculate the 95% confidence interval for μ using the formula $\bar{x} \pm 1.96\sigma_{\bar{x}}$.

Section 8.4

8.37 The normal distribution has two **parameters**: μ and σ. Given the values of these parameters for a normal distribution, we can find the area under the normal curve between any two points. The t distribution has only one **parameter**: the degrees of freedom. The shape of the t distribution curve is determined by the degrees of freedom.

8.39 The t distribution is used to make a confidence interval for the population mean μ if the following three assumptions hold true:

1. The population standard deviation σ is not known.
2. The sample size is small (that is, $n < 30$).
3. The population from which the sample is drawn is (approximately) normally distributed.

8.41 a. $df = n - 1 = 21 - 1 = 20$ and $t = -1.325$ b. $df = n - 1 = 14 - 1 = 13$ and $t = 2.160$

c. $t = 3.281$ d. $t = -2.715$

8.43 a. Area in the left tail = .10

b. $df = n - 1 = 25 - 1 = 24$, area in the right tail = .005

c. $df = n - 1 = 9 - 1 = 8$, area in the right tail = .10

d. Area in the left tail = .01

8.45 a. $df = n - 1 = 22 - 1 = 21$, $\alpha/2 = .5 - (.95/2) = .025$, and $t = 2.080$

b. $\alpha/2 = .5 - (.90/2) = .05$ and $t = 1.671$

c. $df = n - 1 = 24 - 1 = 23$, $\alpha/2 = .5 - (.99/2) = .005$, and $t = 2.807$

8.47 $n = 10$, $\sum x = 441$, and $\sum x^2 = 19{,}977$

$\bar{x} = (\sum x)/n = 441/10 = 44.10$

$$s = \sqrt{\frac{\sum x^2 - \frac{(\sum x)^2}{n}}{n-1}} = \sqrt{\frac{19{,}977 - \frac{(441)^2}{10}}{10-1}} = 7.66594199$$

$s_{\bar{x}} = s/\sqrt{n} = 7.66594199/\sqrt{10} = 2.42418371$

a. $\bar{x} = 44.10$

b. $df = n - 1 = 10 - 1 = 9$, $\alpha/2 = .5 - (.95/2) = .025$, and $t = 2.262$

The 95% confidence interval for μ is

$\bar{x} \pm t s_{\bar{x}} = 44.10 \pm 2.262(2.42418371) = 44.10 \pm 5.48 = 38.62$ to 49.58

c. $E = t s_{\bar{x}} = 2.262(2.42418371) = 5.48$

8.49 $n = 47$, $\bar{x} = 25.5$, $s = 4.9$, $s_{\bar{x}} = s/\sqrt{n} = 4.9/\sqrt{47} = .71473846$, and $df = n - 1 = 47 - 1 = 46$

a. $\alpha/2 = .5 - (.95/2) = .025$ and $t = 2.013$

The 95% confidence interval for μ is

$\bar{x} \pm t s_{\bar{x}} = 25.5 \pm 2.013(.71473846) = 25.5 \pm 1.44 = 24.06$ to 26.94

b. $\alpha/2 = .5 - (.99/2) = .005$ and $t = 2.687$

The 99% confidence interval for μ is

$\bar{x} \pm ts_{\bar{x}} = 25.5 \pm 2.687(.71473846) = 25.5 \pm 1.92 = 23.58$ to 27.42

The width of the 99% confidence interval for μ is larger than that of the 95% confidence interval. This is so because the value of t for a 99% confidence level is larger than that for a 95% confidence level.

c. $s_{\bar{x}} = s/\sqrt{n} = 4.9/\sqrt{32} = .86620581$

$df = n - 1 = 32 - 1 = 31$, $\alpha/2 = .5 - (.95/2) = .025$ and $t = 2.040$

The 95% confidence interval for μ is

$\bar{x} \pm ts_{\bar{x}} = 25.5 \pm 2.040(.86620581) = 25.5 \pm 1.77 = 23.73$ to 27.27

The width of the 95% confidence interval for μ is larger with $n = 32$ than that of the 95% confidence interval with $n = 47$. This is so because the value of the standard deviation of the sample mean increases as sample size decreases.

8.51 In each of the following parts, since $n = 400$ is very large, we use the normal distribution to approximate the t distribution, and construct the confidence interval using $\bar{x} \pm zs_{\bar{x}}$.

a. $n = 400$, $\bar{x} = 92.45$, $s = 12.20$, and $s_{\bar{x}} = s/\sqrt{n} = 12.20/\sqrt{400} = .61$

The 98% confidence interval for μ is

$\bar{x} \pm zs_{\bar{x}} = 92.45 \pm 2.33(.61) = 92.45 \pm 1.42 = 91.03$ to 93.87

b. $n = 400$, $\bar{x} = 91.75$, $s = 14.50$, and $s_{\bar{x}} = s/\sqrt{n} = 14.50/\sqrt{400} = .725$

The 98% confidence interval for μ is

$\bar{x} \pm zs_{\bar{x}} = 91.75 \pm 2.33(.725) = 91.75 \pm 1.69 = 90.06$ to 93.44

c. $n = 400$, $\bar{x} = 89.63$, $s = 13.40$, and $s_{\bar{x}} = s/\sqrt{n} = 13.40/\sqrt{400} = .67$

The 98% confidence interval for μ is

$\bar{x} \pm zs_{\bar{x}} = 89.63 \pm 2.33(.67) = 89.63 \pm 1.56 = 88.07$ to 91.19

d. The confidence intervals of parts b and c cover μ but the confidence interval of part a does not.

8.53 $n = 20$, $\bar{x} = 41.2$ bushels per acre, $s = 3$ bushels per acre, and $s_{\bar{x}} = s/\sqrt{n} = 3/\sqrt{20} = .67082039$ bushel per acre

$df = n - 1 = 20 - 1 = 19$, $\alpha/2 = .5 - (.90/2) = .05$, and $t = 1.729$

The 90% confidence interval for μ is

$\bar{x} \pm ts_{\bar{x}} = 41.2 \pm 1.729(.67082039) = 41.2 \pm 1.16 = 40.04$ to 42.36 bushels per acre

8.55 $n = 36$, $\bar{x} = 31.98$ ounces, $s = .26$ ounce, and $s_{\bar{x}} = s/\sqrt{n} = .26/\sqrt{36} = .04333333$ ounce

$df = n - 1 = 36 - 1 = 35$, $\alpha/2 = .5 - (.95/2) = .025$, and $t = 2.030$

The 95% confidence interval for μ is

$\bar{x} \pm ts_{\bar{x}} = 31.98 \pm 2.030(.04333333) = 31.98 \pm .09 = 31.89$ to 32.07 ounces

8.57 $n = 25$, $\bar{x} = 22$ minutes, $s = 6$ minutes, and $s_{\bar{x}} = s/\sqrt{n} = 6/\sqrt{25} = 1.2$ minutes

$df = n - 1 = 25 - 1 = 24$, $\alpha/2 = .5 - (.99/2) = .005$, and $t = 2.797$

The 99% confidence interval for μ is $\bar{x} \pm ts_{\bar{x}} = 22 \pm 2.797(1.2) = 22 \pm 3.36 = 18.64$ to 25.36 minutes

8.59 a. $n = 40$, $\bar{x} = 23$ hours, $s = 3.75$ hours, and $s_{\bar{x}} = s/\sqrt{n} = 3.75/\sqrt{40} = .59292706$ hour

$df = n - 1 = 40 - 1 = 39$, $\alpha/2 = .5 - (.98/2) = .01$, and $t = 2.426$

The 98% confidence interval for μ is

$\bar{x} \pm ts_{\bar{x}} = 23 \pm 2.426(.59292706) = 23 \pm 1.44 = 21.56$ to 24.44 hours

b. The width of the confidence interval obtained in part a can be reduced by:

1. Lowering the confidence level
2. Increasing the sample size

The second alternative is better because lowering the confidence level gives a less reliable estimate for μ.

8.61 $n = 9$, $\sum x = 72$, and $\sum x^2 = 708$

$\bar{x} = (\sum x)/n = 72/9 = 8$ hours

$$s = \sqrt{\frac{\sum x^2 - \frac{(\sum x)^2}{n}}{n-1}} = \sqrt{\frac{708 - \frac{(72)^2}{9}}{9-1}} = 4.06201920 \text{ hours}$$

$s_{\bar{x}} = s/\sqrt{n} = 4.06201920/\sqrt{9} = 1.35400640$ hours

$df = n - 1 = 9 - 1 = 8$, $\alpha/2 = .5 - (.95/2) = .025$, and $t = 2.306$

The 95% confidence interval for μ is

$\bar{x} \pm ts_{\bar{x}} = 8 \pm 2.306(1.354006401) = 8 \pm 3.12 = 4.88$ to 11.12 hours

8.63 Since $n = 90$ is very large, we use the normal distribution to approximate the t distribution, and construct the confidence interval using $\bar{x} \pm zs_{\bar{x}}$.

$n = 90$, $\bar{x} = \$1250$, $s = \$270$, and $s_{\bar{x}} = s/\sqrt{n} = 270/\sqrt{90} = \$28.46049894$

The 95% confidence interval for μ is

$\bar{x} \pm z s_{\bar{x}} = 1250 \pm 1.96(28.46049894) = 1250 \pm 55.78 = \$1194.22$ to $\$1305.78$

8.65 Since $n = 1100$ is very large, we use the normal distribution to approximate the t distribution, and construct the confidence interval using $\bar{x} \pm z s_{\bar{x}}$.

$n = 1100$, $\bar{x} = \$21{,}213$, $s = \$5100$, and $s_{\bar{x}} = s/\sqrt{n} = 5100/\sqrt{1100} = \$153.7707857$

The 98% confidence interval for μ is

$\bar{x} \pm z s_{\bar{x}} = 21{,}213 \pm 2.33(153.7707857) = 21{,}213 \pm 358.29 = \$20{,}854.71$ to $\$21{,}571.29$

8.67 To estimate the mean waiting time, μ (assuming that waiting times are normally distributed):

1. Take a random sample of 45 customers at the bank.
2. Record the waiting time for each of the 45 customers.
3. Calculate the sample mean, the standard deviation, and $s_{\bar{x}}$.
4. Choose the confidence level and determine the t value for $df = n - 1 = 45 - 1 = 44$.
5. Construct the confidence interval for μ using $\bar{x} \pm t s_{\bar{x}}$.

Section 8.5

8.69 The sample proportion $\hat{p}$ is the **point estimator of *p***.

8.71

a. $n = 120$, $\hat{p} = .04$, $\hat{q} = 1 - \hat{p} = 1 - .04 = .96$, $n\hat{p} = (120)(.04) = 4.8$, and $n\hat{q} = (120)(.96) = 115.2$

Since $n\hat{p} < 5$, the sample size is not large enough to use the normal distribution.

b. $n = 60$, $\hat{p} = .08$, $\hat{q} = 1 - \hat{p} = 1 - .08 = .92$, $n\hat{p} = (60)(.08) = 4.8$, and $n\hat{q} = (60)(.92) = 55.2$

Since $n\hat{p} < 5$, the sample size is not large enough to use the normal distribution.

c. $n = 40$, $\hat{p} = .50$, $\hat{q} = 1 - \hat{p} = 1 - .50 = .50$, $n\hat{p} = (40)(.50) = 20$, and $n\hat{q} = (40)(.50) = 20$

Since $n\hat{p} > 5$ and $n\hat{q} > 5$, the sample size is large enough to use the normal distribution.

d. $n = 900$, $\hat{p} = .15$, $\hat{q} = 1 - \hat{p} = 1 - .15 = .85$, $n\hat{p} = (900)(.15) = 135$, and $n\hat{q} = (900)(.85) = 765$

Since $n\hat{p} > 5$ and $n\hat{q} > 5$, the sample size is large enough to use the normal distribution.

8.73 a. $n = 900$, $\hat{p} = .32$, $\hat{q} = 1 - \hat{p} = 1 - .32 = .68$, and

$s_{\hat{p}} = \sqrt{\hat{p}\hat{q}/n} = \sqrt{(.32)(.68)/900} = .01554921$

The 90% confidence interval for p is

$\hat{p} \pm z s_{\hat{p}} = .32 \pm 1.65(.01554921) = .32 \pm .026 = .294 \text{ to } .346$

b. $n = 900$, $\hat{p} = .36$, $\hat{q} = 1 - \hat{p} = 1 - .36 = .64$, and $s_{\hat{p}} = \sqrt{\hat{p}\hat{q}/n} = \sqrt{(.36)(.64)/900} = .0160$

The 90% confidence interval for p is $\hat{p} \pm z s_{\hat{p}} = .36 \pm 1.65(.0160) = .36 \pm .026 = .334 \text{ to } .386$

c. $n = 900$, $\hat{p} = .30$, $\hat{q} = 1 - \hat{p} = 1 - .30 = .70$, and

$s_{\hat{p}} = \sqrt{\hat{p}\hat{q}/n} = \sqrt{(.30)(.70)/900} = .01527525$

The 90% confidence interval for p is

$\hat{p} \pm z s_{\hat{p}} = .30 \pm 1.65(.01527525) = .30 \pm .025 = .275 \text{ to } .325$

d. The confidence intervals of parts a and b cover p, but the confidence interval of part c does not.

8.75 $n = 200$, $\hat{p} = .27$, $\hat{q} = 1 - \hat{p} = 1 - .27 = .73$, and $s_{\hat{p}} = \sqrt{\hat{p}\hat{q}/n} = \sqrt{(.27)(.73)/200} = .03139267$

a. The 99% confidence interval for p is

$\hat{p} \pm z s_{\hat{p}} = .27 \pm 2.58(.03139267) = .27 \pm .081 = .189 \text{ to } .351$

b. The 97% confidence interval for p is

$\hat{p} \pm z s_{\hat{p}} = .27 \pm 2.17(.03139267) = .27 \pm .068 = .202 \text{ to } .338$

c. The 90% confidence interval for p is

$\hat{p} \pm z s_{\hat{p}} = .27 \pm 1.65(.03139267) = .27 \pm .052 = .218 \text{ to } .322$

d. Yes, the width of the confidence intervals decreases as the confidence level decreases. This occurs because as the confidence level decreases, the value of z decreases.

8.77 $\hat{p} = .31$ and $\hat{q} = 1 - \hat{p} = 1 - .31 = .69$

a. $n = 1200$ and $s_{\hat{p}} = \sqrt{\hat{p}\hat{q}/n} = \sqrt{(.31)(.69)/1200} = .01335103$

The 95% confidence interval for p is $\hat{p} \pm z s_{\hat{p}} = .31 \pm 1.96(.01335103) = .31 \pm .026 = .284 \text{ to } .336$

b. $n = 500$ and $s_{\hat{p}} = \sqrt{\hat{p}\hat{q}/n} = \sqrt{(.31)(.69)/500} = .02068333$

The 95% confidence interval for p is $\hat{p} \pm z s_{\hat{p}} = .31 \pm 1.96(.02068333) = .31 \pm .041 = .269 \text{ to } .351$

c. $n = 80$ and $s_{\hat{p}} = \sqrt{\hat{p}\hat{q}/n} = \sqrt{(.31)(.69)/80} = .05170832$

The 95% confidence interval for p is $\hat{p} \pm z s_{\hat{p}} = .31 \pm 1.96(.05170832) = .31 \pm .101 = .209$ to $.411$

d. Yes, the width of the confidence intervals increases as the sample size decreases. This occurs because decreasing the sample size increases the standard deviation of the sample proportion.

8.79 a. $E = .045$, $\hat{p} = .53$, $\hat{q} = 1 - \hat{p} = 1 - .53 = .47$, and $z = 2.33$ for 98% confidence level

$$n = \frac{z^2 \hat{p}\hat{q}}{E^2} = \frac{(2.33)^2(.53)(.47)}{(.045)^2} = 667.82 \approx 668$$

b. $E = .045$, $p = q = .50$ for the most conservative sample size, and $z = 2.33$ for 98% confidence level

$$n = \frac{z^2 pq}{E^2} = \frac{(2.33)^2(.50)(.50)}{(.045)^2} = 670.23 \approx 671$$

8.81 a. $E = .03$, $\hat{p} = .32$, $\hat{q} = 1 - \hat{p} = 1 - .32 = .68$, and $z = 2.58$ for 99% confidence level

$$n = \frac{z^2 \hat{p}\hat{q}}{E^2} = \frac{(2.58)^2(.32)(.68)}{(.03)^2} = 1609.37 \approx 1610$$

b. $E = .04$, $\hat{p} = .78$, $\hat{q} = 1 - \hat{p} = 1 - .78 = .22$, and $z = 1.96$ for 95% confidence level

$$n = \frac{z^2 \hat{p}\hat{q}}{E^2} = \frac{(1.96)^2(.78)(.22)}{(.04)^2} = 412.01 \approx 413$$

c. $E = .02$, $\hat{p} = .64$, $\hat{q} = 1 - \hat{p} = 1 - .64 = .36$, and $z = 1.65$ for 90% confidence level

$$n = \frac{z^2 \hat{p}\hat{q}}{E^2} = \frac{(1.65)^2(.64)(.36)}{(.02)^2} = 1568.16 \approx 1569$$

8.83 $n = 200$, $\hat{p} = 74/200 = .37$, $\hat{q} = 1 - \hat{p} = 1 - .37 = .63$, and

$s_{\hat{p}} = \sqrt{\hat{p}\hat{q}/n} = \sqrt{(.37)(.63)/200} = .03413942$

a. The 98% confidence interval for p is

$\hat{p} \pm z s_{\hat{p}} = .37 \pm 2.33(.03413942) = .37 \pm .080 = .290$ to $.450$

The corresponding interval for the population percentage is 29% to 45%.

b. The width of the confidence interval constructed in part a may be reduced by:

1. Lowering the confidence level
2. Increasing the sample size

The second alternative is better because lowering the confidence level results in a less reliable estimate of p.

8.85 $n = 240$, $\hat{p} = 96/240 = .40$, $\hat{q} = 1 - \hat{p} = 1 - .40 = .60$, and

$s_{\hat{p}} = \sqrt{\hat{p}\hat{q}/n} = \sqrt{(.40)(.60)/240} = .03162278$

a. $\hat{p} = .40$, so the point estimate for the population percentage is 40%

b. The 97% confidence interval for p is

$\hat{p} \pm z s_{\hat{p}} = .40 \pm 2.17(.03162278) = .40 \pm .069 = .331 \text{ to } .469$

The corresponding interval for the population percentage is 33.1% to 46.9%.

8.87 a. $n = 50$, $\hat{p} = 19/50 = .38$, $\hat{q} = 1 - \hat{p} = 1 - .38 = .62$, and

$s_{\hat{p}} = \sqrt{\hat{p}\hat{q}/n} = \sqrt{(.38)(.62)/50} = .06864401$

The 99% confidence interval for p is

$\hat{p} \pm z s_{\hat{p}} = .38 \pm 2.58(.06864401) = .38 \pm .177 = .203 \text{ to } .557$

The corresponding interval for the population percentage is 20.3% to 55.7%.

b. The width of the confidence interval constructed in part a may be reduced by:

1. Lowering the confidence level
2. Increasing the sample size

The second alternative is better because lowering the confidence level gives a less reliable estimate of p.

8.89 $n = 7861$, $\hat{p} = .18$, $\hat{q} = 1 - \hat{p} = 1 - .18 = .82$, and $s_{\hat{p}} = \sqrt{\hat{p}\hat{q}/n} = \sqrt{(.18)(.82)/7861} = .00433316$

a. The 99% confidence interval for p is

$\hat{p} \pm z s_{\hat{p}} = .18 \pm 2.58(.00433316) = .18 \pm .011 = .169 \text{ to } .191$

b. The sample proportion of .18 is an estimate of p based on a random sample. Because of sampling error, this estimate might differ from the true proportion p, so we make an interval estimate to allow for this uncertainty and sampling error.

8.91 $n = 24$, $\hat{p} = 8/24 = .333$, $\hat{q} = 1 - \hat{p} = 1 - .333 = .667$, and

$s_{\hat{p}} = \sqrt{\hat{p}\hat{q}/n} = \sqrt{(.333)(.667)/24} = .09620096$

a. $\hat{p} = .333$

b. The 99% confidence interval for p is

$$\hat{p} \pm z s_{\hat{p}} = .333 \pm 2.58(.09620096) = .333 \pm .248 = .085 \text{ to } .581$$

The corresponding interval for the population percentage is 8.5% to 58.1%.

8.93 $E = .02$, $\hat{p} = .93$, $\hat{q} = 1 - \hat{p} = 1 - .93 = .07$, and $z = 2.58$ for 99% confidence level

$$n = \frac{z^2 \hat{p}\hat{q}}{E^2} = \frac{(2.58)^2(.93)(.07)}{(.02)^2} = 1083.33 \approx 1084$$

8.95 $E = .03$, $p = q = .50$ for the most conservative sample size, and $z = 2.58$ for 99% confidence level

$$n = \frac{z^2 pq}{E^2} = \frac{(2.58)^2(.50)(.50)}{(.03)^2} = 1849$$

8.97 To estimate the percentage of students who are satisfied with campus food services:

1. Take a random sample of 30 students who use campus food services.
2. Determine the number of students in this sample who are satisfied with the campus food services.
3. Calculate $\hat{p}$ and $s_{\hat{p}}$.
4. Choose the confidence level and find the required value of z in the normal distribution table.
5. Construct the confidence interval for p using the formula $\hat{p} \pm z s_{\hat{p}}$.
6. Obtain the corresponding interval for the population percentage by multiplying the upper and lower limits of the confidence interval for p by 100%.

Supplementary Exercises

8.99 $n = 100$, $\bar{x} = \$2640$, $\sigma = \$578$, and $\sigma_{\bar{x}} = \sigma/\sqrt{n} = 578/\sqrt{100} = \$57.80$

a. $\bar{x} = \$2640$

b. The 97% confidence interval for μ is

$$\bar{x} \pm z\sigma_{\bar{x}} = 2640 \pm 2.17(57.80) = 2640 \pm 125.43 = \$2514.57 \text{ to } \$2765.43$$

8.101 $n = 20$, $\bar{x} = 3.99$ inches, $\sigma = .04$ inch, and $\sigma_{\bar{x}} = \sigma/\sqrt{n} = .04/\sqrt{20} = .00894427$ inch

The 98% confidence interval for μ is

$\bar{x} \pm z\sigma_{\bar{x}} = 3.99 \pm 2.33(.00894427) = 3.99 \pm .021 = 3.969 \text{ to } 4.011$ inches

Since the lower limit of the confidence interval is 3.969, which is less than 3.98, the machine needs an adjustment.

8.103 $n = 36$, $\sum x = 547$, and $\sum x^2 = 9265$

$\bar{x} = (\sum x)/n = 547/36 = 15.19$ hours

$$s = \sqrt{\frac{\sum x^2 - \frac{(\sum x)^2}{n}}{n-1}} = \sqrt{\frac{9265 - \frac{(547)^2}{36}}{36-1}} = 5.21984917 \text{ hours}$$

$s_{\bar{x}} = s/\sqrt{n} = 5.21984917/\sqrt{36} = .86997486$ hour

$df = n - 1 = 36 - 1 = 35$, $\alpha/2 = .5 - (.99/2) = .005$, and $t = 2.724$

The 99% confidence interval for μ is

$\bar{x} \pm ts_{\bar{x}} = 15.19 \pm 2.724(.86997486) = 15.19 \pm 2.37 = 12.82 \text{ to } 17.56$ hours

8.105 $n = 18$, $\bar{x} = 24$ minutes, $s = 4.5$ minutes, and $s_{\bar{x}} = s/\sqrt{n} = 4.5/\sqrt{18} = 1.06066017$ minutes

$df = n - 1 = 18 - 1 = 17$, $\alpha/2 = .5 - (.95/2) = .025$, and $t = 2.110$

The 95% confidence interval for μ is

$\bar{x} \pm ts_{\bar{x}} = 24 \pm 2.110(1.06066017) = 24 \pm 2.24 = 21.76 \text{ to } 26.24$ minutes

8.107 Since $n = 300$ is very large, we use the normal distribution to approximate the t distribution, and construct the confidence interval using $\bar{x} \pm zs_{\bar{x}}$.

$n = 300$, $\bar{x} = 4.5$ hours, $s = .75$ hour, and $s_{\bar{x}} = s/\sqrt{n} = .75/\sqrt{300} = .04330127$ hour

The 98% confidence interval for μ is $\bar{x} \pm zs_{\bar{x}} = 4.5 \pm 2.33(.04330127) = 4.5 \pm .10 = 4.40 \text{ to } 4.60$ hours

8.109 $n = 10$, $\sum x = 1514$, and $\sum x^2 = 229,646$

$\bar{x} = (\sum x)/n = 1514/10 = 151.40$ calories

$$s = \sqrt{\frac{\sum x^2 - \frac{(\sum x)^2}{n}}{n-1}} = \sqrt{\frac{229,646 - \frac{(1514)^2}{10}}{10-1}} = 6.88315173 \text{ calories}$$

$s_{\bar{x}} = s/\sqrt{n} = 6.88315173/\sqrt{10} = 2.17664369$ calories

$df = n - 1 = 10 - 1 = 9$, $\alpha/2 = .5 - (.99/2) = .005$, and $t = 3.250$

The 99% confidence interval for μ is

$\bar{x} \pm ts_{\bar{x}} = 151.40 \pm 3.250(2.17664369) = 151.40 \pm 7.07 = 144.33 \text{ to } 158.47$ calories

8.111 $n = 500$, $\hat{p} = .44$, $\hat{q} = 1 - \hat{p} = 1 - .44 = .56$, and $s_{\hat{p}} = \sqrt{\hat{p}\hat{q}/n} = \sqrt{(.44)(.56)/500} = .02219910$

a. $\hat{p} = .44$, so the point estimate for the population percentage is 44%

b. The 90% confidence interval for p is

$\hat{p} \pm zs_{\hat{p}} = .44 \pm 1.65(.02219910) = .44 \pm .037 = .403 \text{ to } .477$

The corresponding interval for the population percentage is 40.3% to 47.7%.

8.113 $n = 16$, $\hat{p} = 5/16 = .313$, $\hat{q} = 1 - \hat{p} = 1 - .313 = .687$, and

$s_{\hat{p}} = \sqrt{\hat{p}\hat{q}/n} = \sqrt{(.313)(.687)/16} = .11592859$

The 97% confidence interval for p is $\hat{p} \pm zs_{\hat{p}} = .313 \pm 2.17(.11592859) = .313 \pm .252 = .061 \text{ to } .565$

The corresponding interval for the population percentage is 6.1% to 56.5%.

8.115 $E = \$3500$, $\sigma = \$31{,}500$, and $z = 1.65$ for 90% confidence level

$$n = \frac{z^2\sigma^2}{E^2} = \frac{(1.65)^2(31{,}500)^2}{(3500)^2} = 220.52 \approx 221$$

8.117 $E = .05$, $\hat{p} = .63$, $\hat{q} = 1 - \hat{p} = 1 - .63 = .37$, and $z = 1.96$ for 95% confidence level

$$n = \frac{z^2\hat{p}\hat{q}}{E^2} = \frac{(1.96)^2(.63)(.37)}{(.05)^2} = 358.19 \approx 359$$

8.119 Let p_1 = proportion of adults who consider job security essential, p_2 = proportion of adults who consider health benefits essential, p_3 = proportion of adults who consider interesting work essential, and p_4 = proportion of adults who consider a good salary essential. Note that for $n = 601$, $n\hat{p}$ and $n\hat{q}$ exceed 5 for all of these proportions, so the sample is considered large.

$\hat{p}_1 = .71$, $\hat{q}_1 = 1 - \hat{p}_1 = 1 - .71 = .29$, and $s_{\hat{p}_1} = \sqrt{\hat{p}_1\hat{q}_1/n} = \sqrt{(.71)(.29)/601} = .01850934$

The 95% confidence interval for p_1 is $\hat{p}_1 \pm zs_{\hat{p}_1} = .71 \pm 1.96(.01850934) = .71 \pm .036 = .674 \text{ to } .746$

The corresponding interval for the population percentage is 67.4% to 74.6%.

$\hat{p}_2 = .63$, $\hat{q}_2 = 1 - \hat{p}_2 = 1 - .63 = .37$, and $s_{\hat{p}_2} = \sqrt{\hat{p}_2\hat{q}_2/n} = \sqrt{(.63)(.37)/601} = .01969400$

The 95% confidence interval for p_2 is $\hat{p}_2 \pm zs_{\hat{p}_2} = .63 \pm 1.96(.01969400) = .63 \pm .039 = .591 \text{ to } .669$

The corresponding interval for the population percentage is 59.1% to 66.9%.

$\hat{p}_3 = .60$, $\hat{q}_3 = 1 - \hat{p}_3 = 1 - .60 = .40$, and $s_{\hat{p}_3} = \sqrt{\hat{p}_3\hat{q}_3 / n} = \sqrt{(.60)(.40)/601} = .01998335$

The 95% confidence interval for p_3 is $\hat{p}_3 \pm z s_{\hat{p}_3} = .60 \pm 1.96(.01998335) = .60 \pm .039 = .561$ to $.639$

The corresponding interval for the population percentage is 56.1% to 63.9%.

$\hat{p}_4 = .56$, $\hat{q}_4 = 1 - \hat{p}_4 = 1 - .56 = .44$, and $s_{\hat{p}_4} = \sqrt{\hat{p}_4\hat{q}_4 / n} = \sqrt{(.56)(.44)/601} = .02024805$

The 95% confidence interval for p_4 is $\hat{p}_4 \pm z s_{\hat{p}_4} = .56 \pm 1.96(.02024805) = .56 \pm .040 = .520$ to $.600$

The corresponding interval for the population percentage is 52.0% to 60.0%.

(1) A confidence interval is a range of numbers (in this particular case proportions or percentages) which gives an estimate for the true value of the population parameter. (2) A single percentage that we assign as an estimate would almost always differ from the true value, hence a range with the associated confidence level is more informative. (3) The 95% means that we are 95% confident that this interval, calculated from this particular sample, actually contains the true value of the population parameter. (4) We assume that the 601 persons selected for the sample constitute a random sample of adults.

8.121 $E = 1.0$ mile, $s = 4.1$ miles, and $z = 1.96$ for 95% confidence level; σ may be estimated by s

$$n = \frac{z^2 s^2}{E^2} = \frac{(1.96)^2 (4.1)^2}{(1.0)^2} = 64.58 \approx 65$$

Thus, an additional $65 - 20 = 45$ observations must be taken.

8.123 a. $E = 100$ cars, $\sigma = 170$ cars, and $z = 2.58$ for 99% confidence level

$$n = \frac{z^2 \sigma^2}{E^2} = \frac{(2.58)^2 (170)^2}{(100)^2} = 19.24 \approx 20$$

Note that since $n < 30$, we must assume that the number of cars passing each day is approximately normally distributed, or we may take a larger $(n \geq 30)$ sample.

b. Since $n = \dfrac{z^2 \sigma^2}{E^2}$, $z = \dfrac{E\sqrt{n}}{\sigma} = \dfrac{100\sqrt{20}}{272} = 1.64$ which corresponds to a confidence level of approximately 90%.

c. Since $n = \dfrac{z^2 \sigma^2}{E^2}$, $E = \dfrac{z\sigma}{\sqrt{n}} = \dfrac{(2.58)(130)}{\sqrt{20}} = 75$.

Thus, they can be 99% confident that their point estimate is within 75 cars of the true average.

8.125 When the sample size is doubled, the margin of error is reduced by a factor of $\sqrt{2}/2$. When the sample size is quadrupled, the margin of error is reduced by a factor of 1/2. This relationship does not hold true when calculating a confidence interval for the population mean with an unknown population standard deviation for the following reasons:

1) While σ is constant, the sample standard deviation, s, is not. The sample standard deviation will change with each sample.
2) If σ is unknown, the confidence interval is calculated with a t value. This value will change as the sample size changes because the degrees of freedom will change.

8.127 Answers will vary. The following provide one possible example for each case.

a. $n = 200$, $\hat{p} = .01$, $\hat{q} = 1 - \hat{p} = 1 - .01 = .99$, and

$$s_{\hat{p}} = \sqrt{\hat{p}\hat{q}/n} = \sqrt{(.01)(.99)/200} = .00703562$$

The 95% confidence interval for p is

$$\hat{p} \pm z s_{\hat{p}} = .01 \pm 1.96(.00703562) = .01 \pm .014 = -.004 \text{ to } .024$$

b. $n = 160$, $\hat{p} = .9875$, $\hat{q} = 1 - \hat{p} = 1 - .9875 = .0125$, and

$$s_{\hat{p}} = \sqrt{\hat{p}\hat{q}/n} = \sqrt{(.9875)(.0125)/160} = .00878342$$

The 95% confidence interval for p is

$$\hat{p} \pm z s_{\hat{p}} = .9875 \pm 1.96(.00878342) = .9875 \pm .0172 = .9703 \text{ to } 1.0047$$

The value of the lower limit in the first interval is less than 0, and the value of the upper limit in the second interval is greater than 1. These are not possible values for proportions.

Self – Review Test

1. a. Estimation means assigning values to a *population parameter* based on the value of a *sample statistic*.

b. An estimator is the *sample statistic* used to estimate a *population parameter*.

c. The value of a *sample statistic* is called the point estimate of the corresponding *population parameter*.

2. b **3.** a **4.** a **5.** c **6.** b

7. $n = 36$, $\bar{x} = \$159{,}000$, $\sigma = \$27{,}000$, and $\sigma_{\bar{x}} = \sigma/\sqrt{n} = 27{,}000/\sqrt{36} = \$4500$

a. $\bar{x} = \$159{,}000$

b. The 99% confidence interval for μ is

$\bar{x} \pm z\sigma_{\bar{x}} = 159{,}000 \pm 2.58(4500) = 159{,}000 \pm 11{,}610 = \$147{,}390$ to $\$170{,}610$

$E = z\sigma_{\bar{x}} = 2.58(4500) = \$11{,}610$

8. $n = 25$, $\bar{x} = \$410{,}425$, $s = \$74{,}820$, and $s_{\bar{x}} = s/\sqrt{n} = 74{,}820/\sqrt{25} = \$14{,}964$

$df = n - 1 = 25 - 1 = 24$, $\alpha/2 = .5 - (.95/2) = .025$, and $t = 2.064$

The 95% confidence interval for μ is

$\bar{x} \pm ts_{\bar{x}} = 410{,}425 \pm 2.064(14{,}964) = 410{,}425 \pm 30{,}885.70 = \$379{,}539.30$ to $\$441{,}310.70$

9. $n = 1590$, $\hat{p} = .508$, $\hat{q} = 1 - \hat{p} = 1 - .508 = .492$, and

$s_{\hat{p}} = \sqrt{\hat{p}\hat{q}/n} = \sqrt{(.508)(.492)/1590} = .01253764$

a. $\hat{p} = .508$

b. The 95% confidence interval for p is $\hat{p} \pm zs_{\hat{p}} = .508 \pm 1.96(.01253764) = .508 \pm .025 = .483$ to $.533$

$E = zs_{\hat{p}} = 1.96(.01253764) = .025$

10. $E = .65$ houses, $\sigma = 2.2$ houses, and $z = 1.96$ for 95% confidence level

$$n = \frac{z^2\sigma^2}{E^2} = \frac{(1.96)^2(2.2)^2}{(.65)^2} = 44.01 \approx 45.$$

11. $E = .05$, $p = q = .50$ for the most conservative sample size, and $z = 1.65$ for 90% confidence level

$$n = \frac{z^2 pq}{E^2} = \frac{(1.65)^2(.50)(.50)}{(.05)^2} = 272.25 \approx 273.$$

12. $E = .05$, $\hat{p} = .70$, $\hat{q} = 1 - \hat{p} = 1 - .70 = .30$, and $z = 1.65$ for 90% confidence level

$$n = \frac{z^2 \hat{p}\hat{q}}{E^2} = \frac{(1.65)^2(.70)(.30)}{(.05)^2} = 228.69 \approx 229.$$

13. The width of the confidence interval can be reduced by:

1. Lowering the confidence level
2. Increasing the sample size

The second alternative is better because lowering the confidence level results in a less reliable estimate for μ.

14. To estimate the mean number of hours that all students at your college work per week:

1. Take a random sample of 12 students from your college who hold jobs.
2. Record the number of hours each of these students worked last week.
3. Calculate $\bar{x}$, s and $s_{\bar{x}}$ from these data.
4. After choosing the confidence level, find the value for the t distribution with 11 *df* and for an area of $\alpha/2$ in the right tail.
5. Construct the confidence interval for μ by using the formula $\bar{x} \pm ts_{\bar{x}}$.

   You are assuming that the hours worked by all students at your college have a normal distribution.

15. To estimate the proportion of people who are happy with their current jobs:

1. Take a random sample of 35 workers.
2. Determine whether or not each worker is happy with his or her job.
3. Calculate $\hat{p}$, $\hat{q}$, and $s_{\hat{p}}$.
4. Choose the confidence level and find the required value of z from the normal distribution table.
5. Construct the confidence interval for p by using the formula $\hat{p} \pm zs_{\hat{p}}$.

Chapter Nine

Section 9.1

9.1

a. The **null hypothesis** is a claim about a population parameter that is assumed to be true until it is declared false.

b. An **alternative hypothesis** is a claim about a population parameter that will be true if the null hypothesis is false.

c. The **critical point(s)** divide(s) the whole area under a distribution curve into rejection and non-rejection regions.

d. The **significance level**, denoted by α, is the probability of making a Type I error, that is, the probability of rejecting the null hypothesis when it is actually true.

e. The **nonrejection region** is the area where the null hypothesis is not rejected.

f. The **rejection region** is the area where the null hypothesis is rejected.

g. A hypothesis test is a **two-tailed test** if the rejection regions are in both tails of the distribution curve; it is a **left-tailed test** if the rejection region is in the left tail; and it is a **right-tailed test** if the rejection region is in the right tail.

h. A **Type I error** occurs when a true null hypothesis is rejected. The probability of committing a Type I error is $\alpha = P(H_0 \text{ is rejected} \mid H_0 \text{ is true})$. A **Type II error** occurs when a false null hypothesis is not rejected. The probability of committing a Type II error is $\beta = P(H_0 \text{ is not rejected} \mid H_0 \text{ is false})$.

9.3 A hypothesis test is a two-tailed test if the sign in the alternative hypothesis is "≠"; it is a left-tailed test if the sign in the alternative hypothesis is "<"; and it is a right-tailed test if the sign in the alternative hypothesis is ">". Table 9.3 on page 385 of the text describes these relationships.

9.5 a. Left-tailed test because the rejection region lies in the left tail of the distribution curve.

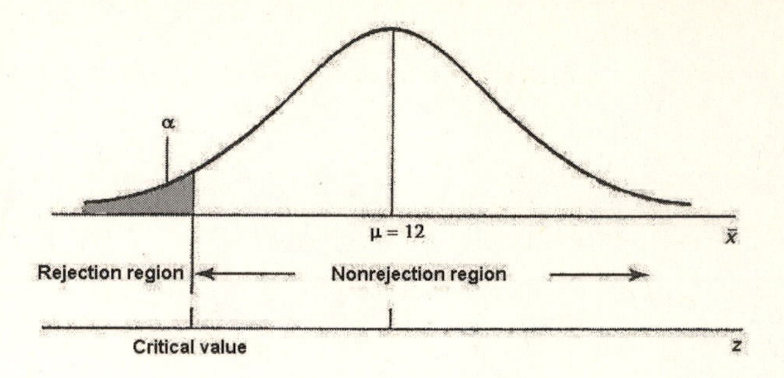

b. Right-tailed test because the rejection region is in the right tail of the distribution curve.

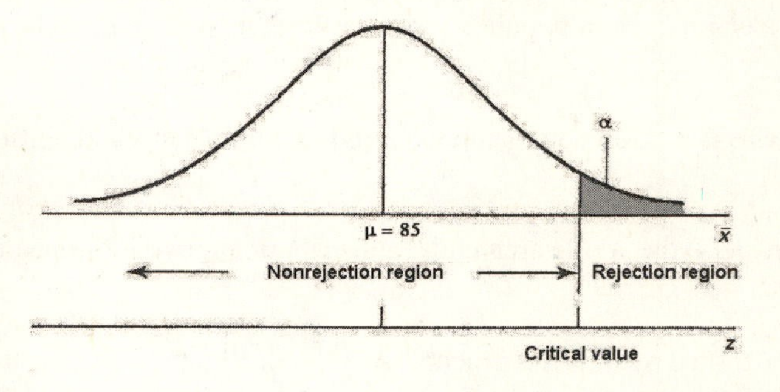

c. Two-tailed test because the rejection region lies in both tails of the distribution curve.

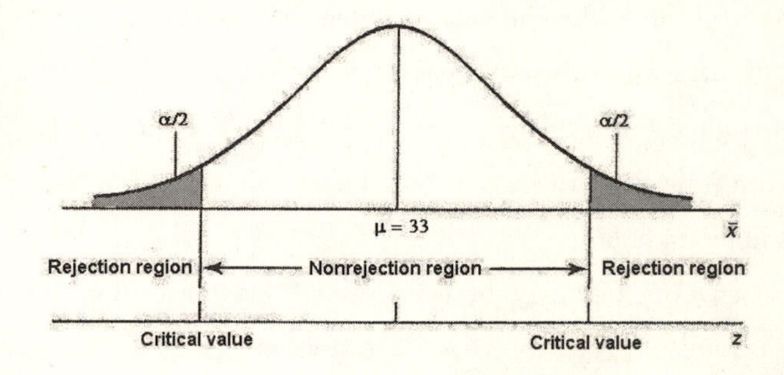

9.7 a. Type II error b. Type I error

9.9 a. H_0: $\mu = 20$ hours, H_1: $\mu \neq 20$ hours, two-tailed test

b. H_0: $\mu = 10$ hours, H_1: $\mu > 10$ hours, right-tailed test

c. H_0: $\mu = 3$ years, H_1: $\mu \neq 3$ years, two-tailed test

d. H_0: $\mu = \$1000$, H_1: $\mu < \$1000$, left-tailed test

e. H_0: $\mu = 12$ minutes, H_1: $\mu > 12$ minutes, right-tailed test

Section 9.2

9.11 The five steps of a test of hypothesis using the **critical value approach** are:

1. State the null and alternative hypotheses.
2. Select the distribution to use.
3. Determine the rejection and nonrejection regions.
4. Calculate the value of the test statistic.
5. Make a decision.

9.13 Rejecting H_0 is equivalent to stating that the evidence from the sample is strong enough to claim that H_1 is true.

9.15 For a two-tailed test, the ***p*-value** is twice the area in the tail of the sampling distribution curve beyond the observed value of the sample test statistic. For a one-tailed test, the p-value is the area in the tail of the sampling distribution curve beyond the observed value of the sample test statistic.

9.17 a. Step 1: H_0: $\mu = 46$, H_1: $\mu \neq 46$

Step 2: Since $n > 30$, use the normal distribution.

Step 3: $\sigma_{\bar{x}} = \sigma/\sqrt{n} = 9.7/\sqrt{40} = 1.53370467$

$z = (\bar{x} - \mu)/\sigma_{\bar{x}} = (49.60 - 46)/1.53370467 = 2.35$

From the normal distribution table, area to the right of $z = 2.35$ is $1 - .9906 = .0094$.

p-value $= 2(.0094) = .0188$

b. Step 1: H_0: $\mu = 26$, H_1: $\mu < 26$

Step 2: Since $n > 30$, use the normal distribution.

Step 3: $\sigma_{\bar{x}} = \sigma/\sqrt{n} = 4.3/\sqrt{33} = .74853392$

$z = (\bar{x} - \mu)/\sigma_{\bar{x}} = (24.3 - 26)/.74853392 = -2.27$

From the normal distribution table, area to the left of $z = -2.27$ is .0116.

p-value $= .0116$

c. Step 1: H_0: $\mu = 18$, H_1: $\mu > 18$

Step 2: Since $n > 30$, use the normal distribution.

Step 3: $\sigma_{\bar{x}} = \sigma/\sqrt{n} = 7.8/\sqrt{55} = 1.05175179$

$z = (\bar{x} - \mu)/\sigma_{\bar{x}} = (20.50 - 18)/1.05175179 = 2.38$

From the normal distribution table, area to the right of $z = 2.38$ is $1 - .9913 = .0087$.

p-value $= .0087$

9.19 a. Step 1: H_0: $\mu = 72$, H_1: $\mu > 72$

Step 2: Since the population is normally distributed, use the normal distribution.

Step 3: $\sigma_{\bar{x}} = \sigma/\sqrt{n} = 6/\sqrt{16} = 1.5$

$z = (\bar{x} - \mu)/\sigma_{\bar{x}} = (75.2 - 72) / 1.5 = 2.13$

The area under the standard normal curve to the right of $z = 2.13$ is $1 - .9834 =$ .0166. p-value = .0166

b. For $\alpha = .01$, do not reject H_0 since $.0166 > .01$.

c. For $\alpha = .025$, reject H_0 since $.0166 < .025$.

9.21 a.

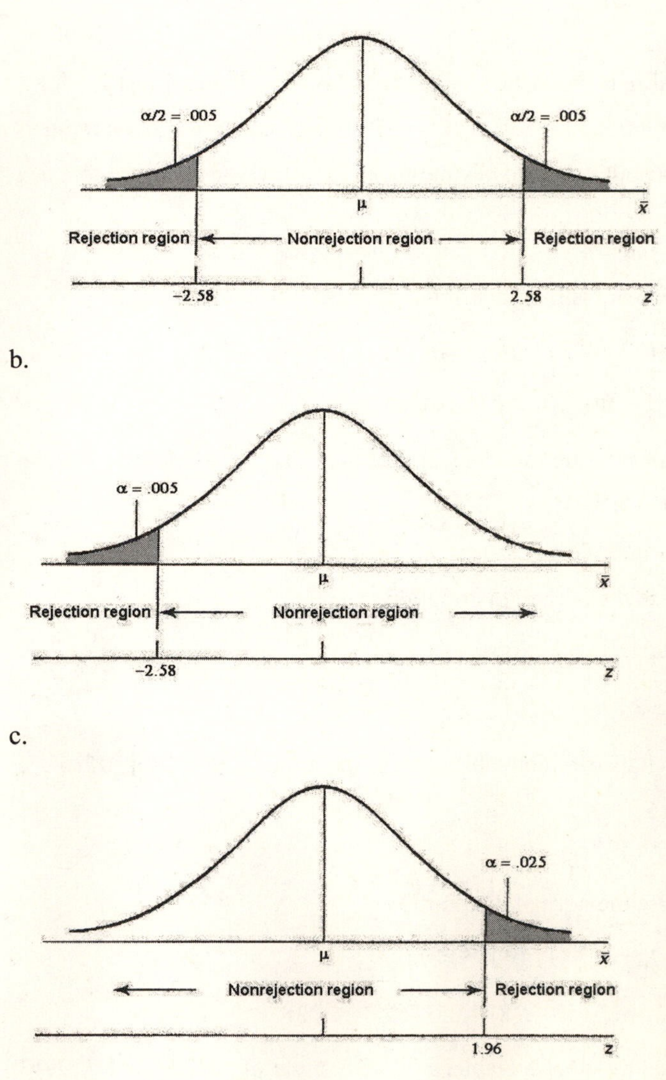

9.23 If H_0 is not rejected, the difference between the hypothesized value of μ and the observed value of $\bar{x}$ is "statistically not significant".

9.25 a. .10 b. .02 c. .005

9.27 a. Step 1: H_0: $\mu = 20$, H_1: $\mu < 20$

Step 2: Since the population is normally distributed, use the normal distribution.

Step 3: For $\alpha = .01$, the critical value of z is –2.33.

Step 4: $\sigma_{\bar{x}} = \sigma/\sqrt{n} = 4/\sqrt{28} = .75592895$

$z = (\bar{x} - \mu)/\sigma_{\bar{x}} = (15 - 20)/.75592895 = -6.61$

b. Step 1: H_0: $\mu = 20$, H_1: $\mu \neq 20$

Step 2: Since the population is normally distributed, use the normal distribution.

Step 3: For $\alpha = .01$, the critical values of z are –2.58 and 2.58.

Step 4: $\sigma_{\bar{x}} = \sigma/\sqrt{n} = 4/\sqrt{28} = .75592895$

$z = (\bar{x} - \mu)/\sigma_{\bar{x}} = (15 - 20)/.75592895 = -6.61$

9.29 a.

b.

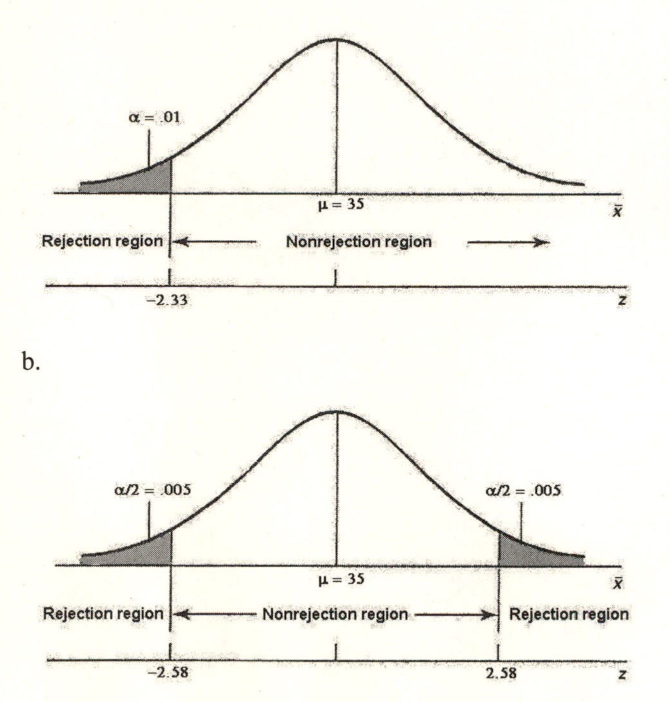

c.

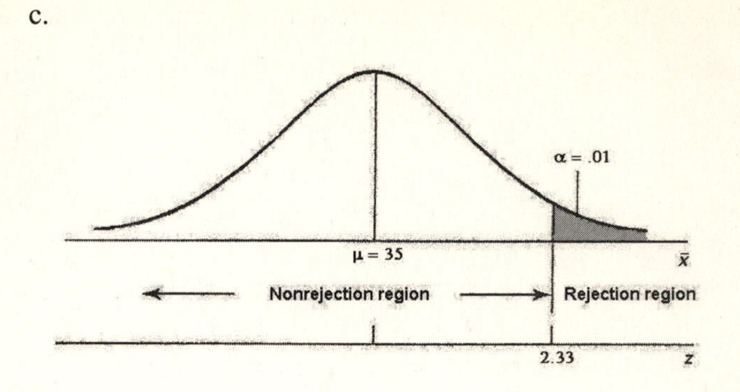

9.31 a. Step 1: H_0: $\mu = 45$, H_1: $\mu < 45$

Step 2: Since the population is normally distributed, use the normal distribution.

Step 3: For $\alpha = .025$, the critical value of z is -1.96.

Step 4: $\sigma_{\bar{x}} = \sigma/\sqrt{n} = 6/\sqrt{25} = 1.20$

$z = (\bar{x} - \mu)/\sigma_{\bar{x}} = (41.8 - 45)/1.20 = -2.67$

Step 5: Reject H_0 since $-2.67 < -1.96$.

b. Step 1: H_0: $\mu = 45$, H_1: $\mu < 45$

Step 2: Since the population is normally distributed, use the normal distribution.

Step 3: For $\alpha = .025$, the critical value of z is -1.96.

Step 4: $\sigma_{\bar{x}} = \sigma/\sqrt{n} = 6/\sqrt{25} = 1.20$

$z = (\bar{x} - \mu)/\sigma_{\bar{x}} = (43.8 - 45)/1.20 = -1.00$

Step 5: Do not reject H_0 since $-1.00 > -1.96$.

Comparing parts a and b shows that two samples selected from the same population can lead to opposite conclusions on the same test of hypothesis. This occurs because the samples may yield different values for the sample mean.

9.33 a. Step 1: H_0: $\mu = 80$, H_1: $\mu \neq 80$

Step 2: Since $n > 30$, use the normal distribution.

Step 3: For $\alpha = .10$, the critical values of z are -1.65 and 1.65.

Step 4: $\sigma_{\bar{x}} = \sigma/\sqrt{n} = 15/\sqrt{33} = 2.61116484$

$z = (\bar{x} - \mu)/\sigma_{\bar{x}} = (76.5 - 80)/2.61116484 = -1.34$

Step 5: Do not reject H_0 since $-1.34 > -1.65$.

b. Step 1: H_0: $\mu = 32$, H_1: $\mu < 32$

Step 2: Since $n > 30$, use the normal distribution.

Step 3: For $\alpha = .01$, the critical value of z is -2.33.

Step 4: $\sigma_{\bar{x}} = \sigma/\sqrt{n} = 7.4/\sqrt{75} = .85447840$

$z = (\bar{x} - \mu)/\sigma_{\bar{x}} = (26.5 - 32)/.85447840 = -6.44$

Step 5: Reject H_0 since $-6.44 < -2.33$.

c. Step 1: H_0: $\mu = 55$, H_1: $\mu > 55$

Step 2: Since $n > 30$, use the normal distribution.

Step 3: For $\alpha = .05$, the critical value of z is 1.65.

Step 4: $\sigma_{\bar{x}} = \sigma/\sqrt{n} = 4/\sqrt{40} = .63245553$

$z = (\bar{x} - \mu)/\sigma_{\bar{x}} = (60.5 - 55)/.63245553 = 8.70$

Step 5: Reject H_0 since $8.70 > 1.65$.

9.35 a. Step 1: H_0: $\mu = 45$ months, H_1: $\mu < 45$ months

Step 2: Since the population is normally distributed, use the normal distribution.

Step 3: $\sigma_{\bar{x}} = \sigma/\sqrt{n} = 4.5/\sqrt{24} = .91855865$ month

$z = (\bar{x} - \mu)/\sigma_{\bar{x}} = (43.05 - 45)/.91855865 = -2.12$

From the normal distribution table, area to the left of $z = -2.12$ is .0170.

p-value = .0170

Step 4: Reject H_0 since $.0170 < .025$.

b. Step 1: H_0: $\mu = 45$ months, H_1: $\mu < 45$ months

Step 2: Since the population is normally distributed, use the normal distribution.

Step 3: For $\alpha = .025$, the critical value of z is -1.96.

Step 4: $\sigma_{\bar{x}} = \sigma/\sqrt{n} = 4.5/\sqrt{24} = .91855865$ month

$z = (\bar{x} - \mu)/\sigma_{\bar{x}} = (43.05 - 45)/.91855865 = -2.12$

Step 5: Reject H_0 since $-2.12 < -1.96$.

For both parts a and b, conclude the mean life of these batteries is less than 45 months.

9.37 a. Step 1: H_0: $\mu = 2320$ square feet, H_1: $\mu > 2320$ square feet

Step 2: Since $n > 30$, use the normal distribution.

Step 3: $\sigma_{\bar{x}} = \sigma/\sqrt{n} = 312/\sqrt{400} = 15.6$ square feet

$z = (\bar{x} - \mu)/\sigma_{\bar{x}} = (2365 - 2320)/15.6 = 2.88$

From the normal distribution table, area to the right of $z = 2.88$ is $1 - .9980 = .0020$.

p-value = .0020

Step 4: Reject H_0 since $.0020 < .02$.

b. Step 1: H_0: $\mu = 2320$ square feet, H_1: $\mu > 2320$ square feet

Step 2: Since $n > 30$, use the normal distribution.

Step 3: For $\alpha = .02$, the critical value of z is 2.05.

Step 4: $\sigma_{\bar{x}} = \sigma/\sqrt{n} = 312/\sqrt{400} = 15.6$ square feet

$z = (\bar{x} - \mu)/\sigma_{\bar{x}} = (2365 - 2320)/15.6 = 2.88$

Step 5: Reject H_0 since $2.88 > 2.05$.

For both parts a and b, conclude the mean size of new homes in the United States exceeds 2320 square feet.

9.39 a. Step 1: H_0: $\mu = 10$ minutes, H_1: $\mu \neq 10$ minutes

Step 2: Since $n > 30$, use the normal distribution.

Step 3: $\sigma_{\bar{x}} = \sigma/\sqrt{n} = 3.80/\sqrt{100} = .38$ minute

$z = (\bar{x} - \mu)/\sigma_{\bar{x}} = (9.20 - 10)/.38 = -2.11$

From the normal distribution table, area to the left of $z = -2.11$ is .0174.

p-value = 2(.0174) = .0348

Step 4: For $\alpha = .02$, do not reject H_0 since $.0348 > .02$.

For $\alpha = .05$, reject H_0 since $.0348 < .05$.

b. Step 1: H_0: $\mu = 10$ minutes, H_1: $\mu \neq 10$ minutes

Step 2: Since $n > 30$, use the normal distribution.

Step 3: For $\alpha = .02$, the critical values of z are -2.33 and 2.33.

For $\alpha = .05$, the critical values of z are -1.96 and 1.96.

Step 4: $\sigma_{\bar{x}} = \sigma/\sqrt{n} = 3.80/\sqrt{100} = .38$ minute

$z = (\bar{x} - \mu)/\sigma_{\bar{x}} = (9.20 - 10)/.38 = -2.11$

Step 5: For $\alpha = .02$, do not reject H_0 since $-2.11 > -2.33$.

For $\alpha = .05$, reject H_0 since $-2.11 < -1.96$.

For both parts a and b, conclude the mean duration of long-distance calls made by residential customers does not differ from 10 minutes at $\alpha = .02$, but does differ from 10 minutes at $\alpha = .05$.

9.41 a. Step 1: H_0: $\mu = 32$ ounces, H_1: $\mu \neq 32$ ounces

Step 2: Since the population is normally distributed, use the normal distribution.

Step 3: $\sigma_{\bar{x}} = \sigma/\sqrt{n} = .15/\sqrt{25} = .03$ ounce

$z = (\bar{x} - \mu)/\sigma_{\bar{x}} = (31.93 - 32)/.03 = -2.33$

From the normal distribution table, area to the left of $z = -2.33$ is .01.

p-value = 2(.01) = .02

Step 4: For $\alpha = .01$, do not reject H_0 since .02 > .01.

For $\alpha = .05$, reject H_0 since .02 < .05.

b. Step 1: H_0: $\mu = 32$ ounces, H_1: $\mu \neq 32$ ounces

Step 2: Since the population is normally distributed, use the normal distribution.

Step 3: For $\alpha = .01$, the critical values of z are –2.58 and 2.58.

For $\alpha = .05$, the critical values of z are –1.96 and 1.96.

Step 4: $\sigma_{\bar{x}} = \sigma/\sqrt{n} = .15/\sqrt{25} = .03$ ounce

$z = (\bar{x} - \mu)/\sigma_{\bar{x}} = (31.93 - 32)/.03 = -2.33$

Step 5: For $\alpha = .01$, do not reject H_0 since –2.33 > –2.58.

For $\alpha = .05$, reject H_0 since –2.33 < –1.96.

For both parts a and b, the inspector will not stop this machine at $\alpha = .01$, but will stop this machine and adjust it at $\alpha = .05$.

9.43 a. Step 1: H_0: $\mu \geq \$35,000$, H_1: $\mu < \$35,000$

Step 2: Since $n > 30$, use the normal distribution.

Step 3: For $\alpha = .01$, the critical value of z is –2.33.

Step 4: $\sigma_{\bar{x}} = \sigma/\sqrt{n} = 5400/\sqrt{150} = \$440.90815370$

$z = (\bar{x} - \mu)/\sigma_{\bar{x}} = (33,400 - 35,000)/\$440.90815370 = -3.63$

Step 5: Reject H_0 since –3.63 < –2.33.

Conclude that the company should not open a restaurant in this area.

b. If $\alpha = 0$, there can be no rejection region, and we cannot reject H_0. Therefore, the decision would be "do not reject H_0."

9.45 a. The first two steps and the calculation of the observed value of the test statistic are the same for both the p-value approach and critical value approach.

Step 1: H_0: $\mu \geq 8$ hours, H_1: $\mu < 8$ hours

Step 2: Since the population is normally distributed, use the normal distribution.

$\sigma_{\bar{x}} = \sigma/\sqrt{n} = 2.1/\sqrt{20} = .46957428$ hour

$z = (\bar{x} - \mu)/\sigma_{\bar{x}} = (7.68 - 8)/.46957428 = -.68$

<u>p-value approach</u>

Step 3: From above, $z = -.68$.

From the normal distribution table, area to the left of $z = -.68$ is .2483.

p-value = .2483

Step 4: Do not reject H_0 since .2483 > .01.

critical value approach

Step 3: For $\alpha = .01$, the critical value of z is -2.33.

Step 4: From above, $z = -.68$.

Step 5: Do not reject H_0 since $-.68 > -2.33$.

Conclude that the claim that all homeowners spend an average of 8 hours on such chores is false.

b. For $\alpha = .025$, reject H_0 using p-value approach since .2483 > .025. The critical value of $z = -1.96$; do not reject H_0 since $-.68 > -1.96$.

The decisions in parts a and b are the same. The results of this sample are rather conclusive, since increasing the significance level from .01 to .025 does not change the decision.

Section 9.3

9.47 To use the ***t* distribution** in Case I, the following conditions must hold:

1. The population standard deviation σ is unknown.
2. The sample size is small (i.e., $n < 30$).
3. The population from which the sample is selected is normally distributed.

To use the t distribution in Case II, the following conditions must hold:

1. The population standard deviation σ is unknown.
2. The sample size is large (i.e., $n > 30$).

9.49 a. $df = n - 1 = 15 - 1 = 14$

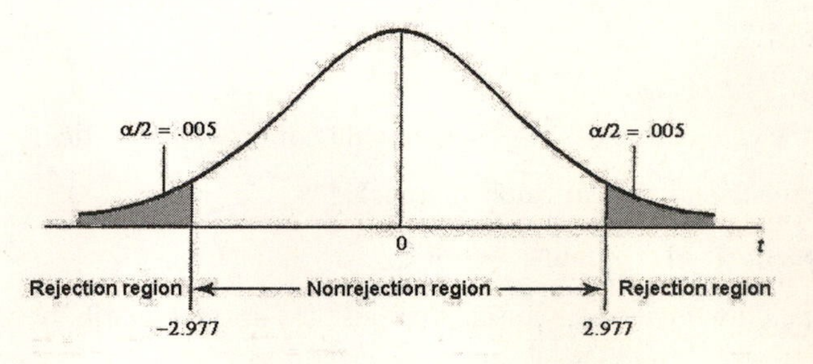

b. $df = n - 1 = 25 - 1 = 24$

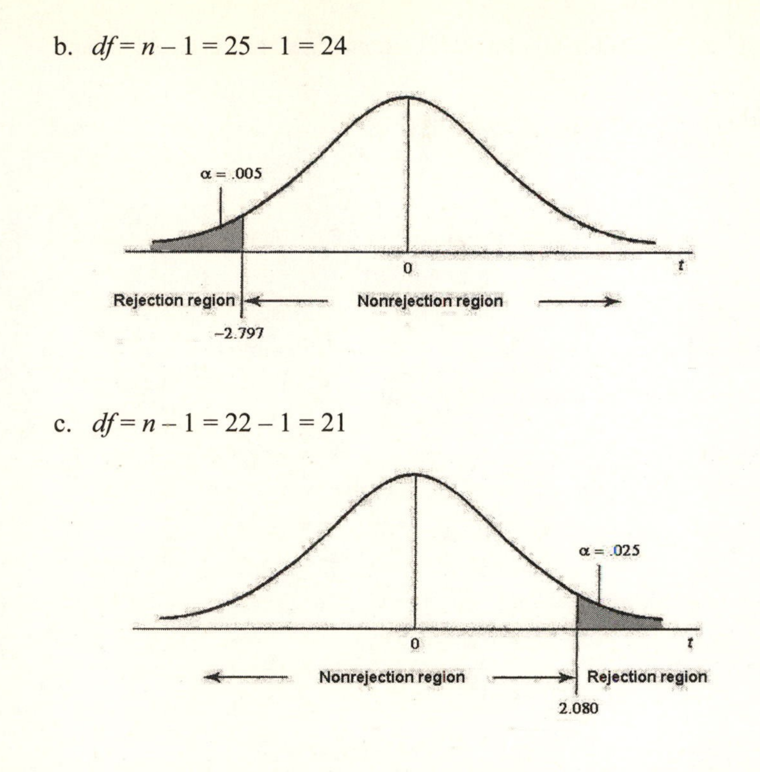

c. $df = n - 1 = 22 - 1 = 21$

9.51 a. The first two steps, the calculation of the observed value of the test statistic, and the number of degrees of freedom are the same for both the *p*-value approach and critical value approach.

Step 1: H_0: $\mu = 46$, H_1: $\mu < 46$

Step 2: Since the population standard deviation is unknown, the sample size is small ($n < 30$), and the population is normally distributed, we use the *t* distribution.

$s_{\bar{x}} = s/\sqrt{n} = 8/\sqrt{16} = 2.00$

$t = (\bar{x} - \mu)/s_{\bar{x}} = (42.4 - 46)/2.00 = -1.800$

$df = n - 1 = 16 - 1 = 15$

<u>*p*-value approach</u>

Step 3: From above, $t = -1.800$ and $df = 15$.

From the *t* distribution table, $-2.131 < -1.800 < -1.753$; hence, $.05 < p\text{-value} < .10$.

<u>critical value approach</u>

Step 3: For $\alpha = .05$ with $df = 15$, the critical value of *t* is -1.753.

b. The first two steps, the calculation of the observed value of the test statistic, and the number of degrees of freedom are the same for both the *p*-value approach and critical value approach.

Step 1: H_0: $\mu = 46$, H_1: $\mu \neq 46$

Step 2: Since the population standard deviation is unknown, the sample size is small

($n < 30$), and the population is normally distributed, we use the t distribution.

$s_{\bar{x}} = s/\sqrt{n} = 8/\sqrt{16} = 2.00$

$t = (\bar{x} - \mu)/s_{\bar{x}} = (42.4 - 46)/2.00 = -1.800$

$df = n - 1 = 16 - 1 = 15$

<u>p-value approach</u>

Step 3: From above, $t = -1.800$ and $df = 15$.

From the t distribution table, $-2.131 < -1.800 < -1.753$ and

$2(.05) < p\text{-value} < 2(.10)$; hence, $.10 < p\text{-value} < .20$.

<u>critical value approach</u>

Step 3: For $\alpha = .05$ with $df = 15$, the critical values of t are -2.131 and 2.131.

9.53 $df = n - 1 = 52 - 1 = 51$

a.

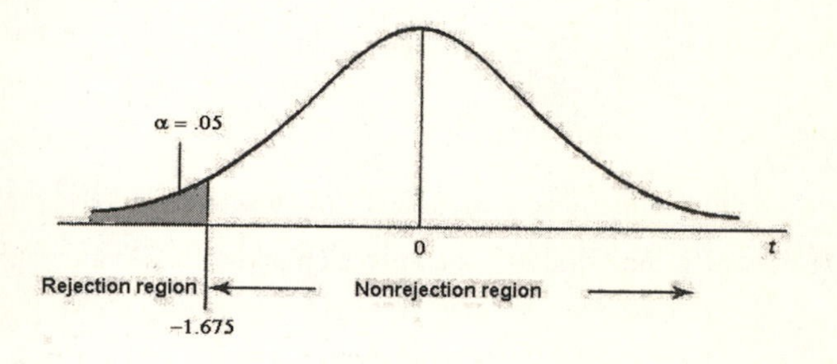

b.

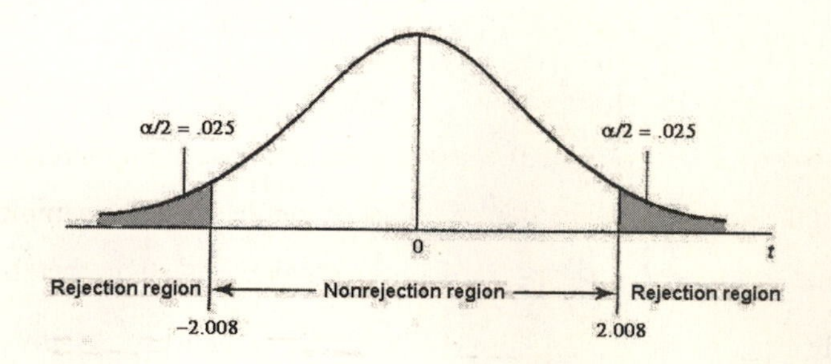

c.

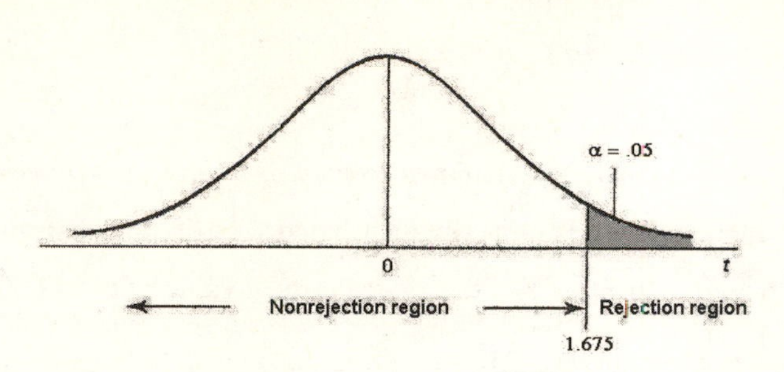

9.55 a. Step 1: H_0: $\mu = 40$, H_1: $\mu > 40$

Step 2: Since the population standard deviation is unknown and the sample size is large ($n > 30$), we use the t distribution.

Step 3: For $\alpha = .025$ with $df = n - 1 = 64 - 1 = 63$, the critical value of t is 1.998.

Step 4: $s_{\bar{x}} = s/\sqrt{n} = 5/\sqrt{64} = .625$

$t = (\bar{x} - \mu)/s_{\bar{x}} = (43 - 40)/.625 = 4.800$

Step 5: Reject H_0 since $4.800 > 1.998$.

b. Step 1: H_0: $\mu = 40$, H_1: $\mu > 40$

Step 2: Since the population standard deviation is unknown and the sample size is large ($n > 30$), we use the t distribution.

Step 3: For $\alpha = .025$ with $df = n - 1 = 64 - 1 = 63$, the critical value of t is 1.998.

Step 4: $s_{\bar{x}} = s/\sqrt{n} = 7/\sqrt{64} = .875$

$t = (\bar{x} - \mu)/s_{\bar{x}} = (41 - 40)/.875 = 1.143$

Step 5: Do not reject H_0 since $1.143 < 1.998$.

Comparing parts a and b shows that two samples selected from the same population can lead to opposite conclusions on the same test of hypothesis. This occurs because the samples may yield different values for the sample mean and the sample standard deviation, thereby affecting the value of the test statistic.

9.57 a. Step 1: H_0: $\mu = 60$, H_1: $\mu \neq 60$

Step 2: Since the population standard deviation is unknown, the sample size is small ($n < 30$), and the population is normally distributed, we use the t distribution.

Step 3: For $\alpha = .05$ with $df = n - 1 = 14 - 1 = 13$, the critical values of t are –2.160 and 2.160.

Step 4: $s_{\bar{x}} = s/\sqrt{n} = 9/\sqrt{14} = 2.40535118$

$t = (\bar{x} - \mu)/s_{\bar{x}} = (57 - 60)/2.40535118 = -1.247$

Step 5: Do not reject H_0 since $-1.247 > -2.160$.

b. Step 1: H_0: $\mu = 35$, H_1: $\mu < 35$

Step 2: Since the population standard deviation is unknown, the sample size is small ($n < 30$), and the population is normally distributed, we use the t distribution.

Step 3: For $\alpha = .005$ with $df = n - 1 = 24 - 1 = 23$, the critical value of t is -2.807.

Step 4: $s_{\bar{x}} = s/\sqrt{n} = 5.4/\sqrt{24} = 1.10227038$

$t = (\bar{x} - \mu)/s_{\bar{x}} = (28 - 35)/1.10227038 = -6.351$

Step 5: Reject H_0 since $-6.351 < -2.807$.

c. Step 1: H_0: $\mu = 47$, H_1: $\mu > 47$

Step 2: Since the population standard deviation is unknown, the sample size is small ($n < 30$), and the population is normally distributed, we use the t distribution.

Step 3: For $\alpha = .001$ with $df = n - 1 = 18 - 1 = 17$, the critical value of t is 3.646.

Step 4: $s_{\bar{x}} = s/\sqrt{n} = 6/\sqrt{18} = 1.41421356$

$t = (\bar{x} - \mu)/s_{\bar{x}} = (50 - 47)/1.41421356 = 2.121$

Step 5: Do not reject H_0 since $2.121 < 3.646$.

9.59 Step 1: H_0: $\mu = \$220,680$, H_1: $\mu > \$220,680$

Step 2: Since the population standard deviation is unknown, the sample size is small ($n < 30$), and the population is normally distributed, we use the t distribution.

Step 3: $s_{\bar{x}} = s/\sqrt{n} = 34,256/\sqrt{16} = \$8564$

$t = (\bar{x} - \mu)/s_{\bar{x}} = (262,570 - 220,680)/8564 = 4.891$

For $\alpha = .01$ with $df = n - 1 = 16 - 1 = 15$,

$3.788 < 4.891 < \infty$; hence, $0 < p\text{-value} < .001$.

Step 4: Reject H_0 since p-value $< .01$.

Conclude that the average jury award in vehicular liability cases currently exceeds \$220,680.

9.61 Step 1: H_0: $\mu = \$850$, H_1: $\mu < \$850$

Step 2: Since the population standard deviation is unknown and the sample size is large ($n > 30$), we use the t distribution.

Step 3: For $\alpha = .01$ with $df = n - 1 = 55 - 1 = 54$, the critical value of t is -2.397.

Step 4: $s_{\bar{x}} = s/\sqrt{n} = 230/\sqrt{55} = \$31.01319367$

$t = (\bar{x} - \mu)/s_{\bar{x}} = (780 - 850)/31.01319367 = -2.257$

Step 5: Do not reject H_0 since $-2.257 > -2.397$.

Conclude that the mean balance of checking accounts at the bank has not decreased during this period.

If $\alpha = .025$, the critical value for t is -2.005. We would reject H_0 since $-2.257 < -2.005$ and conclude that the mean balance of checking accounts at the bank has decreased during this period.

9.63 The first two steps, the calculation of the observed value of the test statistic, and the number of degrees of freedom are the same for both the p-value approach and critical value approach.

Step 1: H_0: $\mu = \$17{,}989$, H_1: $\mu \neq 17{,}989$

Step 2: Since the population standard deviation is unknown and the sample size is large ($n > 30$), we use the t distribution.

$s_{\bar{x}} = s/\sqrt{n} = 4650/\sqrt{75} = \$536.9357503$

$t = (\bar{x} - \mu)/s_{\bar{x}} = (16{,}450 - 17{,}989)/536.9357503 = -2.866$

$df = n - 1 = 75 - 1 = 74$

<u>p-value approach</u>

Step 3: From above, $t = -2.866$ and $df = 74$.

From the t distribution table, $-3.204 < -2.866 < -2.644$ and $2(.001) < p\text{-value} < 2(.005)$; hence, $.002 < p\text{-value} < .01$.

Step 4: Reject H_0 since p-value $< .02$.

<u>critical value approach</u>

Step 3: For $\alpha = .02$ with $df = 74$, the critical values of t are -2.378 and 2.378.

Step 4: From above, $t = -2.866$.

Step 5: Reject H_0 since $-2.866 < -2.378$.

Conclude that the current mean personal debt for all households in New Hampshire is different from \$17,989.

9.65 a. Step 1: H_0: $\mu \geq \$150$, H_1: $\mu < \$150$

Step 2: Since the population standard deviation is unknown, the sample size is small ($n < 30$), and the population is normally distributed, we use the t distribution.

Step 3: For $\alpha = .01$ with $df = n - 1 = 25 - 1 = 24$, the critical value of t is -2.492.

Step 4: $s_{\bar{x}} = s/\sqrt{n} = 28/\sqrt{25} = \$5.60$

$t = (\bar{x} - \mu)/s_{\bar{x}} = (139 - 150)/5.60 = -1.964$

Step 5: Do not reject H_0 since $-1.964 > -2.492$.

Conclude that the manager's claim is true: the average weekly tip is \$150 or more per week.

b. The Type I error would occur if the mean waiter's earnings from tips was \$150 or more but we concluded that it was less than \$150. The probability of making such an error is $\alpha = .01$.

9.67 a. If $\alpha = 0$, there is no rejection region, so we could not reject H_0. Thus, we could not conclude that the current mean age of all CEOs of medium-sized companies in the United States is different from 58 years.

b. The first two steps, the calculation of the observed value of the test statistic, and the number of degrees of freedom are the same for both the *p*-value approach and critical value approach.

Step 1: H_0: $\mu = 58$ years, H_1: $\mu \neq 58$ years

Step 2: Since the population standard deviation is unknown and the sample size is large ($n > 30$), we use the *t* distribution.

$$s_{\bar{x}} = s/\sqrt{n} = 6/\sqrt{70} = .71713717 \text{ years}$$

$$t = (\bar{x} - \mu)/s_{\bar{x}} = (55 - 58)/.71713717 = -4.183$$

$$df = n - 1 = 70 - 1 = 69$$

<u>*p*-value approach</u>

Step 3: From above, $t = -4.183$ and $df = 69$.

From the *t* distribution table, $-\infty < -4.183 < -3.213$ and $2(0) < p\text{-value} < 2(.001)$; hence, $0 < p\text{-value} < .002$.

Step 4: Reject H_0 since $p\text{-value} < .01$.

<u>critical value approach</u>

Step 3: For $\alpha = .01$ with $df = 69$, the critical values of *t* are -3.213 and 3.213.

Step 4: From above, $t = -4.183$.

Step 5: Reject H_0 since $-4.183 < -3.213$.

Conclude that the current mean age of all CEOs of medium-sized companies in the United States is different from 58 years.

9.69 $n = 12$, $\sum x = 885$, and $\sum x^2 = 78{,}045$

$\bar{x} = (\sum x)/n = 885/12 = \$73.75$

$$s = \sqrt{\frac{\sum x^2 - \frac{(\sum x)^2}{n}}{n-1}} = \sqrt{\frac{78{,}045 - \frac{(885)^2}{12}}{12-1}} = \$34.08045294$$

Step 1: H_0: $\mu = \$65$, H_1: $\mu > \$65$

Step 2: Since the population standard deviation is unknown, the sample size is small ($n < 30$), and the population is normally distributed, we use the *t* distribution.

Step 3: For $\alpha = .01$ with $df = n - 1 = 12 - 1 = 11$, the critical value of t is 2.718.

Step 4: $s_{\bar{x}} = s/\sqrt{n} = 34.08045294/\sqrt{12} = \$9.83817934$

$t = (\bar{x} - \mu)/s_{\bar{x}} = (73.75 - 65)/9.83817934 = .889$

Step 5: Do not reject H_0 since $.889 < 2.718$.

Conclude that the mean amount of money spent by all customers at this supermarket after the campaign was started is not more than \$65.

9.71 The first two steps, the calculation of the observed value of the test statistic, and the number of degrees of freedom are the same for both the p-value approach and critical value approach.

Step 1: H_0: $\mu = 231$ minutes, H_1: $\mu \neq 231$ minutes

Step 2: Since the population standard deviation is unknown and the sample size is large ($n > 30$), we use the t distribution. However, since $n = 800$ is very large, we use the normal distribution to approximate the t distribution.

$s_{\bar{x}} = s/\sqrt{n} = 55/\sqrt{800} = 1.94454365$ minutes

$z = (\bar{x} - \mu)/s_{\bar{x}} = (250 - 231)/1.94454365 = 9.77$

<u>p-value approach</u>

Step 3: From above, $z = 9.77$.

From the normal distribution table, area to the right of $z = 9.77$ is 1 – 1 . 0.

p-value . 2(0) = 0

Step 4: Reject H_0 since $0 < .02$.

<u>critical value approach</u>

Step 3: For $\alpha = .02$, the critical value of z is 2.05.

Step 4: From above, $z = 9.77$.

Step 5: Reject H_0 since $9.77 > 2.05$.

Conclude that the current mean time spent watching television per day by children in this age group differs from 231 minutes.

9.73 To make a hypothesis test about the mean number of bolts manufactured per hour by this machine (assuming the number of bolts manufactured is normally distributed):

1. Take a sample of less than 30 one hour periods.
2. Record the number of bolts produced during each one hour period.
3. Calculate $\bar{x}$, s, and $s_{\bar{x}} = s/\sqrt{n}$.

4. Determine the significance level and find the critical value of t from the t distribution table for $df = n - 1$.
5. Make the test of hypothesis for H_0: $\mu \geq 88$ bolts versus H_1: $\mu < 88$ bolts using $t = (\bar{x} - \mu)/s_{\bar{x}}$.

Section 9.4

9.75 a. $np = (40)(.11) = 4.4 < 5$

Since $np < 5$, the sample size is not large enough to use the normal distribution.

b. $np = (100)(.73) = 73 > 5$ and $nq = (100)(.27) = 27 > 5$

Since $np > 5$ and $nq > 5$, the sample size is large enough to use the normal distribution.

c. $np = (80)(.05) = 4 < 5$

Since $np < 5$, the sample size is not large enough to use the normal distribution.

d. $np = (50)(.14) = 7 > 5$ and $nq = (50)(.86) = 43 > 5$

Since $np > 5$ and $nq > 5$, the sample size is large enough to use the normal distribution.

9.77 a.

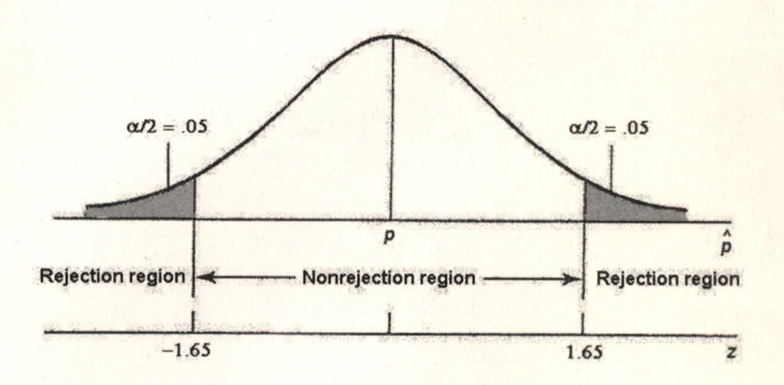

b.

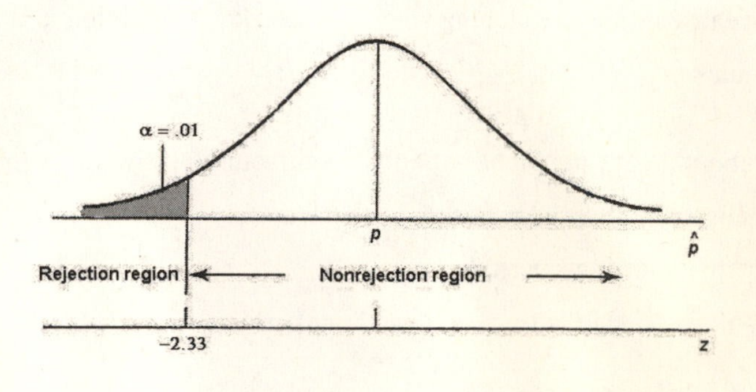

c.

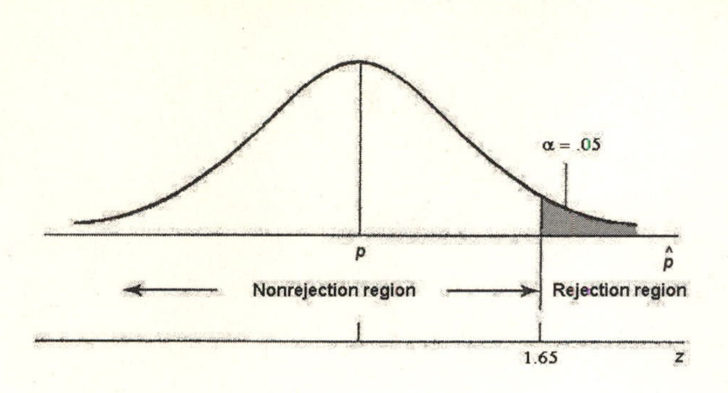

9.79 a. Step 1: H_0: $p = .30$, H_1: $p > .30$

Step 2: $np = (500)(.30) = 150 > 5$ and $nq = (500)(.70) = 350 > 5$

Since $np > 5$ and $nq > 5$, use the normal distribution.

Step 3: For $\alpha = .05$, the critical value of z is 1.65.

Step 4: $\sigma_{\hat{p}} = \sqrt{pq/n} = \sqrt{(.30)(.70)/500} = .02049390$

$z = (\hat{p} - p)/\sigma_{\hat{p}} = (.38 - .30)/.02049390 = 3.90$

b. Step 1: H_0: $p = .30$, H_1: $p \neq .30$

Step 2: $np = (500)(.30) = 150 > 5$ and $nq = (500)(.70) = 350 > 5$

Since $np > 5$ and $nq > 5$, use the normal distribution.

Step 3: For $\alpha = .05$, the critical values of z are –1.96 and 1.96.

Step 4: $\sigma_{\hat{p}} = \sqrt{pq/n} = \sqrt{(.30)(.70)/500} = .02049390$

$z = (\hat{p} - p)/\sigma_{\hat{p}} = (.38 - .30)/.02049390 = 3.90$

9.81 a.

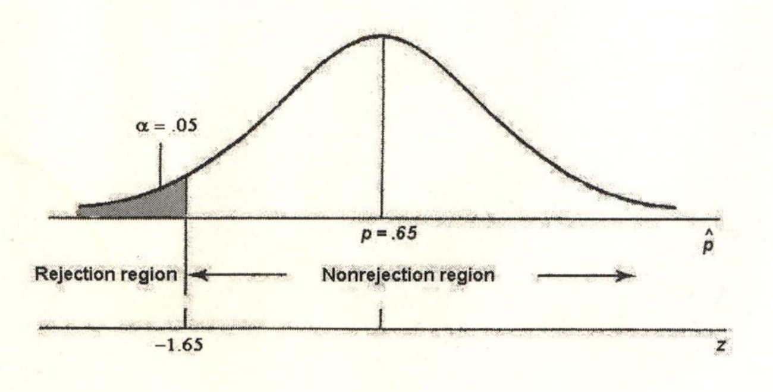

b.

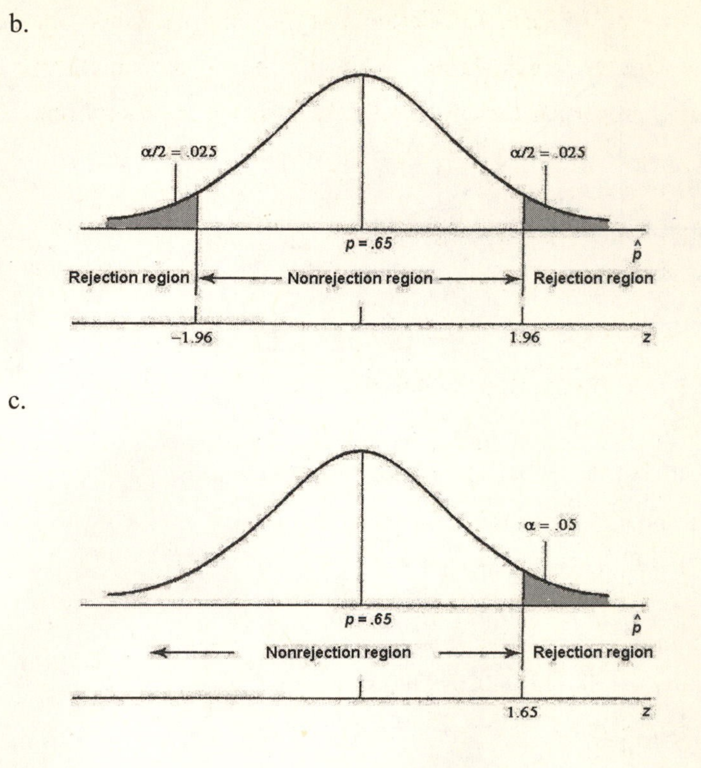

9.83 a. Step 1: H_0: $p = .70, H_1$: $p \neq .70$

Step 2: $np = (600)(.70) = 420 > 5$ and $nq = (600)(.30) = 180 > 5$

Since $np > 5$ and $nq > 5$, use the normal distribution.

Step 3: For $\alpha = .01$, the critical values of z are -2.58 and 2.58.

Step 4: $\sigma_{\hat{p}} = \sqrt{pq/n} = \sqrt{(.70)(.30)/600} = .01870829$

$z = (\hat{p} - p)/\sigma_{\hat{p}} = (.68 - .70)/.01870829 = -1.07$

Step 5: Do not reject H_0 since $-1.07 > -2.58$.

b. Step 1: H_0: $p = .70, H_1$: $p \neq .70$

Step 2: $np = (600)(.70) = 420 > 5$ and $nq = (600)(.30) = 180 > 5$

Since $np > 5$ and $nq > 5$, use the normal distribution.

Step 3: For $\alpha = .01$, the critical values of z are -2.58 and 2.58.

Step 4: $\sigma_{\hat{p}} = \sqrt{pq/n} = \sqrt{(.70)(.30)/600} = .01870829$

$z = (\hat{p} - p)/\sigma_{\hat{p}} = (.76 - .70)/.01870829 = 3.21$

Step 5: Reject H_0 since $3.21 > 2.58$.

Comparing parts a and b shows that two samples selected from the same population can lead to opposite conclusions on the same test of hypothesis. This occurs because the samples may yield different values for the sample proportion, thereby affecting the value of the test statistic.

9.85 a. Step 1: H_0: $p = .45$, H_1: $p \neq .45$

Step 2: $np = (100)(.45) = 45 > 5$ and $nq = (100)(.55) = 55 > 5$

Since $np > 5$ and $nq > 5$, use the normal distribution.

Step 3: For $\alpha = .10$, the critical values of z are -1.65 and 1.65.

Step 4: $\sigma_{\hat{p}} = \sqrt{pq/n} = \sqrt{(.45)(.55)/100} = .04974937$

$z = (\hat{p} - p)/\sigma_{\hat{p}} = (.49 - .45)/.04974937 = .80$

Step 5: Do not reject H_0 since $.80 < 1.65$.

b. Step 1: H_0: $p = .72$, H_1: $p < .72$

Step 2: $np = (700)(.72) = 504 > 5$ and $nq = (700)(.28) = 196 > 5$

Since $np > 5$ and $nq > 5$, use the normal distribution.

Step 3: For $\alpha = .05$, the critical value of z is -1.65.

Step 4: $\sigma_{\hat{p}} = \sqrt{pq/n} = \sqrt{(.72)(.28)/700} = .01697056$

$z = (\hat{p} - p)/\sigma_{\hat{p}} = (.64 - .72)/.01697056 = -4.71$

Step 5: Reject H_0 since $-4.71 < -1.65$.

c. Step 1: H_0: $p = .30$, H_1: $p > .30$

Step 2: $np = (200)(.30) = 60 > 5$ and $nq = (200)(.70) = 140 > 5$

Since $np > 5$ and $nq > 5$, use the normal distribution.

Step 3: For $\alpha = .01$, the critical value of z is 2.33.

Step 4: $\sigma_{\hat{p}} = \sqrt{pq/n} = \sqrt{(.30)(.70)/200} = .03240370$

$z = (\hat{p} - p)/\sigma_{\hat{p}} = (.33 - .30)/.03240370 = .93$

Step 5: Do not reject H_0 since $.93 < 2.33$.

9.87 The first two steps and the calculation of the observed value of the test statistic are the same for both the p-value approach and critical value approach.

Step 1: H_0: $p = .22$, H_1: $p \neq .22$

Step 2: $np = (500)(.22) = 110 > 5$ and $nq = (500)(.78) = 390 > 5$

Since $np > 5$ and $nq > 5$, use the normal distribution.

$\sigma_{\hat{p}} = \sqrt{pq/n} = \sqrt{(.22)(.78)/500} = .01852566$

$z = (\hat{p} - p)/\sigma_{\hat{p}} = (.29 - .22)/.01852566 = 3.78$

<u>p-value approach</u>

Step 3: From above, $z = 3.78$.

From the normal distribution table, area to the right of $z = 3.78$ is $1 - 1 . 0$.

p-value . $2(0) = 0$

Step 4: Reject H_0 since $0 < .05$.

<u>critical value approach</u>

Step 3: For $\alpha = .05$, the critical values of z are -1.96 and 1.96.

Step 4: From above, $z = 3.78$.

Step 5: Reject H_0 since $3.78 > 1.96$.

Conclude that the current percentage of U.S. high school students who smoke is different from 22%.

9.89 The first two steps and the calculation of the observed value of the test statistic are the same for both the p-value approach and critical value approach.

Step 1: H_0: $p = .41$, H_1: $p < .41$

Step 2: $np = (900)(.41) = 369 > 5$ and $nq = (900)(.59) = 531 > 5$

Since $np > 5$ and $nq > 5$, use the normal distribution.

$\sigma_{\hat{p}} = \sqrt{pq/n} = \sqrt{(.41)(.59)/900} = .01639444$

$\hat{p} = 324/900 = .36$

$z = (\hat{p} - p)/\sigma_{\hat{p}} = (.36 - .41)/.01639444 = -3.05$

<u>p-value approach</u>

Step 3: From above, $z = -3.05$.

From the normal distribution table, area to the left of $z = -3.05$ is .0011.

p-value = .0011

Step 4: Reject H_0 since $.0011 < .025$.

<u>critical value approach</u>

Step 3: For $\alpha = .025$, the critical value of z is -1.96.

Step 4: From above, $z = -3.05$.

Step 5: Reject H_0 since $-3.05 < -1.96$.

Conclude that the current percentage of men who hold this opinion is less than 41%.

9.91 a. Step 1: H_0: $p = .60$, H_1: $p > .60$

Step 2: $np = (300)(.60) = 180 > 5$ and $nq = (300)(.40) = 120 > 5$

Since $np > 5$ and $nq > 5$, use the normal distribution.

Step 3: For $\alpha = .02$, the critical value of z is 2.05.

Step 4: $\sigma_{\hat{p}} = \sqrt{pq/n} = \sqrt{(.60)(.40)/300} = .02828427$

$\hat{p} = 186/300 = .62$

$z = (\hat{p} - p)/\sigma_{\hat{p}} = (.62 - .60)/.02828427 = .71$

Step 5: Do not reject H_0 since $.71 < 2.05$.

Conclude that the percentage of all U.S. companies who paid no federal income taxes last year is not higher than 60%.

b. The Type I error would occur if the percentage of all U.S. companies who paid no federal income taxes last year is 60%, but we concluded this percentage is more than 60%. The probability of this error is $\alpha = .02$.

c. From the normal distribution table, area to the right of $z = .71$ is $1 - .7611 = .2389$.

p-value = .2389

Since $.2389 > .02$, conclude that the percentage of all U.S. companies who paid no federal income taxes last year is not higher than 60%.

9.93 a. Step 1: H_0: $p \geq .35$, H_1: $p < .35$

Step 2: $np = (400)(.35) = 140 > 5$ and $nq = (400)(.65) = 260 > 5$

Since $np > 5$ and $nq > 5$, use the normal distribution.

Step 3: For $\alpha = .025$, the critical value of z is -1.96.

Step 4: $\sigma_{\hat{p}} = \sqrt{pq/n} = \sqrt{(.35)(.65)/400} = .02384848$

$\hat{p} = 112/400 = .28$

$z = (\hat{p} - p)/\sigma_{\hat{p}} = (.28 - .35)/.02384848 = -2.94$

Step 5: Reject H_0 since $-2.94 < -1.96$.

Conclude that the percentage of people who like the yogurt is less than 35%, and the company should not market this product.

b. If $\alpha = 0$, there can be no rejection region, and we cannot reject H_0. Therefore, the decision would be "do not reject H_0."

c. Step 1: H_0: $p \geq .35$, H_1: $p < .35$

Step 2: $np = (400)(.35) = 140 > 5$ and $nq = (400)(.65) = 260 > 5$

Since $np > 5$ and $nq > 5$, use the normal distribution.

Step 3: $\sigma_{\hat{p}} = \sqrt{pq/n} = \sqrt{(.35)(.65)/400} = .02384848$

$\hat{p} = 112/400 = .28$

$z = (\hat{p} - p)/\sigma_{\hat{p}} = (.28 - .35)/.02384848 = -2.94$

From the normal distribution table, area to the left of $z = -2.94$ is .0016.

p-value = .001

Step 4: Reject H_0 since $.0016 < .025$.

9.95 a. Step 1: H_0: $p \leq .05$, H_1: $p > .05$

Step 2: $np = (200)(.05) = 10 > 5$ and $nq = (200)(.95) = 190 > 5$

Since $np > 5$ and $nq > 5$, use the normal distribution.

Step 3: For $\alpha = .025$, the critical value of z is 1.96.

Step 4: $\sigma_{\hat{p}} = \sqrt{pq/n} = \sqrt{(.05)(.95)/200} = .01541104$

$\hat{p} = 17/200 = .085$

$z = (\hat{p} - p)/\sigma_{\hat{p}} = (.085 - .05)/.01541104 = 2.27$

Step 5: Reject H_0 since $2.27 > 1.96$.

Conclude that the percentage of defective DVDs is greater than 5%, and the production process should be stopped to make necessary adjustments.

b. If $\alpha = .01$, the critical value of z is 2.33. Since the value of the test statistic is 2.27, we do not reject H_0. Thus, our decision differs from that of part a; we conclude that the process should not be stopped. The results of this sample are not very conclusive, since lowering the significance level from 2.5% to 1% reverses the decision.

9.97 To make a hypothesis test about the percentage of bank customers who are satisfied with the services provided:

1. Take a large sample of the bank's customers such that $np > 5$ and $nq > 5$.
2. Ask these customers if they are satisfied with the services provided by the bank.
3. Calculate $\hat{p}$ and $\sigma_{\hat{p}} = \sqrt{pq/n}$.
4. Determine the significance level and find the critical value of z from the standard normal distribution table.
5. Make the test of hypothesis for H_0: $p = .75$ versus H_1: $p \neq .75$ using $z = (\hat{p} - p)/\sigma_{\hat{p}}$.

Supplementary Exercises

9.99 a. Step 1: H_0: $\mu = 120$, H_1: $\mu > 120$

Step 2: Since $n > 30$, use the normal distribution.

Step 3: For $\alpha = .025$, the critical value of z is 1.96.

Step 4: $\sigma_{\bar{x}} = \sigma/\sqrt{n} = 15/\sqrt{81} = 1.66666667$

$z = (\bar{x} - \mu)/\sigma_{\bar{x}} = (123.5 - 120)/1.6666667 = 2.10$

Step 5: Reject H_0 since $2.10 > 1.96$.

b. $P(\text{Type I error}) = \alpha = .025$

c. From the normal distribution table, area to the right of $z = 2.10$ is $1 - .9821 = .0179$.

p-value = .0179

For $\alpha = .01$, do not reject H_0 since $.0179 > .01$.

For $\alpha = .05$, reject H_0 since $.0179 < .05$.

9.101 a. Step 1: H_0: $p = .82$, H_1: $p \neq .82$

Step 2: $np = (600)(.82) = 492 > 5$ and $nq = (600)(.18) = 108 > 5$

Since $np > 5$ and $nq > 5$, use the normal distribution.

Step 3: For $\alpha = .02$, the critical values of z are –2.33 and 2.33.

Step 4: $\sigma_{\hat{p}} = \sqrt{pq/n} = \sqrt{(.82)(.18)/600} = .01568439$

$z = (\hat{p} - p)/\sigma_{\hat{p}} = (.86 - .82)/.01568439 = 2.55$

Step 5: Reject H_0 since $2.55 > 2.33$.

b. $P(\text{Type I error}) = \alpha = .02$

c. From the normal distribution table, area to the right of $z = 2.55$ is $1 - .9946 = .0054$

p-value = 2(.0054) = .0108

For $\alpha = .025$, reject H_0 since $.0108 < .025$.

For $\alpha = .005$, do not reject H_0 since $.0108 > .005$.

9.103 a. Step 1: H_0: $\mu = 377$ minutes, H_1: $\mu \neq 377$ minutes

Step 2: Since $n > 30$, use the normal distribution.

Step 3: $\sigma_{\bar{x}} = \sigma/\sqrt{n} = 85/\sqrt{100} = 8.5$ minutes

$z = (\bar{x} - \mu)/\sigma_{\bar{x}} = (422 - 377)/8.5 = 5.29$

From the normal distribution table, area to the right of $z = 5.29$ is 1 – 1 . 0.

p-value . 2(0) = 0

Step 4: Reject H_0 since $0 < .05$.

b. Step 1: H_0: $\mu = 377$ minutes, H_1: $\mu \neq 377$ minutes

Step 2: Since $n > 30$, use the normal distribution.

Step 3: For $\alpha = .05$, the critical values of z are -1.96 and 1.96.

Step 4: $\sigma_{\bar{x}} = \sigma/\sqrt{n} = 85/\sqrt{100} = 8.5$ minutes

$z = (\bar{x} - \mu)/\sigma_{\bar{x}} = (422 - 377)/8.5 = 5.29$

Step 5: Reject H_0 since $5.29 > 1.96$.

Conclude that the mean time taken by all tax payers in this state to prepare form 1040 was different from 377 minutes.

9.105 a. Step 1: H_0: $\mu \geq 50$ mph, H_1: $\mu < 50$ mph

Step 2: Since $n > 30$, use the normal distribution.

Step 3: For $\alpha = .025$, the critical value of z is -1.96.

Step 4: $\sigma_{\bar{x}} = \sigma/\sqrt{n} = 4/\sqrt{36} = .66666667$ mph

$z = (\bar{x} - \mu)/\sigma_{\bar{x}} = (48 - 50)/.66666667 = -3.00$

Step 5: Reject H_0 since $-3.00 < -1.96$.

Conclude that the mean speed of vehicles passing through this construction zone is less than 50 mph and the worker's claim is false.

b. The Type I error would be to conclude that the site worker's claim is false when it is actually true. The probability of making such an error is .025.

c. If $\alpha = 0$, there is no rejection region so we cannot reject H_0. Thus, our conclusion of part a is changed.

d. From the normal distribution table, area to the left of $z = -3.00$ is .0013

p-value = .0013

For $\alpha = .025$, reject H_0 since $.0013 < .025$.

9.107 a. Step 1: H_0: $\mu \leq 2400$ square feet, H_1: $\mu > 2400$ square feet

Step 2: Since the population standard deviation is unknown and the sample size is large ($n > 30$), we use the t distribution.

Step 3: For $\alpha = .05$ with $df = n - 1 = 50 - 1 = 49$, the critical value of t is 1.677.

Step 4: $s_{\bar{x}} = s/\sqrt{n} = 472/\sqrt{50} = 66.75088014$

$t = (\bar{x} - \mu)/s_{\bar{x}} = (2540 - 2400)/66.75088014 = 2.097$

Step 5: Reject H_0 since $2.097 > 1.677$.

Conclude that the mean living area of all single-family homes in this county exceeds 2400 square feet, and the realtor's claim is false.

b. If $\alpha = .01$, the critical value of t is 2.405. Since the value of the test statistic is 2.097, we do not reject H_0. Thus, our decision differs from that of part a; we conclude that the mean living area of all single-family homes in this county is at most 2400 square feet, and the realtor's claim is true.

9.109 Step 1: H_0: $\mu \le 15$ minutes, H_1: $\mu > 15$ minutes

Step 2: Since the population standard deviation is unknown and the sample size is large ($n > 30$), we use the t distribution.

Step 3: For $\alpha = .01$ with $df = n - 1 = 36 - 1 = 35$, the critical value of t is 2.438.

Step 4: $s_{\bar{x}} = s/\sqrt{n} = 2.4/\sqrt{36} = .4$ minute

$t = (\bar{x} - \mu)/s_{\bar{x}} = (15.75 - 15)/.4 = 1.875$

Step 5: Do not reject H_0 since $1.875 < 2.438$.

Conclude that the average amount of time between placing an order and service of food is not more than 15 minutes and the restaurant's claim is true. The journalist's conclusion is unfair to the restaurant.

9.111 Step 1: H_0: $\mu = 25$ minutes, H_1: $\mu \ne 25$ minutes

Step 2: Since the population standard deviation is unknown, the sample size is small ($n < 30$), and the population is normally distributed, we use the t distribution.

Step 3: For $\alpha = .01$ with $df = n - 1 = 16 - 1 = 15$, the critical values of t are -2.947 and 2.947.

Step 4: $s_{\bar{x}} = s/\sqrt{n} = 4.8/\sqrt{16} = 1.2$ minutes

$t = (\bar{x} - \mu)/s_{\bar{x}} = (27.5 - 25)/1.2 = 2.083$

Step 5: Do not reject H_0 since $2.083 < 2.947$.

Conclude that the mean waiting time at the emergency ward is not different than 25 minutes.

9.113 $n = 12$, $\sum x = 26.8$, and $\sum x^2 = 62.395$

$\bar{x} = (\sum x)/n = 26.8/12 = 2.23333333$ hours

$$s = \sqrt{\frac{\sum x^2 - \frac{(\sum x)^2}{n}}{n-1}} = \sqrt{\frac{62.395 - \frac{(26.8)^2}{12}}{12-1}} = .48068764 \text{ hour}$$

Step 1: H_0: $\mu \le 2$ hours, H_1: $\mu > 2$ hours

Step 2: Since the population standard deviation is unknown, the sample size is small ($n < 30$), and the population is normally distributed, we use the t distribution.

Step 3: For $\alpha = .01$ with $df = n - 1 = 12 - 1 = 11$, the critical value of t is 2.718.

Step 4: $s_{\bar{x}} = s/\sqrt{n} = .48068764/\sqrt{12} = .13876257$ hour

$t = (\bar{x} - \mu)/s_{\bar{x}} = (2.23333333 - 2)/.13876257 = 1.682$

Step 5: Do not reject H_0 since $1.682 < 2.718$.

Conclude that the mean time it takes to learn how to use this software is not more than two hours and the company's claim is true.

9.115 a. Step 1: H_0: $p = .156$, H_1: $p > .156$

Step 2: $np = (1200)(.156) = 187.2 > 5$ and $nq = (1200)(.844) = 1012.8 > 5$

Since $np > 5$ and $nq > 5$, use the normal distribution.

Step 3: For $\alpha = .02$, the critical value of z is 2.05.

Step 4: $\sigma_{\hat{p}} = \sqrt{pq/n} = \sqrt{(.156)(.844)/1200} = .01047473$

$\hat{p} = 216/1200 = .18$

$z = (\hat{p} - p)/\sigma_{\hat{p}} = (.18 - .156)/.01047473 = 2.29$

Step 5: Reject H_0 since $2.29 > 2.05$.

Conclude that the current percentage of Americans who have no health insurance coverage exceeds 15.6%.

b. The Type I error would occur if the current percentage of Americans who have no health insurance coverage does not exceed 15.6%, but we concluded this percentage does exceed 15.6%. The probability of this error is $\alpha = .02$.

c. From the normal distribution table, area to the right of $z = 2.29$ is $1 - .9890 = .0110$.

p-value = .0110

For $\alpha = .02$, reject H_0 since $.0110 < .02$.

9.117 The first two steps and the calculation of the observed value of the test statistic are the same for both the p-value approach and critical value approach.

Step 1: H_0: $p = .40$, H_1: $p \neq .40$

Step 2: $np = (700)(.40) = 280 > 5$ and $nq = (700)(.60) = 420 > 5$

Since $np > 5$ and $nq > 5$, use the normal distribution.

$\sigma_{\hat{p}} = \sqrt{pq/n} = \sqrt{(.40)(.60)/700} = .01851640$

$\hat{p} = 259/700 = .37$

$z = (\hat{p} - p)/\sigma_{\hat{p}} = (.37 - .40)/.01851640 = -1.62$

p-value approach

Step 3: From above, $z = -1.62$.

From the normal distribution table, area to the left of $z = -1.62$ is .0526

p-value = 2(.0526) = .1052

Step 4: Do not reject H_0 since .1052 > .01.

critical value approach

Step 3: For $\alpha = .01$, the critical values of z are –2.58 and 2.58.

Step 4: From above, $z = -1.62$.

Step 5: Do not reject H_0 since $-1.62 > -2.58$.

Conclude that the percentage of people who buy national brand coffee is not different from 40%.

9.119 a. Step 1: H_0: $p = .80$, H_1: $p \neq .80$

Step 2: $np = (40)(.80) = 32 > 5$ and $nq = (40)(.20) = 8 > 5$

Since $np > 5$ and $nq > 5$, use the normal distribution.

Step 3: For $\alpha = .01$, the critical values of z are –2.58 and 2.58.

Step 4: $\sigma_{\hat{p}} = \sqrt{pq/n} = \sqrt{(.80)(.20)/40} = .06324555$

$z = (\hat{p} - p)/\sigma_{\hat{p}} = (.75 - .80)/.06324555 = -.79$

Step 5: Do not reject H_0 since $-.79 > -2.58$.

Conclude that the percentage of all such batteries that are good for 70 months or longer is not different from 80%, and the companies claim is true.

b. If $\alpha = 0$, there can be no rejection region, and we cannot reject H_0. Therefore, the decision would be "do not reject H_0."

9.121 a. P(a student guessing randomly wins all five bets) = $\left(\frac{18}{38}\right)^5 = .0238$

b. The significance level of α represents the probability of a Type I error, i.e., concluding that a student is not guessing when in fact the student is guessing. From part a, we know that the probability that we reject H_0 despite the fact that the student is guessing is .0238.

c. $\mu = np = (100)(.0238) = 2.38$. So, we would expect 2 students to win all five bets even if all the students are guessing.

9.123 H_0: $p \leq .15$, H_1: $p > .15$

Let x be the number of late arrivals in a random sample of 50 flights. The significance level, α, is the probability of rejecting H_0 when H_0 is true. In this case, $\alpha = P(x > 9 \mid p = .15)$. If H_0 is true, x is a binomial random variable with $n = 50$ and $p = .15$. Since np and nq are both greater than 5, we may use a normal approximation.

$\mu = np = 50(.15) = 7.5$ and $\sqrt{npq} = \sqrt{50(.15)(.85)} = 2.52487623$

Correcting for continuity, $\alpha = P(x > 8.5)$, and $z = (8.5 - 7.5) / 2.52487623 = .40$. Thus, $\alpha = P(x > 8.5) = P(z > .40) = 1 - .6554 = .3446$. Hence, the significance level for this test is approximately .34.

9.125 Let μ = mean life of these light bulbs

H_0: $\mu = 750$ hours, H_1: $\mu < 750$ hours

$\sigma_{\bar{x}} = \sigma/\sqrt{n} = 50/\sqrt{64} = 6.25$

First decision rule: Reject H_0 if $\bar{x} < 735$

$z = (\bar{x} - \mu)/\sigma_{\bar{x}} = (735 - 750)/6.25 = -2.40$

Thus, $\alpha = P(\bar{x} < 735 \mid \mu = 750) = P(z < -2.40) = .0082$

Using this decision rule, the probability of concluding that GE has printed too high an average length of life on the package when in fact they have not is .0082.

Second decision rule: Reject H_0 if $\bar{x} < 700$

$z = (\bar{x} - \mu)/\sigma_{\bar{x}} = (700 - 750)/6.25 = -8.00$

Thus, $\alpha = P(\bar{x} < 700 \mid \mu = 750) = P(z < -8.00) = 0$ approximately

Using this decision rule, there is virtually no chance of concluding that GE has printed too high an average length of life on the package when in fact they have not.

9.127 The following are two possible experiments we might conduct to investigate the effectiveness of middle taillights.

I. Let p_1 = proportion of all collisions involving cars built since 1984 that were rear–end collisions. Let p_2 = proportion of all collisions involving cars built before 1984 that were rear–end collisions. (p_2 would be known)

We would test H_0: $p_1 = p_2$ versus H_1: $p_1 < p_2$.

We would take a random sample of collisions involving cars built since 1984 and determine the number that were rear–end collisions. We would have to assume the following:

1. The only change in cars built since 1984 that would reduce rear–end collisions is the middle taillight.

2. None of the cars built before 1984 had middle taillights.
3. People's driving habits, traffic volume, and other variables that might affect rear–end collisions have not changed appreciably since 1984.

II. Let μ_1 = mean number of rear–end collisions per 1000 cars built since 1984.

Let μ_2 = mean number of rear–end collisions per 1000 cars built before 1984.

(μ_2 would be known)

We would test H_0: $\mu_1 = \mu_2$ versus H_1: $\mu_1 < \mu_2$.

We could take several random samples of 1000 cars built since 1984. We would determine the number of rear–end collisions in each sample of 1000 cars. We would find the mean and standard deviation of these numbers and use them to form the test statistic. We would require the same assumptions as those listed for the test in part I. Also, if we took less than 30 samples, we would have to assume that the number of rear–end collisions per 1000 cars has a normal distribution.

9.129 Since $np > 5$ and $nq > 5$, the distribution of $\hat{p}$ is approximately normal.

For $\alpha = .05$, the critical values of z are -1.96 and 1.96.

a. $\sigma_{\hat{p}} = \sqrt{pq/n} = \sqrt{(1/6)(5/6)/120} = .03402069$

Now $z = (\hat{p} - p)/\sigma_{\hat{p}}$, and this is a two-tailed test, so

$-1.96 = (\hat{p} - .166666667)/.03402069$ and $1.96 = (\hat{p} - .166666667)/.03402069$. Then,

$\hat{p} = (-1.96)(.03402069) + .166666667 = .09998611$ or

$\hat{p} = (1.96)(.03402069) + .166666667 = .23334722$.

Thus, we would reject H_0 if $\hat{p} < .09998611$ or $\hat{p} > .23334722$.

Then $.09998611 = a/n = a/120$ and $.23334722 = b/n = b/120$.

Hence, $a = (120)(.09998611) = 11.998 \approx 12$, and $b = (120)(.23334722) = 28.002 \approx 28$.

Therefore, reject H_0 and conclude that the die is not balanced if you obtain fewer than 12 or more than 28 2-spots in 120 rolls of the die.

b. $\sigma_{\hat{p}} = \sqrt{pq/n} = \sqrt{(1/6)(5/6)/1200} = .01075829$

Now $z = (\hat{p} - p)/\sigma_{\hat{p}}$, and this is a two-tailed test, so

$-1.96 = (\hat{p} - .166666667)/.01075829$ and $1.96 = (\hat{p} - .166666667)/.01075829$. Then,

$\hat{p} = (-1.96)(.01075829) + .166666667 = .014558042$ or

$\hat{p} = (1.96)(.01075829) + .166666667 = .18775292$.

Thus, we would reject H_0 if $\hat{p} < .014558042$ or $\hat{p} > .18775292$.

Then $.014558042 = a/n = a/120$ and $.18775292 = b/n = b/120$.

Hence, $a = (1200)(.014558042) = 174.697 \approx 175$, and $b = (1200)(.18775292) = 225.304 \approx 225$.

Therefore, reject H_0 and conclude that the die is not balanced if you obtain fewer than 175 or more than 225 2-spots in 1200 rolls of the die.

c. $\sigma_{\hat{p}} = \sqrt{pq/n} = \sqrt{(1/6)(5/6)/12000} = .00340207$

Now $z = (\hat{p} - p)/\sigma_{\hat{p}}$, and this is a two-tailed test, so

$-1.96 = (\hat{p} - .166666667)/.00340207$ and $1.96 = (\hat{p} - .166666667)/.000340207$. Then,

$\hat{p} = (-1.96)(.00340207) + .166666667 = .15999861$ or

$\hat{p} = (1.96)(.00340207) + .166666667 = .17333472$.

Thus, we would reject H_0 if $\hat{p} < .15999861$ or $\hat{p} > .1733347217333472$.

Then $.15999861 = a/n = a/120$ and $.17333472 = b/n = b/120$.

Hence, $a = (12{,}000)(.15999861) = 1919.983 \approx 1920$, and

$b = (12{,}000)(.17333472) = 2080.017 \approx 2080$.

Therefore, reject H_0 and conclude that the die is not balanced if you obtain fewer than 1920 or more than 2080 2-spots in 120 rolls of the die.

9.131 Because we know the value of the population mean, we know that the null hypothesis is true. If the null hypothesis is rejected in this circumstance, we have made a Type I error. Since $\alpha = .05$, we expect to make a Type I error $(50)(.05) = 2.5$ times when we test using 50 different samples. Therefore, 47 tests failing to reject the null hypothesis is not surprising.

Self–Review Test

1. a **2.** b **3.** a **4.** b **5.** a **6.** a **7.** a **8.** b

9. c **10.** a **11.** c **12.** b **13.** c **14.** a **15.** b

16. a. Step 1: H_0: $\mu = 52$ minutes, H_1: $\mu \neq 52$ minutes

Step 2: Since $n > 30$, use the normal distribution.

Step 3: For $\alpha = .01$, the critical values of z are –2.58 and 2.58.

Step 4: $\sigma_{\bar{x}} = \sigma/\sqrt{n} = 12/\sqrt{1400} = .32071349$ minute

$z = (\bar{x} - \mu)/\sigma_{\bar{x}} = (58 - 52)/.32071349 = 18.71$

Step 5: Reject H_0 since $18.71 > 2.58$.

Conclude that the mean time spent driving per day by this age group differs from 52 minutes.

b. Step 1: H_0: $\mu = 52$ minutes, H_1: $\mu > 52$ minutes

Step 2: Since $n > 30$, use the normal distribution.

Step 3: For $\alpha = .025$, the critical value of z is 1.96.

Step 4: $\sigma_{\bar{x}} = \sigma/\sqrt{n} = 12/\sqrt{1400} = .32071349$ minute

$z = (\bar{x} - \mu)/\sigma_{\bar{x}} = (58 - 52)/.32071349 = 18.71$

Step 5: Reject H_0 since $18.71 > 1.96$.

Conclude that the mean time spent driving per day by this age group exceeds from 52 minutes.

c. The Type I error would occur if the we conclude the mean time spent driving per day by this age group is different from 52 minutes (part a) or exceeds 52 minutes (part b), when the actual mean time spent driving is 52 minutes. The probability of making such error is $\alpha = .01$ in part a and $\alpha = .025$ in part b.

d. From the normal distribution table, area to the right of $z = 18.71$ is $1 - 1 = 0$ approximately.

p-value = 2(0) = 0

For $\alpha = .01$, reject H_0 since $0 < .01$.

e. From the normal distribution table, area to the right of $z = 18.71$ is $1 - 1 = 0$ approximately.

p-value = 0

For $\alpha = .025$, reject H_0 since $0 < .025$

17. a. Step 1: H_0: $\mu = 185$ minutes, H_1: $\mu < 185$ minutes

Step 2: Since the population standard deviation is unknown and the sample size is large ($n >$ 30), we use the t distribution.

Step 3: For $\alpha = .01$ with $df = n - 1 = 36 - 1 = 35$, the critical value of t is -2.438.

Step 4: $s_{\bar{x}} = s/\sqrt{n} = 12/\sqrt{36} = 2$ minutes

$t = (\bar{x} - \mu)/s_{\bar{x}} = (179 - 185)/2 = -3.000$

Step 5: Reject H_0 since $-3.000 < -2.448$.

Conclude that the mean duration of nine-inning games has decreased after the meeting.

b. The Type I error would be to conclude that the mean durations of games have decreased after the meeting when they are actually equal to the duration of games before the meeting. The probability of making such an error is .01.

c. If $\alpha = 0$, there can be no rejection region, and we cannot reject H_0. Therefore, the decision would be "do not reject H_0."

d. From the t distribution table, $-3.340 < -3.000 < -2.724$; hence, $.001 < p\text{-value} < .005$.

18. a. Step 1: H_0: $\mu \geq 31$ months, H_1: $\mu < 31$ months

Step 2: Since the population standard deviation is unknown, the sample size is small ($n < 30$), and the population is normally distributed, we use the t distribution.

Step 3: For $\alpha = .025$ with $df = n - 1 = 16 - 1 = 15$, the critical value of t is -2.131.

Step 4: $s_{\bar{x}} = s/\sqrt{n} = 7.2/\sqrt{16} = 1.8$

$t = (\bar{x} - \mu)/s_{\bar{x}} = (25 - 31)/1.8 = -3.333$

Step 5: Reject H_0 since $-3.333 < -2.131$.

Conclude that the mean time it takes to write a textbook is less than 31 months, and the editor's claim is false.

b. The Type I error would be to conclude that the editor's claim is false when it is actually true. The probability of making such an error is .025.

c. If $\alpha = .001$, the critical value of t is -3.733. Since the value of the test statistic is -3.333, we do not reject H_0.

19. a. Step 1: H_0: $p = .50$, H_1: $p < .50$

Step 2: $np = (1000)(.50) = 500 > 5$ and $nq = (1000)(.50) = 500 > 5$

Since $np > 5$ and $nq > 5$, use the normal distribution.

Step 3: For $\alpha = .05$, the critical value of z is -1.65.

Step 4: $\sigma_{\hat{p}} = \sqrt{pq/n} = \sqrt{(.50)(.50)/1000} = .01581139$

$\hat{p} = 450/1000 = .45$

$z = (\hat{p} - p)/\sigma_{\hat{p}} = (.45 - .50)/.01581139 = -3.16$

Step 5: Reject H_0 since $-3.16 < -1.65$.

Conclude that the percentage of people who have a will is less than 50%.

b. The Type I error would be to conclude that the percentage of adults with wills was less than 50 % when it is actually 50%. The probability of making such an error is .05.

c. If $\alpha = 0$, there can be no rejection region, and we cannot reject H_0. Therefore, the decision would be "do not reject H_0."

d. From the normal distribution table, area to the left of $z = -3.16$ is .0008.

p-value = .0008

For $\alpha = .05$, reject H_0 since $.0008 < .05$.

For $\alpha = .01$, reject H_0 since $.0008 < .01$.

Chapter Ten

Section 10.1

10.1 Two samples are **independent** if they are drawn from two different populations and the elements of the two samples are not related. Example 10–1 in the text provides an example of independent samples. In two **dependent** samples, the elements of one sample are related to the elements of the second sample. Example 10–2 in the text provides an example of dependent samples.

10.3 a. $\bar{x}_1 - \bar{x}_2 = 5.56 - 4.80 = .76$

b. $\sigma_{\bar{x}_1-\bar{x}_2} = \sqrt{\dfrac{\sigma_1^2}{n_1} + \dfrac{\sigma_2^2}{n_2}} = \sqrt{\dfrac{(1.65)^2}{24} + \dfrac{(1.58)^2}{27}} = .45375848$

The 99% confidence interval for $\mu_1 - \mu_2$ is

$(\bar{x}_1 - \bar{x}_2) \pm z\sigma_{\bar{x}_1-\bar{x}_2} = .76 \pm 2.58(.45375848) = 0.76 \pm 1.17 = -.41$ to 1.93

$E = z\sigma_{\bar{x}_1-\bar{x}_2} = 2.58(.45375848) = 1.17$

10.5 Step 1: H_0: $\mu_1 - \mu_2 = 0$, H_1: $\mu_1 - \mu_2 \neq 0$

Step 2: Since σ_1 and σ_2 are known, the samples are independent, and the populations are normally distributed, use the normal distribution.

Step 3: For $\alpha = .05$, the critical values of z are -1.96 and 1.96.

Step 4: $\sigma_{\bar{x}_1-\bar{x}_2} = \sqrt{\dfrac{\sigma_1^2}{n_1} + \dfrac{\sigma_2^2}{n_2}} = \sqrt{\dfrac{(1.65)^2}{24} + \dfrac{(1.58)^2}{27}} = .45375848$

$$z = \frac{(\bar{x}_1 - \bar{x}_2) - (\mu_1 - \mu_2)}{\sigma_{\bar{x}_1-\bar{x}_2}} = \frac{(5.56 - 4.80) - 0}{.45375848} = 1.67$$

Step 5: Do not reject H_0 since $1.67 < 1.96$.

10.7 Step 1: H_0: $\mu_1 - \mu_2 = 0$, H_1: $\mu_1 - \mu_2 < 0$

Step 2: Since σ_1 and σ_2 are known, the samples are independent, and the sample sizes are large ($n_1 \geq 30$ and $n_2 \geq 30$), use the normal distribution.

Step 3: For $\alpha = .05$, the critical value of z is –1.65.

Step 4: $$\sigma_{\bar{x}_1-\bar{x}_2} = \sqrt{\frac{\sigma_1^2}{n_1}+\frac{\sigma_2^2}{n_2}} = \sqrt{\frac{(4.9)^2}{300}+\frac{(4.5)^2}{250}} = .40128959$$

$$z = \frac{(\bar{x}_1-\bar{x}_2)-(\mu_1-\mu_2)}{\sigma_{\bar{x}_1-\bar{x}_2}} = \frac{(22.0-27.6)-0}{.40128959} = -13.96$$

Step 5: Reject H_0 since $-13.96 < -1.65$.

10.9 a. $\bar{x}_1 - \bar{x}_2 = 101 - 92 = 9$ hours

b. $$\sigma_{\bar{x}_1-\bar{x}_2} = \sqrt{\frac{\sigma_1^2}{n_1}+\frac{\sigma_2^2}{n_2}} = \sqrt{\frac{(15)^2}{29}+\frac{(10)^2}{27}} = 3.38560547 \text{ hours}$$

The 97% confidence interval for $\mu_1 - \mu_2$ is

$$(\bar{x}_1-\bar{x}_2) \pm z\sigma_{\bar{x}_1-\bar{x}_2} = 9 \pm 2.17(3.38560547) = 9 \pm 7.35 = 1.65 \text{ to } 16.35 \text{ hours}$$

c. The first two steps and the calculation of the observed value of the test statistic are the same for both the p-value approach and critical value approach.

Step 1: H_0: $\mu_1 - \mu_2 = 0$, H_1: $\mu_1 - \mu_2 \neq 0$

Step 2: Since σ_1 and σ_2 are known, the samples are independent, and the populations are normally distributed, use the normal distribution.

$$z = \frac{(\bar{x}_1-\bar{x}_2)-(\mu_1-\mu_2)}{\sigma_{\bar{x}_1-\bar{x}_2}} = \frac{9-0}{3.38560547} = 2.66$$

<u>p-value approach</u>

Step 3: From above, $z = 2.66$.

From the normal distribution table, area to the right of $z = 2.66$ is $1 - .9961 = .0039$.

p-value $= 2(.0039) = .0078$

Step 4: Reject H_0 since $.0078 < .02$.

<u>critical value approach</u>

Step 3: For $\alpha = .02$, the critical values of z are –2.33 and 2.33.

Step 4: From above, $z = 2.66$.

Step 5: Reject H_0 since $2.66 > 2.33$.

Conclude that the claim that the mean elapsed times for repellent A and B are different.

10.11 a. $\sigma_{\bar{x}_1-\bar{x}_2} = \sqrt{\frac{\sigma_1^2}{n_1}+\frac{\sigma_2^2}{n_2}} = \sqrt{\frac{(1.20)^2}{45}+\frac{(1.85)^2}{50}} = .31693848$ day

The 98% confidence interval for $\mu_1 - \mu_2$ is

$(\bar{x}_1 - \bar{x}_2) \pm z\sigma_{\bar{x}_1-\bar{x}_2} = (6.4 - 9.3) \pm 2.33(.31693848) = -2.9 \pm .74 = -3.64$ to -2.16 days

b. Step 1: H_0: $\mu_1 - \mu_2 = 0$, H_1: $\mu_1 - \mu_2 < 0$

Step 2: Since σ_1 and σ_2 are known, the samples are independent, and the sample sizes are large ($n_1 \geq 30$ and $n_2 \geq 30$), use the normal distribution.

Step 3: For $\alpha = .025$, the critical value of z is -1.96.

Step 4: $z = \frac{(\bar{x}_1 - \bar{x}_2)-(\mu_1 - \mu_2)}{\sigma_{\bar{x}_1-\bar{x}_2}} = \frac{-2.9-0}{.31693848} = -9.15$

Step 5: Reject H_0 since $-9.15 < -1.96$.

Conclude that the mean number of days missed per year by mothers working for companies that provide day-care facilities on premises is less than the mean number of days missed per year by mothers working for companies that do not provide day-care facilities on premises.

c. The Type I error would be to conclude that the mean number of days missed by the first group of mothers is less than that the mean for the second group when the two means are actually equal. The probability of making such an error is $\alpha = .025$.

10.13 a. $\sigma_{\bar{x}_1-\bar{x}_2} = \sqrt{\frac{\sigma_1^2}{n_1}+\frac{\sigma_2^2}{n_2}} = \sqrt{\frac{(800)^2}{45}+\frac{(1000)^2}{51}} = \$183.9295119$

The 99% confidence interval for $\mu_1 - \mu_2$ is

$(\bar{x}_1 - \bar{x}_2) \pm z\sigma_{\bar{x}_1-\bar{x}_2} = (3300 - 3850) \pm 2.58(183.9295119)$

$= -550 \pm 474.54 = -\$1024.54$ to $-\$75.46$

b. Step 1: H_0: $\mu_1 - \mu_2 = 0$, H_1: $\mu_1 - \mu_2 \neq 0$

Step 2: Since σ_1 and σ_2 are known, the samples are independent, and the sample sizes are large ($n_1 \geq 30$ and $n_2 \geq 30$), use the normal distribution.

Step 3: For $\alpha = .01$, the critical values of z are -2.58 and 2.58.

Step 4: $z = \frac{(\bar{x}_1 - \bar{x}_2)-(\mu_1 - \mu_2)}{\sigma_{\bar{x}_1-\bar{x}_2}} = \frac{-550-0}{183.9295119} = -2.99$

Step 5: Reject H_0 since $-2.99 < -2.58$.

Conclude that such mean repair costs are different for these two types of cars.

c. If $\alpha = 0$, there can be no rejection region, and we cannot reject H_0. Therefore, the decision would be "do not reject H_0."

10.15 a. $\sigma_{\bar{x}_1-\bar{x}_2} = \sqrt{\dfrac{\sigma_1^2}{n_1}+\dfrac{\sigma_2^2}{n_2}} = \sqrt{\dfrac{(5.5)^2}{27}+\dfrac{(6.4)^2}{25}} = 1.66095466$ calories

The 98% confidence interval for $\mu_1 - \mu_2$ is

$(\bar{x}_1 - \bar{x}_2) \pm z\sigma_{\bar{x}_1-\bar{x}_2} = (141 - 144) \pm 2.33(1.66095466) = -3 \pm 3.87 = -6.87$ to $.87$ calories

b. Step 1: H_0: $\mu_1 - \mu_2 = 0$, H_1: $\mu_1 - \mu_2 < 0$

Step 2: Since σ_1 and σ_2 are known, the samples are independent, and the populations are normally distributed, use the normal distribution.

Step 3: For $\alpha = .01$, the critical value of z is -2.33.

Step 4: $z = \dfrac{(\bar{x}_1 - \bar{x}_2) - (\mu_1 - \mu_2)}{\sigma_{\bar{x}_1-\bar{x}_2}} = \dfrac{-3-0}{1.66095466} = -1.81$

Step 5: Do not reject H_0 since $-1.81 > -2.33$.

Conclude that the mean number of calories in the eight-ounce low-fat yogurt cups produced by Maine Mountain Dairy is not less than the mean number of calories in the yogurt cups produced by the competitor. Maine Mountain Dairy's claim is false.

c. From the normal distribution table, area to the left of $z = -1.81$ is .0352.

p-value = .0352

For $\alpha = .005$, do not reject H_0 since $.0352 > .005$.

For $\alpha = .025$, do not reject H_0 since $.0352 > .025$.

Section 10.2

10.17 a. $\bar{x}_1 - \bar{x}_2 = 12.50 - 14.60 = -2.10$

b. $s_p = \sqrt{\dfrac{(n_1-1)s_1^2 + (n_2-1)s_2^2}{n_1+n_2-2}} = \sqrt{\dfrac{(25-1)(3.75)^2 + (20-1)(3.10)^2}{25+20-2}} = 3.47780337$

$s_{\bar{x}_1-\bar{x}_2} = s_p\sqrt{\dfrac{1}{n_1}+\dfrac{1}{n_2}} = 3.47780337\sqrt{\dfrac{1}{25}+\dfrac{1}{20}} = 1.04334101$

$df = n_1 + n_2 - 2 = 25 + 20 - 2 = 43$

The 95% confidence interval for $\mu_1 - \mu_2$ is

$(\bar{x}_1 - \bar{x}_2) \pm ts_{\bar{x}_1-\bar{x}_2} = -2.10 \pm 2.017(1.04334101) = -2.10 \pm 2.10 = -4.20$ to 0

10.19 Step 1: H_0: $\mu_1 - \mu_2 = 0$, H_1: $\mu_1 - \mu_2 \neq 0$

Step 2: Since σ_1 and σ_2 are unknown but assumed to be equal, the samples are independent, the sample sizes are small, and the populations are normally distributed, use the t distribution.

Step 3: From the solution to Exercise 10.17, $df = 43$
For $\alpha = .05$, the critical values of t are –2.017 and 2.017.

Step 4: From the solution to Exercise 10.17, $s_{\bar{x}_1-\bar{x}_2} = 1.04334101$

$$t = \frac{(\bar{x}_1 - \bar{x}_2) - (\mu_1 - \mu_2)}{s_{\bar{x}_1-\bar{x}_2}} = \frac{(12.50 - 14.60) - 0}{1.04334101} = -2.013$$

Step 5: Do not reject H_0 since $-2.013 > -2.017$.

10.21 Step 1: H_0: $\mu_1 - \mu_2 = 0$, H_1: $\mu_1 - \mu_2 < 0$

Step 2: Since σ_1 and σ_2 are unknown but assumed to be equal, the samples are independent, the sample sizes are small, and the populations are normally distributed, use the t distribution.

Step 3: From the solution to Exercise 10.17, $df = 43$
For $\alpha = .01$, the critical value of t is –2.416.

Step 4: From the solution to Exercise 10.17, $s_{\bar{x}_1-\bar{x}_2} = 1.04334101$

$$t = \frac{(\bar{x}_1 - \bar{x}_2) - (\mu_1 - \mu_2)}{s_{\bar{x}_1-\bar{x}_2}} = \frac{(12.50 - 14.60) - 0}{1.04334101} = -2.013$$

Step 5: Do not reject H_0 since $-2.013 > -2.416$.

10.23 From Sample 1: $n_1 = 13$, $\Sigma x = 385$, $\Sigma x^2 = 11{,}701$

$\bar{x}_1 = \Sigma x / n_1 = 385/13 = 29.62$

$$s_1 = \sqrt{\frac{\Sigma x^2 - \frac{(\Sigma x)^2}{n}}{n-1}} = \sqrt{\frac{11{,}701 - \frac{(385)^2}{13}}{13-1}} = 4.99230177$$

From Sample 2: $n_2 = 12$, $\Sigma x = 306$, $\Sigma x^2 = 7918$

$\bar{x}_2 = \Sigma x / n_2 = 306/12 = 25.50$

$$s_2 = \sqrt{\frac{\Sigma x^2 - \frac{(\Sigma x)^2}{n}}{n-1}} = \sqrt{\frac{7918 - \frac{(306)^2}{12}}{12-1}} = 3.23334895$$

a. $\bar{x}_1 - \bar{x}_2 = 29.62 - 25.50 = 4.12$

b. $s_p = \sqrt{\dfrac{(n_1 - 1)s_1^2 + (n_2 - 1)s_2^2}{n_1 + n_2 - 2}}$

$= \sqrt{\dfrac{(13-1)(4.99230177)^2 + (12-1)(3.23334895)^2}{13+12-2}} = 4.24303482$

$s_{\bar{x}_1 - \bar{x}_2} = s_p \sqrt{\dfrac{1}{n_1} + \dfrac{1}{n_2}} = 4.24303482 \sqrt{\dfrac{1}{13} + \dfrac{1}{12}} = 1.69857333$

$df = n_1 + n_2 - 2 = 13 + 12 - 2 = 23$

The 98% confidence interval for $\mu_1 - \mu_2$ is

$(\bar{x}_1 - \bar{x}_2) \pm t s_{\bar{x}_1 - \bar{x}_2} = 4.12 \pm 2.500(1.69857333) = 4.12 \pm 4.25 = -.13$ to 8.37

c. Step 1: H_0: $\mu_1 - \mu_2 = 0$, H_1: $\mu_1 - \mu_2 > 0$

Step 2: Since σ_1 and σ_2 are unknown but assumed to be equal, the samples are independent, the sample sizes are small, and the populations are normally distributed, use the *t* distribution.

Step 3: For $\alpha = .01$ with $df = 23$, the critical value of t is 2.500.

Step 4: $t = \dfrac{(\bar{x}_1 - \bar{x}_2) - (\mu_1 - \mu_2)}{s_{\bar{x}_1 - \bar{x}_2}} = \dfrac{4.12 - 0}{1.69857333} = 2.426$

Step 5: Do not reject H_0 since $2.426 > 2.500$.

10.25 a. $s_p = \sqrt{\dfrac{(n_1 - 1)s_1^2 + (n_2 - 1)s_2^2}{n_1 + n_2 - 2}} = \sqrt{\dfrac{(35-1)(17.50)^2 + (40-1)(14.40)^2}{35+40-2}} = \$15.9191192$

$s_{\bar{x}_1 - \bar{x}_2} = s_p \sqrt{\dfrac{1}{n_1} + \dfrac{1}{n_2}} = 15.9191192 \sqrt{\dfrac{1}{35} + \dfrac{1}{40}} = \$3.68456013$

$df = n_1 + n_2 - 2 = 35 + 40 - 2 = 73$

The 99% confidence interval for $\mu_1 - \mu_2$ is

$(\bar{x}_1 - \bar{x}_2) \pm t s_{\bar{x}_1 - \bar{x}_2} = (80 - 96) \pm 2.645(3.68456013) = -16 \pm 9.75 = -\$25.75$ to $-\$6.25$

b. Step 1: H_0: $\mu_1 - \mu_2 = 0$, H_1: $\mu_1 - \mu_2 < 0$

Step 2: Since σ_1 and σ_2 are unknown but assumed to be equal, the samples are independent, and both samples are large ($n_1 \geq 30$ and $n_2 \geq 30$), use the *t* distribution.

Step 3: For $\alpha = .025$ with $df = 73$, the critical value of t is -1.993.

Step 4: $t = \dfrac{(\bar{x}_1 - \bar{x}_2) - (\mu_1 - \mu_2)}{s_{\bar{x}_1 - \bar{x}_2}} = \dfrac{-16 - 0}{3.68456013} = -4.342$

Step 5: Reject H_0 since $-4.342 < -1.993$.

Conclude that the mean amount spent by all male customers at this supermarket is less than that by all female customers.

10.27 a. $s_p = \sqrt{\dfrac{(n_1 - 1)s_1^2 + (n_2 - 1)s_2^2}{n_1 + n_2 - 2}} = \sqrt{\dfrac{(27-1)(2.2)^2 + (18-1)(2.5)^2}{27 + 18 - 2}} = 2.32323952$ mph

$s_{\bar{x}_1 - \bar{x}_2} = s_p \sqrt{\dfrac{1}{n_1} + \dfrac{1}{n_2}} = 2.32323952\sqrt{\dfrac{1}{27} + \dfrac{1}{18}} = .70693927$ mph

$df = n_1 + n_2 - 2 = 27 + 18 - 2 = 43$

The 98% confidence interval for $\mu_1 - \mu_2$ is

$(\bar{x}_1 - \bar{x}_2) \pm t s_{\bar{x}_1 - \bar{x}_2} = (72 - 68) \pm 2.416(.70693927) = 4 \pm 1.71 = 2.29$ to 5.71 mph

b. Step 1: H_0: $\mu_1 - \mu_2 = 0$, H_1: $\mu_1 - \mu_2 > 0$

Step 2: Since σ_1 and σ_2 are unknown but assumed to be equal, the samples are independent, the sample sizes are small, and the populations are normally distributed, use the t distribution.

Step 3: For $\alpha = .01$ with $df = 43$, the critical value of t is 2.416.

Step 4: $t = \dfrac{(\bar{x}_1 - \bar{x}_2) - (\mu_1 - \mu_2)}{s_{\bar{x}_1 - \bar{x}_2}} = \dfrac{4 - 0}{.70693927} = 5.658$

Step 5: Reject H_0 since $5.658 > 2.416$.

Conclude that the mean speeds of cars driven by all men drivers on this highway is greater than that of cars driven by all women drivers.

10.29 a. $s_p = \sqrt{\dfrac{(n_1 - 1)s_1^2 + (n_2 - 1)s_2^2}{n_1 + n_2 - 2}} = \sqrt{\dfrac{(25-1)(11)^2 + (22-1)(9)^2}{25 + 22 - 2}} = 10.11599394$ minutes

$s_{\bar{x}_1 - \bar{x}_2} = s_p \sqrt{\dfrac{1}{n_1} + \dfrac{1}{n_2}} = 10.11599394\sqrt{\dfrac{1}{25} + \dfrac{1}{22}} = 2.95716900$ minutes

$df = n_1 + n_2 - 2 = 25 + 22 - 2 = 45$

The 99% confidence interval for $\mu_1 - \mu_2$ is

$(\bar{x}_1 - \bar{x}_2) \pm t s_{\bar{x}_1 - \bar{x}_2} = (44 - 49) \pm 2.690(2.95716900) = -5 \pm 7.95 = -12.95$ to 2.95 minutes

b. Step 1: H_0: $\mu_1 - \mu_2 = 0$, H_1: $\mu_1 - \mu_2 < 0$

Step 2: Since σ_1 and σ_2 are unknown but assumed to be equal, the samples are independent, the sample sizes are small, and the populations are normally distributed, use the t distribution.

Step 3: For $\alpha = .01$ with $df = 45$, the critical value of t is -2.412.

Step 4: $$t = \frac{(\bar{x}_1 - \bar{x}_2) - (\mu_1 - \mu_2)}{s_{\bar{x}_1 - \bar{x}_2}} = \frac{-5 - 0}{2.95716900} = -1.691$$

Step 5: Do not reject H_0 since $-1.691 > -2.412$.

Conclude that the mean relief time for Brand A is not less than that for Brand B.

10.31 a. $$s_p = \sqrt{\frac{(n_1 - 1)s_1^2 + (n_2 - 1)s_2^2}{n_1 + n_2 - 2}} = \sqrt{\frac{(380 - 1)(.75)^2 + (370 - 1)(.59)^2}{380 + 370 - 2}} = .67582036$$

$$s_{\bar{x}_1 - \bar{x}_2} = s_p \sqrt{\frac{1}{n_1} + \frac{1}{n_2}} = .67582036 \sqrt{\frac{1}{380} + \frac{1}{370}} = .04935933$$

Since $n_1 = 380$ and $n_2 = 370$ are very large, we use the normal distribution to approximate the t distribution.

The 98% confidence interval for $\mu_1 - \mu_2$ is

$(\bar{x}_1 - \bar{x}_2) \pm z s_{\bar{x}_1 - \bar{x}_2} = (7.6 - 8.1) \pm 2.33(.04935933) = -.5 \pm .12 = -.62$ to $-.38$

b. Step 1: H_0: $\mu_1 - \mu_2 = 0$, H_1: $\mu_1 - \mu_2 \neq 0$

Step 2: Since σ_1 and σ_2 are unknown but assumed to be equal, the samples are independent, and both samples are large ($n_1 \geq 30$ and $n_2 \geq 30$), use the t distribution. However, since $n_1 = 380$ and $n_2 = 370$ are very large, we use the normal distribution to approximate the t distribution.

Step 3: For $\alpha = .01$, the critical values of z are -2.58 and 2.58.

Step 4: $$z = \frac{(\bar{x}_1 - \bar{x}_2) - (\mu_1 - \mu_2)}{s_{\bar{x}_1 - \bar{x}_2}} = \frac{-.5 - 0}{.04935933} = -10.13$$

Step 5: Reject H_0 since $-10.13 < -2.58$.

Conclude that the mean satisfaction indexes for all customers for the two supermarkets are different.

Section 10.3

10.33 $s_{\bar{x}_1-\bar{x}_2} = \sqrt{\frac{s_1^2}{n_1}+\frac{s_2^2}{n_2}} = \sqrt{\frac{(3.90)^2}{24}+\frac{(5.15)^2}{16}} = 1.51373916$

$$df = \frac{\left(\frac{s_1^2}{n_1}+\frac{s_2^2}{n_2}\right)^2}{\frac{\left(\frac{s_1^2}{n_1}\right)^2}{n_1-1}+\frac{\left(\frac{s_2^2}{n_2}\right)^2}{n_2-1}} = \frac{\left(\frac{3.90^2}{24}+\frac{5.15^2}{16}\right)^2}{\frac{\left(\frac{3.90^2}{24}\right)^2}{23}+\frac{\left(\frac{5.15^2}{16}\right)^2}{15}} = 26.17 \approx 26$$

The 95% confidence interval for $\mu_1 - \mu_2$ is

$(\bar{x}_1 - \bar{x}_2) \pm t s_{\bar{x}_1-\bar{x}_2} = (20.50 - 22.60) \pm 2.056(1.51373916) = -2.10 \pm 3.11 = -5.21$ to 1.01

10.35 Step 1: H_0: $\mu_1 - \mu_2 = 0$, H_1: $\mu_1 - \mu_2 \neq 0$

Step 2: Since σ_1 and σ_2 are unknown but assumed to be unequal, the samples are independent, the sample sizes are small, and the populations are normally distributed, use the t distribution.

Step 3: From the solution to Exercise 10.33, $df = 26$.
For $\alpha = .05$, the critical values of t are –2.056 and 2.056.

Step 4: From the solution to Exercise 10.33, $s_{\bar{x}_1-\bar{x}_2} = 1.51373916$

$$t = \frac{(\bar{x}_1 - \bar{x}_2) - (\mu_1 - \mu_2)}{s_{\bar{x}_1-\bar{x}_2}} = \frac{(20.50 - 22.60) - 0}{1.51373916} = -1.387$$

Step 5: Do not reject H_0 since $-1.387 > -2.056$.

10.37 Step 1: H_0: $\mu_1 - \mu_2 = 0$, H_1: $\mu_1 - \mu_2 < 0$

Step 2: Since σ_1 and σ_2 are unknown but assumed to be unequal, the samples are independent, the sample sizes are small, and the populations are normally distributed, use the t distribution.

Step 3: From the solution to Exercise 10.33, $df = 26$.
For $\alpha = .01$, the critical value of t is –2.479.

Step 4: From the solution to Exercise 10.33, $s_{\bar{x}_1-\bar{x}_2} = 1.51373916$

$$t = \frac{(\bar{x}_1 - \bar{x}_2) - (\mu_1 - \mu_2)}{s_{\bar{x}_1 - \bar{x}_2}} = \frac{(20.50 - 22.60) - 0}{1.51373916} = -1.387$$

Step 5: Do not reject H_0 since $-1.387 > -2.479$.

10.39 a. $s_{\bar{x}_1 - \bar{x}_2} = \sqrt{\frac{s_1^2}{n_1} + \frac{s_2^2}{n_2}} = \sqrt{\frac{(17.5)^2}{35} + \frac{(14.40)^2}{40}} = \$3.73282735$

$$df = \frac{\left(\frac{s_1^2}{n_1} + \frac{s_2^2}{n_2}\right)^2}{\frac{\left(\frac{s_1^2}{n_1}\right)^2}{n_1 - 1} + \frac{\left(\frac{s_2^2}{n_2}\right)^2}{n_2 - 1}} = \frac{\left(\frac{17.5^2}{35} + \frac{14.40^2}{40}\right)^2}{\frac{\left(\frac{17.5^2}{35}\right)^2}{34} + \frac{\left(\frac{14.40^2}{40}\right)^2}{39}} = 66.02 \approx 66$$

The 99% confidence interval for $\mu_1 - \mu_2$ is

$(\bar{x}_1 - \bar{x}_2) \pm t s_{\bar{x}_1 - \bar{x}_2} = (80 - 96) \pm 2.652(3.73282735) = -16 \pm 9.90 = -\$25.90$ to $-\$6.10$

b. Step 1: H_0: $\mu_1 - \mu_2 = 0$, H_1: $\mu_1 - \mu_2 < 0$

Step 2: Since σ_1 and σ_2 are unknown but assumed to be unequal, the samples are independent, and both samples are large ($n_1 \geq 30$ and $n_2 \geq 30$), use the t distribution.

Step 3: For $\alpha = .025$ with $df = 66$, the critical value of t is -1.997.

Step 4: $t = \frac{(\bar{x}_1 - \bar{x}_2) - (\mu_1 - \mu_2)}{s_{\bar{x}_1 - \bar{x}_2}} = \frac{-16 - 0}{3.73282735} = -4.286$

Step 5: Reject H_0 since $-4.286 < -1.997$.

Conclude that the mean amount spent by all male customers at this supermarket is less than that of all female customers.

10.41 a. $s_{\bar{x}_1 - \bar{x}_2} = \sqrt{\frac{s_1^2}{n_1} + \frac{s_2^2}{n_2}} = \sqrt{\frac{(2.2)^2}{27} + \frac{(2.5)^2}{18}} = .72559044$ mph

$$df = \frac{\left(\frac{s_1^2}{n_1} + \frac{s_2^2}{n_2}\right)^2}{\frac{\left(\frac{s_1^2}{n_1}\right)^2}{n_1 - 1} + \frac{\left(\frac{s_2^2}{n_2}\right)^2}{n_2 - 1}} = \frac{\left(\frac{2.2^2}{27} + \frac{2.5^2}{18}\right)^2}{\frac{\left(\frac{2.2^2}{27}\right)^2}{26} + \frac{\left(\frac{2.5^2}{18}\right)^2}{17}} = 33.28 \approx 33$$

The 98% confidence interval for $\mu_1 - \mu_2$ is

$(\bar{x}_1 - \bar{x}_2) \pm t s_{\bar{x}_1 - \bar{x}_2} = (72 - 68) \pm 2.445(.72559044) = 4 \pm 1.77 = 2.23$ to 5.77 mph

b. Step 1: H_0: $\mu_1 - \mu_2 = 0$, H_1: $\mu_1 - \mu_2 > 0$

Step 2: Since σ_1 and σ_2 are unknown but assumed to be unequal, the samples are independent, the sample sizes are small, and the populations are normally distributed, use the t distribution..

Step 3: For $\alpha = .01$ with $df = 33$, the critical value of t is 2.445.

Step 4: $$t = \frac{(\bar{x}_1 - \bar{x}_2) - (\mu_1 - \mu_2)}{s_{\bar{x}_1 - \bar{x}_2}} = \frac{4 - 0}{.72559044} = 5.513$$

Step 5: Reject H_0 since $5.513 > 2.445$.

Conclude that the mean speed of cars driven by all men drivers on this highway is higher than that of cars driven by all women drivers.

10.43 a. $$s_{\bar{x}_1 - \bar{x}_2} = \sqrt{\frac{s_1^2}{n_1} + \frac{s_2^2}{n_2}} = \sqrt{\frac{(11)^2}{25} + \frac{(9)^2}{22}} = 2.91921534 \text{ minutes}$$

$$df = \frac{\left(\frac{s_1^2}{n_1} + \frac{s_2^2}{n_2}\right)^2}{\frac{\left(\frac{s_1^2}{n_1}\right)^2}{n_1 - 1} + \frac{\left(\frac{s_2^2}{n_2}\right)^2}{n_2 - 1}} = \frac{\left(\frac{11^2}{25} + \frac{9^2}{22}\right)^2}{\frac{\left(\frac{11^2}{25}\right)^2}{24} + \frac{\left(\frac{9^2}{22}\right)^2}{21}} = 44.78 \approx 44$$

The 99% confidence interval for $\mu_1 - \mu_2$ is

$(\bar{x}_1 - \bar{x}_2) \pm t s_{\bar{x}_1 - \bar{x}_2} = (44 - 49) \pm 2.692(2.91921534) = -5 \pm 7.86 = -12.86$ to 2.86 minutes

b. Step 1: H_0: $\mu_1 - \mu_2 = 0$, H_1: $\mu_1 - \mu_2 < 0$

Step 2: Since σ_1 and σ_2 are unknown but assumed to be unequal, the samples are independent, the sample sizes are small, and the populations are normally distributed, use the t distribution.

Step 3: For $\alpha = .01$ with $df = 44$, the critical value of t is –2.414.

Step 4: $$t = \frac{(\bar{x}_1 - \bar{x}_2) - (\mu_1 - \mu_2)}{s_{\bar{x}_1 - \bar{x}_2}} = \frac{-5 - 0}{2.92921534} = -1.713$$

Step 5: Do not reject H_0 since $-1.713 > -2.414$.

Conclude that the mean relief time for Brand A is not less than that for Brand B.

10.45 a. $s_{\bar{x}_1-\bar{x}_2} = \sqrt{\frac{s_1^2}{n_1}+\frac{s_2^2}{n_2}} = \sqrt{\frac{(.75)^2}{380}+\frac{(.59)^2}{370}} = .04920441$

Since $n_1 = 380$ and $n_2 = 370$ are very large, we use the normal distribution to approximate the t distribution.

The 98% confidence interval for $\mu_1 - \mu_2$ is

$(\bar{x}_1 - \bar{x}_2) \pm z s_{\bar{x}_1-\bar{x}_2} = (7.6 - 8.1) \pm 2.33(.04920441) = -.5 \pm .11 = -.61$ to $-.39$

b. Step 1: H_0: $\mu_1 - \mu_2 = 0$, H_1: $\mu_1 - \mu_2 \neq 0$

Step 2: Since σ_1 and σ_2 are unknown but assumed to be unequal, the samples are independent, and both samples are large ($n_1 \geq 30$ and $n_2 \geq 30$), use the t distribution. However, since $n_1 = 380$ and $n_2 = 370$ are very large, we use the normal distribution to approximate the t distribution.

Step 3: For $\alpha = .01$, the critical values of z are -2.58 and 2.58.

Step 4: $z = \frac{(\bar{x}_1 - \bar{x}_2) - (\mu_1 - \mu_2)}{s_{\bar{x}_1-\bar{x}_2}} = \frac{-.5 - 0}{.04920441} = -10.16$

Step 5: Reject H_0 since $-10.16 < -2.58$.

Conclude that the mean satisfaction indexes for all customers for the two supermarkets are different.

Section 10.4

10.47 The **paired samples procedure** is used when two samples are drawn in such a way that for each data value collected from one sample there is a corresponding data value collected from the second sample, and both of these data values are collected from the same source.

10.49 a. $s_{\bar{d}} = s_d / \sqrt{n} = 6.3 / \sqrt{12} = 1.81865335$

$df = n - 1 = 12 - 1 = 11$

The 99% confidence interval for μ_d is

$\bar{d} \pm t s_{\bar{d}} = 17.5 \pm 3.106(1.81865335) = 17.5 \pm 5.65 = 11.85$ to 23.15

b. $s_{\bar{d}} = s_d / \sqrt{n} = 14.7 / \sqrt{27} = 2.82901632$

$df = n - 1 = 27 - 1 = 26$

The 95% confidence interval for μ_d is

$\bar{d} \pm t s_{\bar{d}} = 55.9 \pm 2.056(2.82901632) = 55.9 \pm 5.82 = 50.08$ to 61.72

c. $s_{\bar{d}} = s_d / \sqrt{n} = 8.3 / \sqrt{16} = 2.075$

$df = n - 1 = 16 - 1 = 15$

The 90% confidence interval for μ_d is

$\bar{d} \pm t s_{\bar{d}} = 29.3 \pm 1.753(2.075) = 29.3 \pm 3.64 = 25.66$ to 32.94

10.51 a. Step 1: H_0: $\mu_d = 0$, H_1: $\mu_d \neq 0$

Step 2: Since the samples are dependent, σ_d is unknown, the sample is small, and the population of paired differences is normally distributed, use the t distribution.

Step 3: $df = n - 1 = 26 - 1 = 25$

For $\alpha = .05$, the critical values of t are -2.060 and 2.060.

Step 4: $s_{\bar{d}} = s_d / \sqrt{n} = 3.9 / \sqrt{26} = .76485293$

$t = (\bar{d} - \mu_d) / s_{\bar{d}} = (9.6 - 0)/.76485293 = 12.551$

Step 5: Reject H_0 since $12.551 > 2.060$.

b. Step 1: H_0: $\mu_d = 0$, H_1: $\mu_d > 0$

Step 2: Since the samples are dependent, σ_d is unknown, the sample is small, and the population of paired differences is normally distributed, use the t distribution.

Step 3: $df = n - 1 = 15 - 1 = 14$

For $\alpha = .01$, the critical value of t is 2.624.

Step 4: $s_{\bar{d}} = s_d / \sqrt{n} = 4.7 / \sqrt{15} = 1.21353478$

$t = (\bar{d} - \mu_d) / s_{\bar{d}} = (8.8 - 0)/1.21353478 = 7.252$

Step 5: Reject H_0 since $7.252 > 2.624$.

c. Step 1: H_0: $\mu_d = 0$, H_1: $\mu_d < 0$

Step 2: Since the samples are dependent, σ_d is unknown, the sample is small, and the population of paired differences is normally distributed, use the t distribution.

Step 3: $df = n - 1 = 20 - 1 = 19$

For $\alpha = .10$, the critical value of t is -1.328.

Step 4: $s_{\bar{d}} = s_d / \sqrt{n} = 2.3 / \sqrt{20} = .51429563$

$t = (\bar{d} - \mu_d) / s_{\bar{d}} = (-7.4 - 0)/.51429563 = -14.389$

Step 5: Reject H_0 since $-14.389 < -1.328$.

10.53

Before	After	d	d^2
103	100	3	9
97	95	2	4
111	104	7	49
95	101	–6	36
102	96	6	36
96	91	5	25
108	101	7	49
		$\Sigma d = 24$	$\Sigma d^2 = 208$

$\bar{d} = \Sigma d / n = 24/7 = 3.43$ minutes

$$s_d = \sqrt{\frac{\Sigma d^2 - \frac{(\Sigma d)^2}{n}}{n-1}} = \sqrt{\frac{208 - \frac{(24)^2}{7}}{7-1}} = 4.57737708 \text{ minutes}$$

$s_{\bar{d}} = s_d / \sqrt{n} = 4.57737708 / \sqrt{7} = 1.73008592$ minutes

$df = n - 1 = 7 - 1 = 6$

a. The 99% confidence interval for μ_d is

$\bar{d} \pm t s_{\bar{d}} = 3.43 \pm 3.707(1.73008592) = 3.43 \pm 6.41 = -2.98$ to 9.84 minutes

b. Step 1: H_0: $\mu_d = 0$, H_1: $\mu_d > 0$

Step 2: Since the samples are dependent, σ_d is unknown, the sample is small, and the population of paired differences is normally distributed, use the t distribution.

Step 3: For $\alpha = .025$ with $df = 6$, the critical value of t is 2.447.

Step 4: $t = (\bar{d} - \mu_d) / s_{\bar{d}} = (3.43 - 0)/1.73008592 = 1.983$

Step 5: Do not reject H_0 since $1.983 < 2.447$.

Conclude that taking the dietary supplement does not result in faster mean times in the time trials.

10.55

Before	After	d	d^2
180	183	–3	9
195	187	8	64
177	161	16	256
221	204	17	289
208	197	11	121
199	189	10	100
		$\Sigma d = 59$	$\Sigma d^2 = 839$

$\bar{d} = \Sigma d / n = 59/6 = 9.83$ pounds

$$s_d = \sqrt{\frac{\Sigma d^2 - \frac{(\Sigma d)^2}{n}}{n-1}} = \sqrt{\frac{839 - \frac{(59)^2}{6}}{6-1}} = 7.19490561 \text{ pounds}$$

$s_{\bar{d}} = s_d / \sqrt{n} = 7.19490561/\sqrt{6} = 2.93730791$ pounds

$df = n - 1 = 6 - 1 = 5$

a. The 95% confidence interval for μ_d is

$\bar{d} \pm t s_{\bar{d}} = 9.83 \pm 2.571(2.93730791) = 9.83 \pm 7.55 = 2.28$ to 17.38 pounds

b. Step 1: H_0: $\mu_d = 0$, H_1: $\mu_d > 0$

Step 2: Since the samples are dependent, σ_d is unknown, the sample is small, and the population of paired differences is normally distributed, use the *t* distribution.

Step 3: For $\alpha = .01$ with $df = 5$, the critical value of t is 3.365.

Step 4: $t = (\bar{d} - \mu_d)/s_{\bar{d}} = (9.83 - 0)/2.93730791 = 3.347$

Step 5: Do not reject H_0 since $3.347 < 3.365$.

Conclude that the mean weight loss for all persons due to this special exercise program is not greater than zero.

10.57

Before	After	d	d^2
23	21	2	4
26	24	2	4
19	23	–4	16
24	25	–1	1
27	24	3	9
22	28	–6	36
20	24	–4	16
18	23	–5	25
		$\Sigma d = -13$	$\Sigma d^2 = 111$

$\bar{d} = \Sigma d / n = -13/8 = -1.63$ minutes

$$s_d = \sqrt{\frac{\Sigma d^2 - \frac{(\Sigma d)^2}{n}}{n-1}} = \sqrt{\frac{111 - \frac{(-13)^2}{8}}{8-1}} = 3.58319490 \text{ minutes}$$

$s_{\bar{d}} = s_d / \sqrt{n} = 3.58319490/\sqrt{8} = 1.26685071$ minutes

$df = n - 1 = 8 - 1 = 7$

a. The 98% confidence interval for μ_d is

$\bar{d} \pm t s_{\bar{d}} = -1.63 \pm 2.998(1.26685071) = -1.63 \pm 3.80 = -5.43$ to 2.17 minutes

b. Step 1: H_0: $\mu_d = 0$, H_1: $\mu_d \neq 0$

Step 2: Since the samples are dependent, σ_d is unknown, the sample is small, and the population of paired differences is normally distributed, use the t distribution.

Step 3: For $\alpha = .05$ with $df = 7$, the critical values of t are –2.365 and 2.365.

Step 4: $t = (\bar{d} - \mu_d)/s_{\bar{d}} = (-1.63 - 0)/1.26685071 = -1.287$

Step 5: Do not reject H_0 since $-1.287 > -2.365$.

Conclude that the mean times taken to assemble a unit of the product are not different for the two types of machines.

Section 10.5

10.59 Two samples are considered large enough for the sampling distribution of the difference between two sample proportions to be approximately normal when $n_1 p_1$, $n_1 q_1$, $n_2 p_2$, and $n_2 q_2$ are all greater than 5.

10.61 $$s_{\hat{p}_1 - \hat{p}_2} = \sqrt{\frac{\hat{p}_1 \hat{q}_1}{n_1} + \frac{\hat{p}_2 \hat{q}_2}{n_2}} = \sqrt{\frac{(.81)(.19)}{100} + \frac{(.77)(.23)}{150}} = .05215042$$

The 95% confidence interval for $p_1 - p_2$ is

$$(\hat{p}_1 - \hat{p}_2) \pm z s_{\hat{p}_1 - \hat{p}_2} = (.81 - .77) \pm 1.96(.05215042) = .04 \pm .102 = -.062 \text{ to } .142$$

10.63 Step 1: H_0: $p_1 - p_2 = 0$, H_1: $p_1 - p_2 \neq 0$

Step 2: Since $n_1 \hat{p}_1$, $n_1 \hat{q}_1$, $n_2 \hat{p}_2$, and $n_2 \hat{q}_2$ are all greater than 5, use the normal distribution.

Step 3: For $\alpha = .05$, the critical values of z are –1.96 and 1.96.

Step 4: $$\bar{p} = \frac{n_1 \hat{p}_1 + n_2 \hat{p}_2}{n_1 + n_2} = \frac{100(.81) + 150(.77)}{100 + 150} = .786 \text{ and } \bar{q} = 1 - \bar{p} = 1 - .786 = .214$$

$$s_{\hat{p}_1 - \hat{p}_2} = \sqrt{\bar{p}\bar{q}\left(\frac{1}{n_1} + \frac{1}{n_2}\right)} = \sqrt{(.786)(.214)\left(\frac{1}{100} + \frac{1}{150}\right)} = .05294714$$

$$z = \frac{(\hat{p}_1 - \hat{p}_2) - (p_1 - p_2)}{s_{\hat{p}_1 - \hat{p}_2}} = \frac{(.81 - .77) - 0}{.05294714} = .76$$

Step 5: Do not reject H_0 since $.76 < 1.96$.

10.65 Step 1: H_0: $p_1 - p_2 = 0$, H_1: $p_1 - p_2 < 0$

Step 2: Since $n_1 \hat{p}_1$, $n_1 \hat{q}_1$, $n_2 \hat{p}_2$, and $n_2 \hat{q}_2$ are all greater than 5, use the normal distribution.

Step 3: For $\alpha = .02$, the critical value of z is 2.05.

Step 4: From the solution to Exercise 10.63, $s_{\hat{p}_1-\hat{p}_2} = .05294714$.

$$z = \frac{(\hat{p}_1 - \hat{p}_2) - (p_1 - p_2)}{s_{\hat{p}_1-\hat{p}_2}} = \frac{(.81-.77)-0}{.05294714} = .76$$

Step 5: Do not reject H_0 since $.76 < 2.05$.

10.67 $\hat{p}_1 = x_1 / n_1 = 290/1000 = .29$ and $\hat{p}_2 = x_2 / n_2 = 396/1200 = .33$

a. $\hat{p}_1 - \hat{p}_2 = .29 - .33 = -.04$

b. $s_{\hat{p}_1-\hat{p}_2} = \sqrt{\frac{\hat{p}_1\hat{q}_1}{n_1} + \frac{\hat{p}_2\hat{q}_2}{n_2}} = \sqrt{\frac{(.29)(.71)}{1000} + \frac{(.33)(.67)}{1200}} = .01975222$

The 98% confidence interval for $p_1 - p_2$ is

$(\hat{p}_1 - \hat{p}_2) \pm z s_{\hat{p}_1-\hat{p}_2} = -.04 \pm 2.33(.01975222) = -.04 \pm .046 = -.086$ to $.006$

c. H_0: $p_1 - p_2 = 0$; H_1: $p_1 - p_2 < 0$

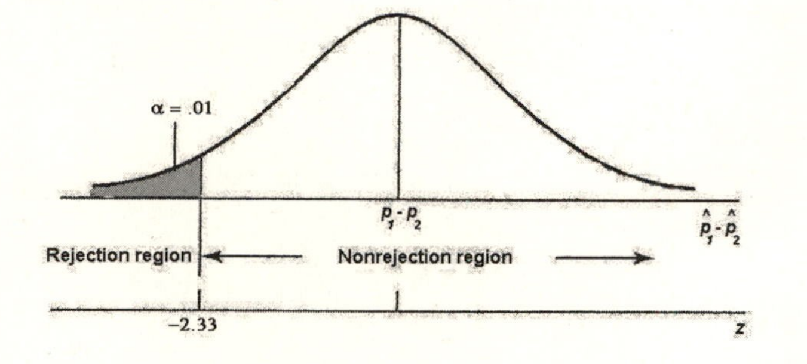

d. $\bar{p} = \frac{x_1 + x_2}{n_1 + n_2} = \frac{290 + 396}{1000 + 1200} = .312$ and $\bar{q} = 1 - \bar{p} = 1 - .312 = .688$

$$s_{\hat{p}_1-\hat{p}_2} = \sqrt{\bar{p}\,\bar{q}\left(\frac{1}{n_1} + \frac{1}{n_2}\right)} = \sqrt{(.312)(.688)\left(\frac{1}{1000} + \frac{1}{1200}\right)} = .01983774$$

$$z = \frac{(\hat{p}_1 - \hat{p}_2) - (p_1 - p_2)}{s_{\hat{p}_1-\hat{p}_2}} = \frac{-.04 - 0}{.01983774} = -2.02$$

e. Do not reject H_0 since $-2.02 > -2.33$.

10.69 a. $s_{\hat{p}_1-\hat{p}_2} = \sqrt{\frac{\hat{p}_1\hat{q}_1}{n_1} + \frac{\hat{p}_2\hat{q}_2}{n_2}} = \sqrt{\frac{(.685)(.315)}{1000} + \frac{(.72)(.28)}{1050}} = .02019344$

The 98% confidence interval for $p_1 - p_2$ is

$(\hat{p}_1 - \hat{p}_2) \pm z s_{\hat{p}_1 - \hat{p}_2} = (.685 - .72) \pm 2.33(.02019344) = -.035 \pm .047 = -.082$ to $.012$

b. Step 1: H_0: $p_1 - p_2 = 0, H_1$: $p_1 - p_2 < 0$

Step 2: Since $n_1\hat{p}_1$, $n_1\hat{q}_1$, $n_2\hat{p}_2$, and $n_2\hat{q}_2$ are all greater than 5, use the normal distribution.

Step 3: For $\alpha = .01$, the critical value of z is -2.33.

Step 4: $\bar{p} = \dfrac{n_1\hat{p}_1 + n_2\hat{p}_2}{n_1 + n_2} = \dfrac{1000(.685) + 1050(.72)}{1000 + 1050} = .703$ and

$\bar{q} = 1 - \bar{p} = 1 - .703 = .297$

$$s_{\hat{p}_1 - \hat{p}_2} = \sqrt{\bar{p}\bar{q}\left(\frac{1}{n_1} + \frac{1}{n_2}\right)} = \sqrt{(.703)(.297)\left(\frac{1}{1000} + \frac{1}{1050}\right)} = .00219009$$

$$z = \frac{(\hat{p}_1 - \hat{p}_2) - (p_1 - p_2)}{s_{\hat{p}_1 - \hat{p}_2}} = \frac{-.035 - 0}{.02019009} = -1.73$$

Step 5: Do not reject H_0 since $-1.73 > -2.33$.

Conclude that the proportion of all drivers in Tennessee who used seat belts in 2003 is not lower than that for 2004.

c. Step 1: H_0: $p_1 - p_2 = 0, H_1$: $p_1 - p_2 < 0$

Step 2: Since $n_1\hat{p}_1$, $n_1\hat{q}_1$, $n_2\hat{p}_2$, and $n_2\hat{q}_2$ are all greater than 5, use the normal distribution.

Step 3: From above, $z = -1.73$.

From the normal distribution table, area to the left of $z = -1.73$ is .0418.

p-value = .0418

Step 4: Do not reject H_0 since $.0418 > .01$.

10.71 a. $\hat{p}_1 = x_1 / n_1 = 53/400 = .133$ and $\hat{p}_2 = x_2 / n_2 = 51/470 = .109$

$\hat{p}_1 - \hat{p}_2 = .133 - .109 = .024$

b. $$s_{\hat{p}_1 - \hat{p}_2} = \sqrt{\frac{\hat{p}_1\hat{q}_1}{n_1} + \frac{\hat{p}_2\hat{q}_2}{n_2}} = \sqrt{\frac{(.133)(.867)}{400} + \frac{(.109)(.891)}{470}} = .02224666$$

The 95% confidence interval for $p_1 - p_2$ is

$(\hat{p}_1 - \hat{p}_2) \pm z s_{\hat{p}_1 - \hat{p}_2} = (.133 - .109) \pm 1.96(.02224666) = .024 \pm .044 = -.020$ to $.068$

c. The first two steps and the calculation of the observed value of the test statistic are the same for both the p-value approach and critical value approach.

Step 1: H_0: $p_1 - p_2 = 0$, H_1: $p_1 - p_2 \neq 0$

Step 2: Since $n_1\hat{p}_1$, $n_1\hat{q}_1$, $n_2\hat{p}_2$, and $n_2\hat{q}_2$ are all greater than 5, use the normal distribution.

$$\bar{p} = \frac{x_1 + x_2}{n_1 + n_2} = \frac{53+51}{400+470} = .120 \text{ and } \bar{q} = 1 - \bar{p} = 1 - .120 = .880$$

$$s_{\hat{p}_1-\hat{p}_2} = \sqrt{\bar{p}\bar{q}\left(\frac{1}{n_1}+\frac{1}{n_2}\right)} = \sqrt{(.120)(.880)\left(\frac{1}{400}+\frac{1}{470}\right)} = .02210613$$

$$z = \frac{(\hat{p}_1 - \hat{p}_2) - (p_1 - p_2)}{s_{\hat{p}_1-\hat{p}_2}} = \frac{.024-0}{.02210613} = 1.09$$

<u>*p*-value approach</u>

Step 3: From above, $z = 1.09$.

From the normal distribution table, area to the right of $z = 1.09$ is $1 - .8621 = .1379$.

p-value $= 2(.1379) = .2758$

Step 4: Do not reject H_0 since $.2758 > .05$.

<u>critical value approach</u>

Step 3: For $\alpha = .05$, the critical values of z are -1.96 and 1.96.

Step 4: From above, $z = 1.09$.

Step 5: Do not reject H_0 since $1.09 < 1.96$.

Conclude that the proportions of cars that fail the emissions test at the two stations are not different.

10.73 a. $\hat{p}_1 = x_1 / n_1 = 160/500 = .32$ and $\hat{p}_2 = x_2 / n_2 = 66/300 = .22$

$\hat{p}_1 - \hat{p}_2 = .32 - .22 = .10$

b. $$s_{\hat{p}_1-\hat{p}_2} = \sqrt{\frac{\hat{p}_1\hat{q}_1}{n_1} + \frac{\hat{p}_2\hat{q}_2}{n_2}} = \sqrt{\frac{(.32)(.68)}{500} + \frac{(.22)(.78)}{300}} = .03173641$$

The 99% confidence interval for $p_1 - p_2$ is

$$(\hat{p}_1 - \hat{p}_2) \pm z s_{\hat{p}_1-\hat{p}_2} = (.32 - .22) \pm 2.58(.03173641) = .10 \pm .082 = .018 \text{ to } .182$$

c. Step 1: H_0: $p_1 - p_2 = 0$, H_1: $p_1 - p_2 \neq 0$

Step 2: Since $n_1\hat{p}_1$, $n_1\hat{q}_1$, $n_2\hat{p}_2$, and $n_2\hat{q}_2$ are all greater than 5, use the normal distribution.

Step 3: For $\alpha = .01$, the critical values of z are -2.58 and 2.58.

Step 4: $\bar{p} = \frac{x_1 + x_2}{n_1 + n_2} = \frac{160 + 66}{500 + 300} = .283$ and $\bar{q} = 1 - \bar{p} = 1 - .283 = .717$

$$s_{\hat{p}_1 - \hat{p}_2} = \sqrt{\bar{p}\bar{q}\left(\frac{1}{n_1} + \frac{1}{n_2}\right)} = \sqrt{(.283)(.717)\left(\frac{1}{500} + \frac{1}{300}\right)} = .03289669$$

$$z = \frac{(\hat{p}_1 - \hat{p}_2) - (p_1 - p_2)}{s_{\hat{p}_1 - \hat{p}_2}} = \frac{.10 - 0}{.03289669} = 3.04$$

Step 5: Reject H_0 since $3.04 > 2.58$.

Conclude that the proportions of all men and all women who play the lottery often are different.

10.75 a. $\hat{p}_1 = x_1 / n_1 = 280/800 = .35$ and $\hat{p}_2 = x_2 / n_2 = 279/900 = .31$

$$s_{\hat{p}_1 - \hat{p}_2} = \sqrt{\frac{\hat{p}_1\hat{q}_1}{n_1} + \frac{\hat{p}_2\hat{q}_2}{n_2}} = \sqrt{\frac{(.35)(.65)}{800} + \frac{(.31)(.69)}{900}} = .02284823$$

The 98% confidence interval for $p_1 - p_2$ is

$(\hat{p}_1 - \hat{p}_2) \pm z s_{\hat{p}_1 - \hat{p}_2} = (.35 - .31) \pm 2.33(.02284823) = .04 \pm .053 = -.013$ to $.093$

b. Step 1: H_0: $p_1 - p_2 = 0$, H_1: $p_1 - p_2 > 0$

Step 2: Since $n_1\hat{p}_1$, $n_1\hat{q}_1$, $n_2\hat{p}_2$, and $n_2\hat{q}_2$ are all greater than 5, use the normal distribution.

Step 3: For $\alpha = .01$, the critical value of z is 2.33.

Step 4: $\bar{p} = \frac{x_1 + x_2}{n_1 + n_2} = \frac{280 + 279}{800 + 900} = .329$ and $\bar{q} = 1 - \bar{p} = 1 - .329 = .671$

$$s_{\hat{p}_1 - \hat{p}_2} = \sqrt{\bar{p}\bar{q}\left(\frac{1}{n_1} + \frac{1}{n_2}\right)} = \sqrt{(.329)(.671)\left(\frac{1}{800} + \frac{1}{900}\right)} = .02283061$$

$$z = \frac{(\hat{p}_1 - \hat{p}_2) - (p_1 - p_2)}{s_{\hat{p}_1 - \hat{p}_2}} = \frac{(.35 - .31) - 0}{.02283061} = 1.75$$

Step 5: Do not reject H_0 since $1.75 > 2.33$.

Conclude that the proportion of all sales for which at least one item is returned is not higher for Store A than for Store B.

Supplementary Exercises

10.77 a. $\sigma_{\bar{x}_1-\bar{x}_2} = \sqrt{\frac{\sigma_1^2}{n_1}+\frac{\sigma_2^2}{n_2}} = \sqrt{\frac{(30)^2}{1500}+\frac{(35)^2}{2000}} = \$1.10113578$

The 95% confidence interval for $\mu_1 - \mu_2$ is

$(\bar{x}_1 - \bar{x}_2) \pm z\sigma_{\bar{x}_1-\bar{x}_2} = (388 - 505) \pm 1.96(1.10113578) = -117 \pm 2.16 = -\$119.16$ to $-\$114.84$

b. Step 1: H_0: $\mu_1 - \mu_2 = 0$, H_1: $\mu_1 - \mu_2 < 0$

Step 2: Since σ_1 and σ_2 are known, the samples are independent, and the sample sizes are large ($n_1 \geq 30$ and $n_2 \geq 30$), use the normal distribution.

Step 3: For $\alpha = .025$, the critical value of z is -1.96.

Step 4: $z = \frac{(\bar{x}_1 - \bar{x}_2) - (\mu_1 - \mu_2)}{\sigma_{\bar{x}_1-\bar{x}_2}} = \frac{-117 - 0}{1.10113578} = -106.25$

Step 5: Reject H_0 since $-106.25 < -1.96$.

Conclude that the mean weekly earnings of female workers who are not union members are less than those of female workers who are union members.

10.79 a. $s_p = \sqrt{\frac{(n_1-1)s_1^2+(n_2-1)s_2^2}{n_1+n_2-2}} = \sqrt{\frac{(25-1)(14)^2+(20-1)(12)^2}{25+20-2}} = \$13.15383046$

$s_{\bar{x}_1-\bar{x}_2} = s_p\sqrt{\frac{1}{n_1}+\frac{1}{n_2}} = 13.15383046\sqrt{\frac{1}{25}+\frac{1}{20}} = \$3.94614914$

$df = n_1 + n_2 - 2 = 25 + 20 - 2 = 43$

The 99% confidence interval for $\mu_1 - \mu_2$ is

$(\bar{x}_1 - \bar{x}_2) \pm ts_{\bar{x}_1-\bar{x}_2} = (97 - 89) \pm 2.695(3.94614914) = 8 \pm 10.63 = -\$2.63$ to $\$18.63$

b. Step 1: H_0: $\mu_1 - \mu_2 = 0$, H_1: $\mu_1 - \mu_2 > 0$

Step 2: Since σ_1 and σ_2 are unknown but assumed to be equal, the samples are independent, the sample sizes are small, and the populations are normally distributed, use the t distribution.

Step 3: For $\alpha = .01$ with $df = 43$, the critical value of t is 2.416.

Step 4: $t = \frac{(\bar{x}_1 - \bar{x}_2) - (\mu_1 - \mu_2)}{s_{\bar{x}_1-\bar{x}_2}} = \frac{8 - 0}{3.94614914} = 2.027$

Step 5: Do not reject H_0 since $2.027 < 2.416$.

Conclude that the mean monthly insurance premium paid by drivers insured with company A is not higher than that of drivers insured with company B.

10.81 a. $s_p = \sqrt{\frac{(n_1-1)s_1^2+(n_2-1)s_2^2}{n_1+n_2-2}} = \sqrt{\frac{(1000-1)(320)^2+(900-1)(305)^2}{1000+900-2}} = \$312.9847768$

$s_{\bar{x}_1-\bar{x}_2} = s_p\sqrt{\frac{1}{n_1}+\frac{1}{n_2}} = 312.9847768\sqrt{\frac{1}{1000}+\frac{1}{900}} = \$14.38065807$

Since $n_1 = 1000$ and $n_2 = 900$ are very large, we use the normal distribution to approximate the t distribution.

The 99% confidence interval for $\mu_1 - \mu_2$ is

$(\bar{x}_1-\bar{x}_2) \pm z s_{\bar{x}_1-\bar{x}_2} = (1019 - 1101) \pm 2.58(14.38065807) = -82 \pm 37.10 = -\$119.10$ to $-\$44.90$

b. Step 1: H_0: $\mu_1 - \mu_2 = 0$, H_1: $\mu_1 - \mu_2 < 0$

Step 2: Since σ_1 and σ_2 are unknown but assumed to be equal, the samples are independent, and both samples are large ($n_1 \geq 30$ and $n_2 \geq 30$), use the t distribution. However, since $n_1 = 1000$ and $n_2 = 900$ are very large, we use the normal distribution to approximate the t distribution.

Step 3: For $\alpha = .01$, the critical value of z is -2.33.

Step 4: $z = \frac{(\bar{x}_1-\bar{x}_2)-(\mu_1-\mu_2)}{s_{\bar{x}_1-\bar{x}_2}} = \frac{-82-0}{14.38065807} = -5.70$

Step 5: Reject H_0 since $-5.70 < -2.33$.

Conclude that the mean amount Americans expected to spend on their longest vacation in 2005 is lower than that for 2004.

10.83 a. $s_{\bar{x}_1-\bar{x}_2} = \sqrt{\frac{s_1^2}{n_1}+\frac{s_2^2}{n_2}} = \sqrt{\frac{(14)^2}{25}+\frac{(12)^2}{20}} = \$3.87814389$

$$df = \frac{\left(\frac{s_1^2}{n_1}+\frac{s_2^2}{n_2}\right)^2}{\frac{\left(\frac{s_1^2}{n_1}\right)^2}{n_1-1}+\frac{\left(\frac{s_2^2}{n_2}\right)^2}{n_2-1}} = \frac{\left(\frac{14^2}{25}+\frac{12^2}{20}\right)^2}{\frac{\left(\frac{14^2}{25}\right)^2}{24}+\frac{\left(\frac{12^2}{20}\right)^2}{19}} = 42.76 \approx 42$$

The 99% confidence interval for $\mu_1 - \mu_2$ is

$(\bar{x}_1 - \bar{x}_2) \pm ts_{\bar{x}_1-\bar{x}_2} = (97 - 89) \pm 2.698(3.87814389) = 8 \pm 10.46 = -\$2.46$ to $\$18.46$

b. Step 1: H_0: $\mu_1 - \mu_2 = 0$, H_1: $\mu_1 - \mu_2 > 0$

Step 2: Since σ_1 and σ_2 are unknown but assumed to be unequal, the samples are independent, the sample sizes are small, and the populations are normally distributed, use the t distribution.

Step 3: For $\alpha = .01$ with $df = 42$, the critical value of t is 2.418.

Step 4: $$t = \frac{(\bar{x}_1 - \bar{x}_2) - (\mu_1 - \mu_2)}{s_{\bar{x}_1-\bar{x}_2}} = \frac{8 - 0}{3.87814389} = 2.063$$

Step 5: Do not reject H_0 since $2.063 < 2.418$.

Conclude that the mean monthly insurance premium paid by drivers insured with company A is not higher than that of drivers insured with company B.

10.85 a. $$s_{\bar{x}_1-\bar{x}_2} = \sqrt{\frac{s_1^2}{n_1} + \frac{s_2^2}{n_2}} = \sqrt{\frac{320^2}{1000} + \frac{305^2}{900}} = \$14.34437559$$

Since $n_1 = 1000$ and $n_2 = 900$ are very large, we use the normal distribution to approximate the t distribution.

The 99% confidence interval for $\mu_1 - \mu_2$ is

$(\bar{x}_1 - \bar{x}_2) \pm ts_{\bar{x}_1-\bar{x}_2} = (1019 - 1101) \pm 2.58(14.34437559) = -.82 \pm 37.01 = -\$119.01$ to $-\$44.99$

b. Step 1: H_0: $\mu_1 - \mu_2 = 0$, H_1: $\mu_1 - \mu_2 < 0$

Step 2: Since σ_1 and σ_2 are unknown but assumed to be equal, the samples are independent, and both samples are large ($n_1 \geq 30$ and $n_2 \geq 30$), use the t distribution. However, since $n_1 = 1000$ and $n_2 = 900$ are very large, we use the normal distribution to approximate the t distribution.

Step 3: For $\alpha = .01$, the critical value of z is -2.33.

Step 4: $$z = \frac{(\bar{x}_1 - \bar{x}_2) - (\mu_1 - \mu_2)}{s_{\bar{x}_1-\bar{x}_2}} = \frac{-82 - 0}{14.34437559} = -5.72$$

Step 5: Reject H_0 since $-5.72 < -2.33$.

Conclude that the mean amount Americans expected to spend on their longest vacation in 2005 is lower than that for 2004.

10.87

Before	After	d	d^2
43	49	–6	36
57	56	1	1
48	55	–7	49
65	77	–12	144
81	89	–8	64
49	57	–8	64
38	36	2	4
69	64	5	25
58	69	–11	121
		$\Sigma d = -44$	$\Sigma d^2 = 508$

$\bar{d} = \Sigma d / n = -44/9 = -4.89$

$$s_d = \sqrt{\frac{\Sigma d^2 - \frac{(\Sigma d)^2}{n}}{n-1}} = \sqrt{\frac{508 - \frac{(-44)^2}{9}}{9-1}} = 6.05071162$$

$s_{\bar{d}} = s_d / \sqrt{n} = 6.05071162 / \sqrt{9} = 2.01690387$

$df = n - 1 = 9 - 1 = 8$

a. The 95% confidence interval for μ_d is

$\bar{d} \pm t s_{\bar{d}} = -4.89 \pm 2.306(2.01690387) = -4.89 \pm 4.65 = -9.54$ to $-.24$

b. Step 1: H_0: $\mu_d = 0$, H_1: $\mu_d < 0$

Step 2: Since the samples are dependent, σ_d is unknown, the sample is small, and the population of paired differences is normally distributed, use the t distribution.

Step 3: For $\alpha = .01$ with $df = 8$, the critical value of t is –2.896.

Step 4: $t = (\bar{d} - \mu_d) / s_{\bar{d}} = (-4.89 - 0)/2.01690387 = -2.425$

Step 5: Do not reject H_0 since $-2.425 > -2.896$.

Conclude that the course does not make any statistically significant improvement in the memory of all students.

10.89 a. $$s_{\hat{p}_1 - \hat{p}_2} = \sqrt{\frac{\hat{p}_1 \hat{q}_1}{n_1} + \frac{\hat{p}_2 \hat{q}_2}{n_2}} = \sqrt{\frac{(.15)(.85)}{1200} + \frac{(.09)(.92)}{1080}} = .01352467$$

The 99% confidence interval for $p_1 - p_2$ is

$(\hat{p}_1 - \hat{p}_2) \pm z s_{\hat{p}_1 - \hat{p}_2} = (.15 - .09) \pm 2.58(.01352467) = .06 \pm .035 = .025$ to $.095$

b. Step 1: H_0: $p_1 - p_2 = 0$, H_1: $p_1 - p_2 > 0$

Step 2: Since $n_1\hat{p}_1$, $n_1\hat{q}_1$, $n_2\hat{p}_2$, and $n_2\hat{q}_2$ are all greater than 5, use the normal distribution.

Step 3: For $\alpha = .01$, the critical value of z is 2.33.

Step 4: $\bar{p} = \frac{n_1\hat{p}_1 + n_2\hat{p}_2}{n_1 + n_2} = \frac{1200(.15) + 1080(.09)}{1200 + 1080} = .122$ and $\bar{q} = 1 - \bar{p} = 1 - .122 = .878$

$$s_{\hat{p}_1-\hat{p}_2} = \sqrt{\bar{p}\bar{q}\left(\frac{1}{n_1} + \frac{1}{n_2}\right)} = \sqrt{(.122)(.878)\left(\frac{1}{1200} + \frac{1}{1080}\right)} = .01372752$$

$$z = \frac{(\hat{p}_1 - \hat{p}_2) - (p_1 - p_2)}{s_{\hat{p}_1-\hat{p}_2}} = \frac{.06 - 0}{.01372752} = 4.37$$

Step 5: Reject H_0 since $4.37 > 2.33$.

Conclude that the percentage of left-handed men in the United States is higher than the percentage of left-handed women.

c. Step 1: H_0: $p_1 - p_2 = 0$, H_1: $p_1 - p_2 > 0$

Step 2: Since $n_1\hat{p}_1$, $n_1\hat{q}_1$, $n_2\hat{p}_2$, and $n_2\hat{q}_2$ are all greater than 5, use the normal distribution.

Step 3: From above, $z = 4.37$.

From the normal distribution table, area to the right of $z = 4.37$ is $1 - 1 \approx 0$.

p-value ≈ 0

Step 4: Reject H_0 since $0 < .01$.

10.91 a. $s_{\hat{p}_1-\hat{p}_2} = \sqrt{\frac{\hat{p}_1\hat{q}_1}{n_1} + \frac{\hat{p}_2\hat{q}_2}{n_2}} = \sqrt{\frac{(.52)(.48)}{800} + \frac{(.32)(.68)}{800}} = .02416609$

The 95% confidence interval for $p_1 - p_2$ is

$(\hat{p}_1 - \hat{p}_2) \pm z s_{\hat{p}_1-\hat{p}_2} = (.52 - .32) \pm 1.96(.02416609) = .20 \pm .047 = .153$ to $.247$

b. Step 1: H_0: $p_1 - p_2 = 0$, H_1: $p_1 - p_2 > 0$

Step 2: Since $n_1\hat{p}_1$, $n_1\hat{q}_1$, $n_2\hat{p}_2$, and $n_2\hat{q}_2$ are all greater than 5, use the normal distribution.

Step 3: For $\alpha = .025$, the critical value of z is 1.96.

Step 4: $\bar{p} = \frac{n_1\hat{p}_1 + n_2\hat{p}_2}{n_1 + n_2} = \frac{800(.52) + 800(.32)}{800 + 800} = .42$ and $\bar{q} = 1 - \bar{p} = 1 - .42 = .58$

$$s_{\hat{p}_1-\hat{p}_2} = \sqrt{\bar{p}\bar{q}\left(\frac{1}{n_1}+\frac{1}{n_2}\right)} = \sqrt{(.42)(.58)\left(\frac{1}{800}+\frac{1}{800}\right)} = .02467793$$

$$z = \frac{(\hat{p}_1-\hat{p}_2)-(p_1-p_2)}{s_{\hat{p}_1-\hat{p}_2}} = \frac{.20-0}{.02467793} = 8.10$$

Step 5: Reject H_0 since $8.10 > 1.96$.

Conclude that the proportion of 18- to 34-year-olds who expected Social Security to pay all such benefits is lower than the proportion of 45- to 54-year-olds who held this opinion.

10.93 $\mu_1 = 200,\ \sigma_1 = 10,\ n_1 = 1,\ \mu_2 = 210,\ \sigma_2 = 12,\ n_2 = 1$

Since the distances are normally distributed with the above means and standard deviations, we can use the normal distribution despite the fact that the samples are small. Then,

$$\sigma_{\bar{x}_1-\bar{x}_2} = \sqrt{\frac{\sigma_1^2}{n_1}+\frac{\sigma_2^2}{n_2}} = \sqrt{\frac{(10)^2}{1}+\frac{(12)^2}{1}} = 15.62049935 \text{ and } \mu_{\bar{x}_1-\bar{x}_2} = 200-210 = -10.$$

Then, for $\bar{x}_1-\bar{x}_2 = 0$, $z = \frac{(\bar{x}_1-\bar{x}_2)-(\mu_1-\mu_2)}{\sigma_{\bar{x}_1-\bar{x}_2}} = \frac{0-(-10)}{15.62049935} = .64$. Hence,

$P(\bar{x}_1-\bar{x}_2 \geq 0) = P(z \geq .64) = 1-P(z < .64) = 1-.7389 = .2611$.

10.95 $E = ts_{\bar{d}} \leq 2$ mpg, so $t \leq \frac{2}{s_d} = \frac{2}{3/\sqrt{n}} = \frac{2}{3}\sqrt{n}$. Thus, we look for the smallest value of n for which $t \leq \frac{2}{3}\sqrt{n}$ with 90% confidence.

n	2	3	4	5	6	7	8	9
critical values of t	6.314	2.920	2.353	2.132	2.015	1.943	1.895	1.860
$\frac{2}{3}\sqrt{n}$	0.9428	1.1547	1.3333	1.4907	1.6330	1.7638	1.8856	2.000

Nine cars should be tested.

10.97 a. The 90% confidence interval for p_1-p_2 is $(\hat{p}_1-\hat{p}_2) \pm z\sigma_{\hat{p}_1-\hat{p}_2}$ with $E = z\sigma_{\hat{p}_1-\hat{p}_2}$. Then,

$1.65\sigma_{\hat{p}_1-\hat{p}_2} = .05$ and $\sigma_{\hat{p}_1-\hat{p}_2} = \frac{.05}{1.65}$. We also have

$$\sigma_{\hat{p}_1-\hat{p}_2} = \sqrt{\frac{p_1q_1}{n_1}+\frac{p_2q_2}{n_2}} = \sqrt{\frac{(.5)(.5)}{n}+\frac{(.5)(.5)}{n}} = \sqrt{\frac{(2)(.5)^2}{n}} \text{ where } n_1 = n_2 = n. \text{ Hence,}$$

$$n = \frac{(2)(.5)^2}{(\sigma_{\hat{p}_1-\hat{p}_2})^2} = \frac{(2)(.5)^2}{\left(\frac{.05}{1.65}\right)^2} = 544.5 \approx 545$$

b. $p_1 = .30,\ p_2 = .23,\ n_1 = n_2 = 100$

$$\mu_{\hat{p}_1-\hat{p}_2} = p_1 - p_2 = .30 - .23 = .07$$

$$\sigma_{\hat{p}_1-\hat{p}_2} = \sqrt{\frac{(.30)(.70)}{100} + \frac{(.23)(.77)}{100}} = .06221736$$

Then, for $\hat{p}_1 - \hat{p}_2 = 0,\ z = \frac{0 - .07}{.06221736} = -1.13$. Hence,

$$P(\hat{p}_1 - \hat{p}_2 \geq 0) = P(z \geq -1.13) = 1 - P(z < -1.13) = 1 - .1292 = .8708.$$

10.99 We need to obtain sample information about the incidence of brain tumors in people who do not use cellular phones and the incidence of brain tumors in people who do use cellular phones. Since the occurrence of brain tumor is a fairly rare event, very large samples (perhaps tens of thousands) of healthy people from each group (people who use cellular phones and those who do not) must be used. For each group, we would count the number of people that develop a brain tumor over some period of time (say a year or two). The hypotheses would be H_0: $p_1 - p_2 = 0$ versus H_1: $p_1 - p_2 < 0$ where $p_1 = P$(brain tumor if do not use cellular phones) and $p_2 = P$(brain tumor if use cellular phones). If the concern is about product liability, the worst mistake would be to conclude cellular phones do not increase the risk of brain tumor when they really do. This would be a Type II error. A higher α level would be appropriate since a higher α level means a lower level for β, the probability of a Type II error. The test is based on the standard normal distribution with the following test statistic.

$$z = \frac{(\hat{p}_1 - \hat{p}_2) - (p_1 - p_2)}{s_{\hat{p}_1-\hat{p}_2}} \text{ where, } s_{\hat{p}_1-\hat{p}_2} = \sqrt{\bar{p}\bar{q}\left(\frac{1}{n_1} + \frac{1}{n_2}\right)}.$$

10.101 $\sigma_{\bar{x}_1-\bar{x}_2} = \sqrt{\frac{\sigma_1^2}{n_1} + \frac{\sigma_2^2}{n_2}} = \sqrt{\frac{(.4)^2}{25} + \frac{(.7)^2}{36}} = .14146063$

a. For $\bar{x}_1 - \bar{x}_2 = -.15$: $z = \frac{(\bar{x}_1 - \bar{x}_2) - (\mu_1 - \mu_2)}{\sigma_{\bar{x}_1-\bar{x}_2}} = \frac{-.15 - (1.3 - 1.5)}{.14146063} = -2.47$

For $\bar{x}_1 - \bar{x}_2 = .15$: $z = \frac{(\bar{x}_1 - \bar{x}_2) - (\mu_1 - \mu_2)}{\sigma_{\bar{x}_1-\bar{x}_2}} = \frac{.15 - (1.3 - 1.5)}{.14146063} = 2.47$

Then, $P[-.15 < (\bar{x}_1 - \bar{x}_2) < .15] = P(-2.47 < z < 2.47) = P(z < 2.47) - P(z < -2.47)$

$= .9932 - .0068 = .9864.$

b. For $\bar{x}_1 - \bar{x}_2 = 0$: $z = \dfrac{(\bar{x}_1 - x_2) - (\mu_1 - \mu_2)}{\sigma_{x_1 - x_2}} = \dfrac{0 - (1.3 - 1.5)}{.14146063} = 1.41$

$P(\bar{x}_1 - \bar{x}_2 > 0) = P(z > 1.41) = 1 - P(z \le 1.41) = 1 - .9207 = .0793$

c. $P[\{-.15 < (\bar{x}_1 - \bar{x}_2) < .15\} \mid \bar{x}_1 = 2.0] = P[-.15 < (2.0 - \bar{x}_2) < .15] = P[1.85 < \bar{x}_2 < 2.15]$

$\sigma_{\bar{x}_2} = \sigma_2 / \sqrt{n_2} = .7 / \sqrt{36} = .11666667$

For $\bar{x}_2 = 1.85$: $z = \dfrac{\bar{x}_2 - \mu_2}{\sigma_{\bar{x}_2}} = \dfrac{1.85 - 1.5}{.11666667} = 3.00$

For $\bar{x}_2 = 2.15$: $z = \dfrac{\bar{x}_2 - \mu_2}{\sigma_{\bar{x}_2}} = \dfrac{2.15 - 1.5}{.11666667} = 5.57$

Hence, $P(1.85 < \bar{x}_2 < 2.15) = P(3.00 < z < 5.57) = P(z < 5.57) - P(z < 3.00)$

$1 - .9987 = .0013.$

d. If the assumption is reasonable, it means that there is not a lot of difference in the average amount of weight loss between the two diets. Hence, it does not matter which diet is selected.

Self-Review Test

1. a

2. See the solution to Exercise 10.1.

3. a. $\sigma_{\bar{x}_1 - \bar{x}_2} = \sqrt{\dfrac{\sigma_1^2}{n_1} + \dfrac{\sigma_2^2}{n_2}} = \sqrt{\dfrac{(.8)^2}{40} + \dfrac{(1.3)^2}{50}} = .22315914$

The 99% confidence interval for $\mu_1 - \mu_2$ is

$(\bar{x}_1 - \bar{x}_2) \pm z\sigma_{\bar{x}_1 - \bar{x}_2} = (7.6 - 5.4) \pm 2.58(.22315914) = 2.2 \pm .58 = 1.62$ to 2.78

b. Step 1: H_0: $\mu_1 - \mu_2 = 0$, H_1: $\mu_1 - \mu_2 > 0$

Step 2: Since σ_1 and σ_2 are known, the samples are independent, and the sample sizes are large ($n_1 \ge 30$ and $n_2 \ge 30$), use the normal distribution.

Step 3: For $\alpha = .025$, the critical value of z is 1.96.

Step 4: $z = \dfrac{(\bar{x}_1 - \bar{x}_2) - (\mu_1 - \mu_2)}{\sigma_{\bar{x}_1 - \bar{x}_2}} = \dfrac{2.2 - 0}{.22315914} = 9.86$

Step 5: Reject H_0 since 9.86 > 1.96.

Conclude that the mean stress score of all executives is higher than that of all professors.

4. a. $s_p = \sqrt{\frac{(n_1-1)s_1^2+(n_2-1)s_2^2}{n_1+n_2-2}} = \sqrt{\frac{(20-1)(.54)^2+(25-1)(.8)^2}{20+25-2}} = .69717703$ hour

$s_{\bar{x}_1-\bar{x}_2} = s_p\sqrt{\frac{1}{n_1}+\frac{1}{n_2}} = .69717703\sqrt{\frac{1}{20}+\frac{1}{25}} = .20915311$ hour

$df = n_1+n_2-2 = 20+25-2 = 43$

The 95% confidence interval for $\mu_1-\mu_2$ is

$(\bar{x}_1-\bar{x}_2) \pm t s_{\bar{x}_1-\bar{x}_2} = (2.3-4.6) \pm 2.017(.20915311) = -2.3 \pm .42 = -2.72$ to -1.88 hours

b. Step 1: H_0: $\mu_1-\mu_2 = 0$, H_1: $\mu_1-\mu_2 < 0$

Step 2: Since σ_1 and σ_2 are unknown but assumed to be equal, the samples are independent, the sample sizes are small, and the populations are normally distributed, use the t distribution.

Step 3: For $\alpha = .01$ with $df = 43$, the critical value of t is -2.416.

Step 4: $t = \frac{(\bar{x}_1-\bar{x}_2)-(\mu_1-\mu_2)}{s_{\bar{x}_1-\bar{x}_2}} = \frac{-2.3-0}{.20915311} = -10.997$

Step 5: Reject H_0 since $-10.997 < -2.416$.

Conclude that the mean time spent per week playing with their children by all alcoholic fathers is less than that of nonalcoholic fathers.

5. a. $s_{\bar{x}_1-\bar{x}_2} = \sqrt{\frac{s_1^2}{n_1}+\frac{s_2^2}{n_2}} = \sqrt{\frac{(.54)^2}{20}+\frac{(.8)^2}{25}} = .20044950$ hour

$$df = \frac{\left(\frac{s_1^2}{n_1}+\frac{s_2^2}{n_2}\right)^2}{\frac{\left(\frac{s_1^2}{n_1}\right)^2}{n_1-1}+\frac{\left(\frac{s_2^2}{n_2}\right)^2}{n_2-1}} = \frac{\left(\frac{.54^2}{20}+\frac{.8^2}{25}\right)^2}{\frac{\left(\frac{.54^2}{20}\right)^2}{19}+\frac{\left(\frac{.8^2}{25}\right)^2}{24}} = 41.94 \approx 41$$

The 95% confidence interval for $\mu_1-\mu_2$ is

$(\bar{x}_1-\bar{x}_2) \pm t s_{\bar{x}_1-\bar{x}_2} = (2.3-4.6) \pm 2.020(.20044950) = -2.3 \pm .40 = -2.70$ to -1.90 hours

b. Step 1: H_0: $\mu_1-\mu_2 = 0$, H_1: $\mu_1-\mu_2 < 0$

Step 2: Since σ_1 and σ_2 are unknown but assumed to be unequal, the samples are independent, the sample sizes are small, and the populations are normally distributed, use the t distribution.

Step 3: For $\alpha = .01$ with $df = 41$, the critical value of t is -2.421.

Step 4: $t = \dfrac{(\bar{x}_1 - \bar{x}_2) - (\mu_1 - \mu_2)}{s_{\bar{x}_1 - \bar{x}_2}} = \dfrac{-2.3 - 0}{.20044950} = -11.474$

Step 5: Reject H_0 since $-11.474 < -2.421$.

Conclude that the mean time spent per week playing with their children by all alcoholic fathers is less than that of nonalcoholic fathers.

6.

Zeke's	Elmer's	d	d^2
1058	995	63	3969
544	540	4	16
1349	1175	174	30,276
1296	1350	–54	2916
676	605	71	5041
998	970	28	784
1698	1520	178	31,684
		$\Sigma d = 464$	$\Sigma d^2 = 74{,}686$

$\bar{d} = \Sigma d / n = 464/7 = \$66.29$

$$s_d = \sqrt{\frac{\Sigma d^2 - \frac{(\Sigma d)^2}{n}}{n-1}} = \sqrt{\frac{74{,}686 - \frac{(464)^2}{7}}{7-1}} = \$85.56618157$$

$s_{\bar{d}} = s_d / \sqrt{n} = 85.56618157 / \sqrt{7} = \$32.34097672$

$df = n - 1 = 7 - 1 = 6$

a. The 99% confidence interval for μ_d is

$\bar{d} \pm t s_{\bar{d}} = 66.29 \pm 3.707(32.34097672) = 66.29 \pm 119.89 = -\$53.60$ to $\$186.18$

b. Step 1: H_0: $\mu_d = 0$, H_1: $\mu_d \neq 0$

Step 2: Since the samples are dependent, σ_d is unknown, the sample is small, and the population of paired differences is normally distributed, use the t distribution.

Step 3: For $\alpha = .05$ with $df = 7$, the critical values of t are -2.447 and 2.447.

Step 4: $t = (\bar{d} - \mu_d) / s_{\bar{d}} = (66.29 - 0)/32.34097672 = 2.050$

Step 5: Do not reject H_0 since $2.050 < 2.447$.

Conclude that the mean estimate for repair costs at Zeke's is not different than the mean estimate at Elmer's.

7. a. $s_{\hat{p}_1-\hat{p}_2} = \sqrt{\dfrac{\hat{p}_1\hat{q}_1}{n_1}+\dfrac{\hat{p}_2\hat{q}_2}{n_2}} = \sqrt{\dfrac{(.57)(.43)}{500}+\dfrac{(.55)(.45)}{400}} = .03330090$

The 97% confidence interval for $p_1 - p_2$ is

$(\hat{p}_1-\hat{p}_2) \pm z s_{\hat{p}_1-\hat{p}_2} = (.57 - .55) \pm 2.17(.03330090) = .02 \pm .072 = -.052$ to $.092$

b. Step 1: H_0: $p_1 - p_2 = 0$, H_1: $p_1 - p_2 \neq 0$

Step 2: Since $n_1\hat{p}_1$, $n_1\hat{q}_1$, $n_2\hat{p}_2$, and $n_2\hat{q}_2$ are all greater than 5, use the normal distribution.

Step 3: For $\alpha = .01$, the critical values of z are -2.58 and 2.58.

Step 4: $\bar{p} = \dfrac{n_1\hat{p}_1 + n_2\hat{p}_2}{n_1+n_2} = \dfrac{500(.57)+400(.55)}{500+400} = .561$ and $\bar{q} = 1 - \bar{p} = 1 - .561 = .439$

$$s_{\hat{p}_1-\hat{p}_2} = \sqrt{\bar{p}\,\bar{q}\left(\frac{1}{n_1}+\frac{1}{n_2}\right)} = \sqrt{(.561)(.439)\left(\frac{1}{500}+\frac{1}{400}\right)} = .03329047$$

$$z = \frac{(\hat{p}_1-\hat{p}_2)-(p_1-p_2)}{s_{\hat{p}_1-\hat{p}_2}} = \frac{.02-0}{.03329047} = .60$$

Step 5: Do not reject H_0 since $.60 < 2.58$.

Conclude that the proportion of all male voters who voted in the last presidential election is not different from that of all female voters.

Chapter Eleven

Section 11.1

11.1 The **chi-square distribution** has only one parameter, called the **degrees of freedom** (*df*). The shape of a chi-square distribution curve is skewed to the right for small *df* and becomes symmetric for large *df*. The entire chi-square distribution curve lies to the right of the vertical axis. The chi-square distribution assumes nonnegative values only, and these are denoted by the symbol χ^2.

11.3 For $df = 28$ and .05 area in the right tail, $\chi^2 = 41.337$.

11.5 For an area of .990 in the left tail, the area in the right tail is $1 - .990 = .01$. Hence, the value of chi–square for $df = 23$ and .990 area in the left tail is the same as for $df = 23$ and .01 area in the right tail. Thus, the chi-square value is $\chi^2 = 41.638$.

11.7
a. For an area of .025 in the left tail, the area in the right tail is $1 - .025 = .975$. Hence, the value of chi–square for $df = 13$ and .025 area in the left tail is the same as for $df = 13$ and .975 area in the right tail. Thus, the chi-square value is $\chi^2 = 5.009$
b. For $df = 13$ and .995 area in the right tail, $\chi^2 = 3.565$.

Section 11.2

11.9 A **goodness–of–fit test** compares the observed frequencies from a multinomial experiment with expected frequencies derived from a certain pattern or theoretical distribution. The test evaluates how well the observed frequencies fit the expected frequencies.

11.11 The **expected frequency** of a category is given by $E = np$ where n is the sample size and p is the probability that an element belongs to that category if the null hypothesis is true. The **degrees of freedom** for a goodness–of–fit test are $k - 1$, where k denotes the number of possible outcomes (or categories) for the experiment.

11.13 Step 1: H_0: The die is fair, H_1: The die is not fair.

Step 2: Since this is a multinomial experiment, use the chi–square distribution.

Step 3: $k = 6$, $df = k - 1 = 6 - 1 = 5$

For $\alpha = .05$, the critical value of χ^2 is 11.070.

Step 4: Note that the die will be fair if the probability of each of the six outcomes is the same, which is 1/6.

Outcome	O	p	$E = np$	$O - E$	$(O - E)^2$	$(O - E)^2/E$
1	7	1/6	10	–3	9	.900
2	12	1/6	10	2	4	.400
3	8	1/6	10	–2	4	.400
4	15	1/6	10	5	25	2.500
5	11	1/6	10	1	1	.100
6	7	1/6	10	–3	9	.900
	$n = 60$					Sum = 5.200

$\chi^2 = \Sigma\,(O - E)^2/E = 5.200$

Step 5: Do not reject H_0 since 5.200 < 11.070.

Conclude that the die is fair.

11.15 Step 1: H_0: The current distribution of grades assigned to the nation's schools is the same as that of 2004.

H_1: The current distribution of grades assigned to the nation's schools differs from that of 2004.

Step 2: Since this is a multinomial experiment, use the chi-square distribution.

Step 3: $k = 6$, $df = k - 1 = 6 - 1 = 5$

For $\alpha = .05$, the critical value of χ^2 is 11.070.

Step 4:

Grade	O	p	$E = np$	$O - E$	$(O - E)^2$	$(O - E)^2/E$
A	16	.02	14	2	4	.286
B	132	.20	140	–8	64	.457
C	314	.47	329	–15	225	.684
D	121	.15	105	16	256	2.438
F	35	.04	28	7	49	1.750
None	82	.12	84	–2	4	.048
	$n = 700$					Sum = 5.663

$\chi^2 = \Sigma\,(O - E)^2/E = 5.663$

Step 5: Do not reject H_0 since 5.663 < 11.070.

Conclude that the current distribution of grades assigned to the nation's schools does not differ from that of 2004.

11.17 Step 1: H_0: The orders are evenly distributed over all days of the week.

H_1: The orders are not evenly distributed over all days of the week.

Step 2: Since this is a multinomial experiment, use the chi-square distribution.

Step 3: $k = 5$, $df = k - 1 = 5 - 1 = 4$

For $\alpha = .05$, the critical value of χ^2 is 9.488.

Step 4:

Day	O	p	$E = np$	$O - E$	$(O - E)^2$	$(O - E)^2/E$
Monday	92	.20	80	12	144	1.800
Tuesday	71	.20	80	− 9	81	1.013
Wednesday	65	.20	80	− 15	225	2.813
Thursday	83	.20	80	3	9	.113
Friday	89	.20	80	9	81	1.013
	$n = 400$					Sum = 6.752

$\chi^2 = \Sigma (O - E)^2/E = 6.752$

Step 5: Do not reject H_0 since $6.752 < 9.488$.

Conclude that the orders are evenly distributed over all days of the week.

11.19 Step 1: H_0: The number of cars sold is the same for each month.

H_1: The number of cars sold is not the same for each month.

Step 2: Since this is a multinomial experiment, use the chi-square distribution.

Step 3: $k = 12$, $df = k - 1 = 12 - 1 = 11$

For $\alpha = .10$, the critical value of χ^2 is 17.275.

Step 4:

Month	O	p	$E = np$	$O - E$	$(O - E)^2$	$(O - E)^2/E$
January	23	1/12	18.67	4.33	18.749	1.004
February	17	1/12	18.67	−1.67	2.789	.149
March	15	1/12	18.67	−3.67	13.469	.721
April	10	1/12	18.67	−8.67	75.169	4.026
May	14	1/12	18.67	−4.67	21.809	1.168
June	12	1/12	18.67	−6.67	44.489	2.383
July	13	1/12	18.67	−5.67	32.149	1.722
August	15	1/12	18.67	−3.67	13.469	0.721
September	23	1/12	18.67	4.33	18.749	1.004
October	26	1/12	18.67	7.33	53.729	2.878
November	27	1/12	18.67	8.33	69.389	3.717
December	29	1/12	18.67	10.33	106.709	5.716
	$n = 224$					Sum = 25.209

$\chi^2 = \Sigma (O - E)^2/E = 25.209$

Step 5: Reject H_0 since $25.209 > 17.275$.

Conclude that the number of cars sold is not the same for each month.

11.21 Step 1: H_0: The percentage distribution of users' opinions is unchanged since the product was redesigned.

H_1: The percentage distribution of users' opinions has changed since the product was redesigned.

Step 2: Since this is a multinomial experiment, use the chi-square distribution.

Step 3: $k = 4$, $df = k - 1 = 4 - 1 = 3$

For $\alpha = .025$, the critical value of χ^2 is 9.348.

Step 4:

Opinion	O	p	$E = np$	$O - E$	$(O - E)^2$	$(O - E)^2/E$
Excellent	495	.53	424	71	5041	11.889
Satisfactory	255	.31	248	7	49	.198
Unsatisfactory	35	.07	56	–21	441	7.875
No opinion	15	.09	72	–57	3249	45.125
	$n = 800$					Sum = 65.087

$\chi^2 = \Sigma (O - E)^2/E = 65.087$

Step 5: Reject H_0 since $65.087 > 9.348$.

Conclude that the percentage distribution of users' opinions has changed since the product was redesigned.

Sections 11.3 - 11.4

11.23 In a **test of independence**, we test the null hypothesis that two characteristics of the elements in a given population are not related against the alternative hypothesis that the two characteristics are related. See Example 11–6 in the text. In a **test of homogeneity**, we test if two (or more) populations are similar with respect to the distribution of certain characteristics. See Example 11–8 in the text.

11.25 The **minimum expected frequency** for each cell should be 5. If this condition is not satisfied, we may increase the sample size or combine some categories.

11.27 a. H_0: The proportion in each row is the same for all four populations.

H_1: The proportion in each row is not the same for all four populations.

b. The expected frequencies are given in parentheses below the observed frequencies in the table below.

	Column 1	Column 2	Column 3	Column 4	Total
Row 1	24 (31.62)	81 (63.95)	60 (89.95)	121 (100.49)	286
Row 2	46 (30.18)	64 (61.04)	91 (85.86)	72 (95.92)	273
Row 3	20 (28.19)	37 (57.01)	105 (80.20)	93 (89.59)	255
Total	90	182	256	286	814

c. $df = (R - 1)(C - 1) = (3 - 1)(4 - 1) = 6$

For $\alpha = .025$, the critical value of χ^2 is 14.449.

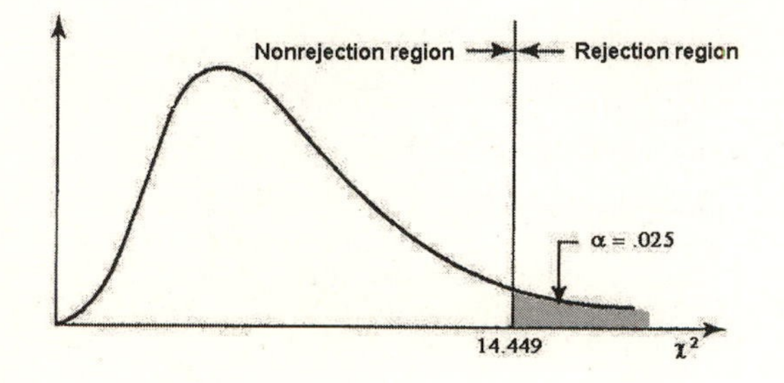

d. $$\chi^2 = \sum \frac{(O-E)^2}{E} = \frac{(24-31.62)^2}{31.62} + \frac{(81-63.95)^2}{63.95} + \frac{(60-89.95)^2}{89.95} + \frac{(121-100.49)^2}{100.49}$$

$$+ \frac{(46-30.18)^2}{30.18} + \frac{(64-61.04)^2}{61.04} + \frac{(91-85.86)^2}{85.86} + \frac{(72-95.92)^2}{95.92} + \frac{(20-28.19)^2}{28.19}$$

$$+ \frac{(37-57.01)^2}{57.01} + \frac{(105-80.20)^2}{80.20} + \frac{(93-89.59)^2}{89.59}$$

$= 1.836 + 4.546 + 9.972 + 4.186 + 8.293 + .144 + .308 + 5.965 + 2.379 + 7.023 + 7.669 + .130$

$= 52.451$

e. Since $52.451 > 14.449$, reject H_0.

11.29 Step 1: H_0: Gender and wearing or not wearing of seat belt are not related.

H_1: Gender and wearing or not wearing of seat belt are related.

Step 2: Since this is a test of independence, use the chi-square distribution.

Step 3: $df = (R - 1)(C - 1) = (2 - 1)(2 - 1) = 1$

For $\alpha = .025$, the critical value of χ^2 is 5.024.

Step 4: The expected frequencies are given in parentheses below the observed frequencies in the table below.

	Wearing seat belt	Not wearing seat belt	Total
Men	34 (36.3)	21 (18.7)	55
Women	32 (29.7)	13 (15.3)	45
Total	66	34	100

$$\chi^2 = \Sigma\frac{(O-E)^2}{E} = \frac{(34-36.3)^2}{36.3} + \frac{(21-18.7)^2}{18.7} + \frac{(32-29.7)^2}{29.7} + \frac{(13-15.3)^2}{15.3}$$

$$= .146 + .283 + .178 + .346 = .953$$

Step 5: Do not reject H_0 since $.953 < 5.024$.

Conclude that gender and wearing or not wearing of seat belt are not related.

11.31 Step 1: H_0: Worries about money and gender are not related.

H_1: Worries about money and gender are related.

Step 2: Since this is a test of independence, use the chi-square distribution.

Step 3: $df = (R - 1)(C - 1) = (2 - 1)(2 - 1) = 1$

For $\alpha = .05$, the critical value of χ^2 is 3.841.

Step 4: The expected frequencies are given in parentheses below the observed frequencies in the table below.

Worries about money	Boys	Girls	Total
Yes	198 (186.63)	181 (192.37)	379
No	127 (138.37)	154 (142.63)	281
Total	325	335	660

$$\chi^2 = \Sigma\frac{(O-E)^2}{E} = \frac{(198-186.63)^2}{186.63} + \frac{(181-192.37)^2}{192.37} + \frac{(127-138.37)^2}{138.37} + \frac{(154-142.63)^2}{142.63}$$

$$= .693 + .672 + .934 + .906 = 3.205$$

Step 5: Do not reject H_0 since $3.205 < 3.841$.

Conclude that worries about money and gender are not related.

11.33 Step 1: H_0: Region and causes of fire are unrelated.

H_1: Region and causes of fire are related.

Step 2: Since this is a test of independence, use the chi-square distribution.

Step 3: $df = (R - 1)(C - 1) = (2 - 1)(4 - 1) = 3$

For $\alpha = .05$, the critical value of χ^2 is 7.815.

Step 4: The expected frequencies are given in parentheses below the observed frequencies in the table below.

Region	Arson	Accident	Lightning	Unknown	Total
A	6 (5.30)	9 (9.38)	6 (8.57)	10 (7.75)	31
B	7 (7.70)	14 (13.62)	15 (12.43)	9 (11.25)	45
Total	13	23	21	19	76

$$\chi^2 = \Sigma\frac{(O-E)^2}{E} = \frac{(6-5.30)^2}{5.30} + \frac{(9-9.38)^2}{9.38} + \frac{(6-8.57)^2}{8.57} + \frac{(10-7.75)^2}{7.75} + \frac{(7-7.70)^2}{7.70} + \frac{(14-13.62)^2}{13.62} + \frac{(15-12.43)^2}{12.43} + \frac{(9-11.25)^2}{11.25}$$

$$= .092 + .015 + .771 + .653 + .064 + .011 + .531 + .450 = 2.587$$

Step 5: Do not reject H_0 since $2.587 < 7.815$.

Conclude that region and causes of fire are unrelated.

11.35 Step 1: H_0: The two drugs are similar in curing the patients.

H_1: The two drugs are not similar in curing the patients.

Step 2: Since this is a test of homogeneity, use the chi-square distribution.

Step 3: $df = (R - 1)(C - 1) = (2 - 1)(2 - 1) = 1$

For $\alpha = .01$, the critical value of χ^2 is 6.635.

Step 4: The expected frequencies are given in parentheses below the observed frequencies in the table below.

	Cured	Not Cured	Total
Drug I	44 (37.2)	16 (22.8)	60
Drug II	18 (24.8)	22 (15.2)	40
Total	62	38	100

$$\chi^2 = \Sigma\frac{(O-E)^2}{E} = \frac{(44-37.2)^2}{37.2} + \frac{(16-22.8)^2}{22.8} + \frac{(18-24.8)^2}{24.8} + \frac{(22-15.2)^2}{15.2}$$

$$= 1.243 + 2.028 + 1.865 + 3.042 = 8.178$$

Step 5: Reject H_0 since $8.178 > 6.635$.

Conclude that the two drugs are not similar in curing the patients.

11.37 Step 1: H_0: The distribution of change in pay is similar for men and women.

H_1: The distribution of change in pay is not similar for men and women.

Step 2: Since this is a test of homogeneity, use the chi-square distribution.

Step 3: $df = (R - 1)(C - 1) = (2 - 1)(3 - 1) = 2$

For $\alpha = .01$, the critical value of χ^2 is 9.210.

Step 4: The expected frequencies are given in parentheses below the observed frequencies in the table below.

	More	Less	Same	Total
Men	140 (116.5)	50 (77.0)	50 (46.5)	240
Women	93 (116.5)	104 (77.0)	43 (46.5)	240
Total	233	154	93	480

$$\chi^2 = \Sigma\frac{(O-E)^2}{E} = \frac{(140-116.5)^2}{116.5} + \frac{(50-77.0)^2}{77.0} + \frac{(50-46.5)^2}{46.5} + \frac{(93-116.5)^2}{116.5} + \frac{(104-77.0)^2}{77.0} + \frac{(43-46.5)^2}{46.5}$$

$$= 4.740 + 9.468 + .263 + 4.740 + 9.468 + .263 = 28.942$$

Step 5: Reject H_0 since $28.942 > 9.210$.

Conclude that the distribution of change in pay is not similar for men and women.

11.39 Step 1: H_0: The distributions of opinions are homogeneous for the two groups of workers.

H_1: The distributions of opinions are not homogeneous for the two groups of workers.

Step 2: Since this is a test of homogeneity, use the chi-square distribution.

Step 3: $df = (R - 1)(C - 1) = (2 - 1)(3 - 1) = 2$

For $\alpha = .025$, the critical value of χ^2 is 7.378.

Step 4: The expected frequencies are given in parentheses below the observed frequencies in the table below.

	Opinion			
	Favor	Oppose	Uncertain	Total
Blue collar Workers	44 (42.59)	39 (42.59)	12 (9.83)	95
White collar Workers	21 (22.41)	26 (22.41)	3 (5.17)	50
Total	65	65	15	145

$$\chi^2 = \Sigma\frac{(O-E)^2}{E} = \frac{(44-42.59)^2}{42.59} + \frac{(39-42.59)^2}{42.59} + \frac{(12-9.83)^2}{9.83} + \frac{(21-22.41)^2}{22.41} + \frac{(26-22.41)^2}{22.41} + \frac{(3-5.17)^2}{5.17}$$

$$= .047 + .303 + .479 + .089 + .575 + .911 = 2.404$$

Step 5: Do not reject H_0 since 2.404 < 7.378.

Conclude that the distributions of opinions are homogeneous for the two groups of workers.

Section 11.5

11.41 $df = n - 1 = 25 - 1 = 24$

a. χ^2 for 24 *df* and .005 area in the right tail = 45.559

χ^2 for 24 *df* and .995 area in the right tail = 9.886

The 99% confidence interval for σ^2 is

$$\frac{(n-1)s^2}{\chi^2_{\alpha/2}} \text{ to } \frac{(n-1)s^2}{\chi^2_{1-\alpha/2}} = \frac{(25-1)(35)}{45.559} \text{ to } \frac{(25-1)(35)}{9.886} = 18.4376 \text{ to } 84.9686$$

b. χ^2 for 24 *df* and .025 area in the right tail = 39.364

χ^2 for 24 *df* and .975 area in the right tail = 12.401

The 95% confidence interval for σ^2 is

$$\frac{(n-1)s^2}{\chi^2_{\alpha/2}} \text{ to } \frac{(n-1)s^2}{\chi^2_{1-\alpha/2}} = \frac{(25-1)(35)}{39.364} \text{ to } \frac{(25-1)(35)}{12.401} = 21.3393 \text{ to } 67.7365$$

c. χ^2 for 24 *df* and .05 area in the right tail = 36.415

χ^2 for 24 *df* and .95 area in the right tail = 13.848

The 90% confidence interval for σ^2 is

$$\frac{(n-1)s^2}{\chi^2_{\alpha/2}} \text{ to } \frac{(n-1)s^2}{\chi^2_{1-\alpha/2}} = \frac{(25-1)(35)}{36.415} \text{ to } \frac{(25-1)(35)}{13.848} = 23.0674 \text{ to } 60.6586$$

As the confidence level decreases, the confidence interval for σ^2 decreases in width.

11.43 a. H_0: $\sigma^2 = .80$, H_1: $\sigma^2 > .80$

b. $df = n - 1 = 16 - 1 = 15$

χ^2 for 15 *df* and .01 area in the right tail = 30.578.

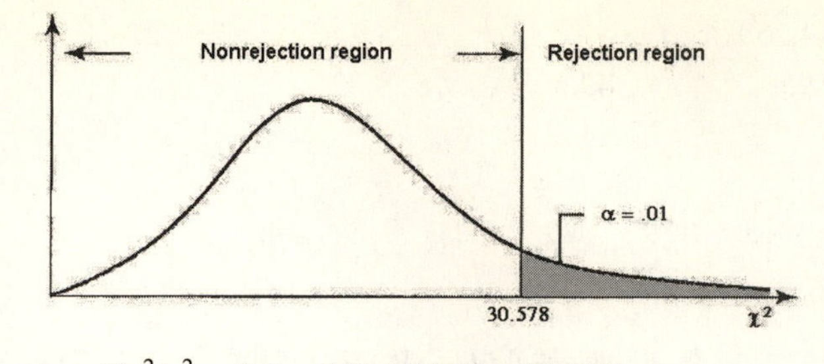

c. $\chi^2 = (n-1)s^2/\sigma^2 = (16-1)(1.10)/.80 = 20.625$

d. Do not reject H_0 since $20.625 < 30.578$.

11.45 a. H_0: $\sigma^2 = 2.2$, H_1: $\sigma^2 \neq 2.2$

b. $df = n - 1 = 18 - 1 = 17$

χ^2 for 17 *df* and .025 area in the right tail = 30.191

χ^2 for 17 *df* and .975 area in the right tail = 7.564

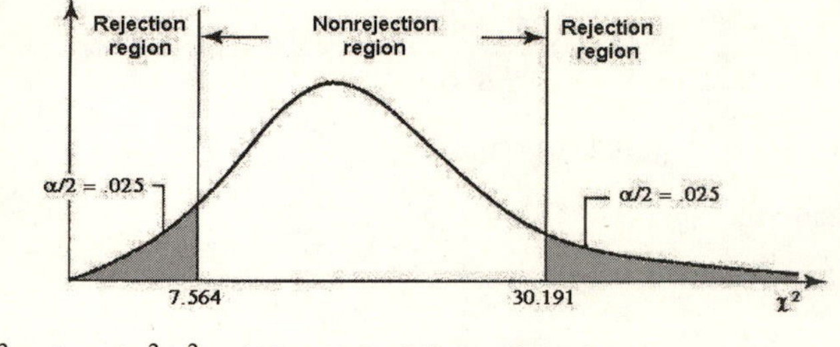

c. $\chi^2 = (n-1)s^2/\sigma^2 = (18-1)(4.6)/2.2 = 35.545$

d. Reject H_0 since $35.545 > 30.191$.

11.47 a. $df = n - 1 = 23 - 1 = 22$

χ^2 for 22 *df* and .01 area in the right tail = 40.289

χ^2 for 22 *df* and .99 area in the right tail = 9.542

The 99% confidence interval for the population variance σ^2 is

$$\frac{(n-1)s^2}{\chi^2_{\alpha/2}} \text{ to } \frac{(n-1)s^2}{\chi^2_{1-\alpha/2}} = \frac{(23-1)(2.7)}{40.289} \text{ to } \frac{(23-1)(2.7)}{9.542} = 1.4743 \text{ to } 6.2251$$

The 99% confidence interval for σ is $\sqrt{1.4743}$ to $\sqrt{6.2251}$ = 1.214 to 2.495

b. Step 1: H_0: $\sigma^2 \leq 2$, H_1: $\sigma^2 > 2$

Step 2: Since this is a test about a population variance, use the chi-square distribution.

Step 3: For $\alpha = .01$ with $df = 22$, the critical value of χ^2 is 40.289.

Step 4: $\chi^2 = (n-1)s^2/\sigma^2 = (23-1)(2.7)/2 = 29.700$

Step 5: Do not reject H_0 since $29.700 < 40.289$.

Conclude that the population variance is not greater than 2 square minutes.

11.49 a. $df = n - 1 = 25 - 1 = 24$

χ^2 for 24 df and .005 area in the right tail = 45.559

χ^2 for 24 df and .995 area in the right tail = 9.886

The 99% confidence interval for σ^2 is

$$\frac{(n-1)s^2}{\chi^2_{\alpha/2}} \text{ to } \frac{(n-1)s^2}{\chi^2_{1-\alpha/2}} = \frac{(25-1)(5200)}{45.559} \text{ to } \frac{(25-1)(5200)}{9.886} = 2739.3051 \text{ to } 12{,}623.9126$$

The 99% confidence interval for σ is $\sqrt{2739.3051}$ to $\sqrt{12{,}623.9126}$ = 52.338 to 112.356

b. Step 1: H_0: $\sigma^2 = 4200$, H_1: $\sigma^2 \neq 4200$

Step 2: Since this is a test about a population variance, use the chi-square distribution.

Step 3: For $\alpha = .05$ with $df = 24$, the critical values of χ^2 are 12.401 and 39.364.

Step 4: $\chi^2 = (n-1)s^2/\sigma^2 = (25-1)(5200)/4200 = 29.714$

Step 5: Do not reject H_0 since $12.401 < 29.714 < 39.364$.

Conclude that the population variance is not different from 4200 square hours.

Supplementary Exercises

11.51 Step 1: H_0: The percentage of people who consume all-bran cereal is the same for all four brands.

H_1: The percentage of people who consume all-bran cereal is not the same for all four brands.

Step 2: Since this is a multinomial experiment, use the chi-square distribution.

Step 3: $k = 4$, $df = k - 1 = 4 - 1 = 3$

For $\alpha = .05$, the critical value of χ^2 is 7.815.

Step 4:

Brand	O	p	$E = np$	$O - E$	$(O - E)^2$	$(O - E)^2/E$
A	212	.25	250	–38	1444	5.776
B	284	.25	250	34	1156	4.624
C	254	.25	250	4	16	.064
D	250	.25	250	0	0	.000
	$n = 1000$					Sum = 10.464

$\chi^2 = \Sigma(O - E)^2/E = 10.464$

Step 5: Reject H_0 since $10.464 > 7.815$.

Conclude that the percentage of people who consume all-bran cereal is not the same for all four brands.

11.53 Step 1: H_0: The current distribution of men's health concerns is the same as that of 2004.

H_1: The current distribution of men's health concerns differs from that of 2004.

Step 2: Since this is a multinomial experiment, use the chi-square distribution.

Step 3: $k = 7$, $df = k - 1 = 7 - 1 = 6$

For $\alpha = .05$, the critical value of χ^2 is 14.067.

step 4:

Health concern	O	p	$E = np$	$O - E$	$(O - E)^2$	$(O - E)^2/E$
No concerns	185	.27	216	−31	961	4.449
Cancer	155	.20	160	−5	25	.156
Heart	138	.16	128	10	100	.781
Diabetes	72	.07	56	16	256	4.571
Blood Pressure	47	.06	48	−1	1	.021
Weight/Diet	52	.05	40	12	144	3.600
Other	151	.19	152	−1	1	.007
	$n = 800$					Sum = 13.585

$$\chi^2 = \Sigma (O - E)^2/E = 13.585$$

Step 5: Do not reject H_0 since $13.585 < 14.067$.

Conclude that the current distribution of men's health concerns is the same as that of 2004.

11.55 Step 1: H_0: The proportions of all allergic persons are the same over the four seasons.

H_1: The proportions of all allergic persons are not the same over the four seasons.

Step 2: Since this is a multinomial experiment, use the chi-square distribution.

Step 3: $k = 4$, $df = k - 1 = 4 - 1 = 3$

For $\alpha = .01$, the critical value of χ^2 is 11.345.

Step 4:

Season	O	p	$E = np$	$O - E$	$(O - E)^2$	$(O - E)^2/E$
Fall	18	.25	25	−7	49	1.960
Winter	13	.25	25	−12	144	5.760
Spring	31	.25	25	6	36	1.440
Summer	38	.25	25	13	169	6.760
	$n = 100$					Sum = 15.920

$$\chi^2 = \Sigma (O - E)^2/E = 15.920$$

Step 5: Reject H_0 since $15.920 > 11.345$.

Conclude that the proportions of all allergic persons are not the same over the four seasons.

11.57 Step 1: H_0: The percentages of insured and uninsured people are the same for all four regions.

H_1: The percentages of insured and uninsured people are not the same for all four regions.

Step 2: Since this is a test of homogeneity, use the chi-square distribution.

Step 3: $df = (R-1)(C-1) = (2-1)(4-1) = 3$

For $\alpha = .025$, the critical value of χ^2 is 9.348.

Step 4: The expected frequencies are given in parentheses below the observed frequencies in the table below.

Coverage	Northeast	Midwest	South	West	Total
Insured	871 (848.75)	880 (848.75)	820 (848.75)	824 (848.75)	3395
Uninsured	129 (151.25)	120 (151.25)	180 (151.25)	176 (151.25)	605
Total	1000	1000	1000	1000	4000

$$\chi^2 = \sum \frac{(O-E)^2}{E} = \frac{(871-848.75)^2}{848.75} + \frac{(880-848.75)^2}{848.75} + \frac{(820-848.75)^2}{848.75} + \frac{(824-848.75)^2}{848.75}$$

$$+ \frac{(129-151.25)^2}{151.25} + \frac{(120-151.25)^2}{151.25} + \frac{(180-151.25)^2}{151.25} + \frac{(176-151.25)^2}{151.25}$$

$$= .583 + 1.151 + .974 + .722 + 3.273 + 6.457 + 5.465 + 4.050 = 22.675$$

Step 5: Reject H_0 since $22.675 > 9.348$.

Conclude that the percentages of insured and uninsured people are not the same for all four regions.

11.59 Step 1: H_0: Gender and marital status are not related for all persons who hold more than one job.

H_1: Gender and marital status are related for all persons who hold more than one job.

Step 2: Since this is a test of independence, use the chi-square distribution.

Step 3: $df = (R-1)(C-1) = (2-1)(3-1) = 2$

For $\alpha = .10$, the critical value of χ^2 is 4.605.

Step 4: The expected frequencies are given in parentheses below the observed frequencies in the table below.

Gender	Single	Married	Other	Total
Male	72 (67.20)	209 (199.04)	39 (53.76)	320
Female	33 (37.80)	102 (111.96)	45 (30.24)	180
Total	105	311	84	500

$$\chi^2 = \Sigma\frac{(O-E)^2}{E} = \frac{(72-67.20)^2}{67.20} + \frac{(209-199.04)^2}{199.04} + \frac{(39-53.76)^2}{53.76} + \frac{(33-37.80)^2}{37.80} + \frac{(102-111.96)^2}{111.96} + \frac{(45-30.24)^2}{30.24}$$

$$= .343 + .498 + 4.052 + .610 + .886 + 7.204 = 13.593$$

Step 5: Reject H_0 since $13.593 > 4.605$.

Conclude that gender and marital status are related for all persons who hold more than one job.

11.61 Step 1: H_0: The percentages of people with different opinions are similar for all four regions.

H_1: The percentages of people with different opinions are not similar for all four regions.

Step 2: Since this is a test of homogeneity, use the chi-square distribution.

Step 3: $df = (R - 1)(C - 1) = (4 - 1)(3 - 1) = 6$

For $\alpha = .01$, the critical value of χ^2 is 16.812.

Step 4: The expected frequencies are given in parentheses below the observed frequencies in the table below.

Region	Favor	Oppose	Uncertain	Total
Northeast	56 (63.75)	33 (29.75)	11 (6.50)	100
Midwest	73 (63.75)	23 (29.75)	4 (6.50)	100
South	67 (63.75)	28 (29.75)	5 (6.50)	100
West	59 (63.75)	35 (29.75)	6 (6.50)	100
Total	255	119	26	400

$$\chi^2 = \Sigma\frac{(O-E)^2}{E} = \frac{(56-63.75)^2}{63.75} + \frac{(33-29.75)^2}{29.75} + \frac{(11-6.50)^2}{6.50} + \frac{(73-63.75)^2}{63.75} + \frac{(23-29.75)^2}{29.75} + \frac{(4-6.50)^2}{6.50} + \frac{(67-63.75)^2}{63.75} + \frac{(28-29.75)^2}{29.75} + \frac{(5-6.50)^2}{6.50} + \frac{(59-63.75)^2}{63.75} + \frac{(35-29.75)^2}{29.75} + \frac{(6-6.50)^2}{6.50}$$

$$= .942 + .355 + 3.115 + 1.342 + 1.532 + .962 + .166 + .103 + .346 + .354 + .926 + .038 = 10.181$$

Step 5: Do not reject H_0 since $10.181 < 16.812$.

Conclude that the percentages of people with different opinions are similar for all four regions.

11.63 a. $df = n - 1 = 10 - 1 = 9$

χ^2 for 9 df and .025 area in the right tail = 19.023

χ^2 for 9 df and .975 area in the right tail = 2.700

The 95% confidence interval for the population variance σ^2 is

$$\frac{(n-1)s^2}{\chi^2_{\alpha/2}} \text{ to } \frac{(n-1)s^2}{\chi^2_{1-\alpha/2}} = \frac{(10-1)(7.2)}{19.023} \text{ to } \frac{(10-1)(7.2)}{2.700} = 3.4064 \text{ to } 24.0000$$

The 95% confidence interval for σ is $\sqrt{3.4064}$ to $\sqrt{24.0000}$ = 1.846 to 4.899

b. $df = n - 1 = 18 - 1 = 17$

χ^2 for 17 df and .025 area in the right tail = 30.191

χ^2 for 17 df and .975 area in the right tail = 7.564

The 95% confidence interval for σ^2 is

$$\frac{(n-1)s^2}{\chi^2_{\alpha/2}} \text{ to } \frac{(n-1)s^2}{\chi^2_{1-\alpha/2}} = \frac{(18-1)(14.8)}{30.191} \text{ to } \frac{(18-1)(14.8)}{7.564} = 8.3336 \text{ to } 33.2628$$

The 95% confidence interval for σ is $\sqrt{8.3336}$ to $\sqrt{33.2628}$ = 2.887 to 5.767

11.65 Step 1: H_0: $\sigma^2 = 1.1$, H_1: $\sigma^2 > 1.1$

Step 2: Since this is a test about a population variance, use the chi-square distribution.

Step 3: $df = n - 1 = 17 - 1 = 16$

For $\alpha = .01$, the critical value of χ^2 is 28.845.

Step 4: $\chi^2 = (n - 1)s^2/\sigma^2 = (17 - 1)(1.7)/1.1 = 24.727$

Step 5: Do not reject H_0 since $24.727 < 28.845$.

Conclude that the population variance is not greater than 1.1.

11.67 Step 1: H_0: $\sigma^2 = 10.4$, H_1: $\sigma^2 \neq 10.4$

Step 2: Since this is a test about a population variance, use the chi-square distribution.

Step 3: $df = n - 1 = 18 - 1 = 17$

For $\alpha = .05$, the critical values of χ^2 are 7.564 and 30.191.

Step 4: $\chi^2 = (n - 1)s^2/\sigma^2 = (18 - 1)(14.8)/10.4 = 24.192$

Step 5: Do not reject H_0 since $7.564 < 24.192 < 30.191$.

Conclude that the population variance is not different from 10.4.

11.69 a. Step 1: H_0: $\sigma^2 = 5000$, H_1: $\sigma^2 < 5000$

Step 2: Since this is a test about a population variance, use the chi-square distribution.

Step 3: $df = n - 1 = 20 - 1 = 19$

For $\alpha = .025$, the critical value of χ^2 is 8.907.

Step 4: $\chi^2 = (n-1)s^2/\sigma^2 = (20 - 1)(3175)/5000 = 12.065$

Step 5: Do not reject H_0 since $12.065 > 8.907$.

Conclude that the population variance is not less than 5000.

b. χ^2 for 19 *df* and .01 area in the right tail = 36.191

χ^2 for 19 *df* and .99 area in the right tail = 7.633

The 98% confidence interval for σ^2 is

$$\frac{(n-1)s^2}{\chi^2_{\alpha/2}} \text{ to } \frac{(n-1)s^2}{\chi^2_{1-\alpha/2}} = \frac{(20-1)(3175)}{36.191} \text{ to } \frac{(20-1)(3175)}{7.633} = 1666.8509 \text{ to } 7903.1835$$

The 98% confidence interval for the population standard deviation σ is

$$\sqrt{1666.8509} \text{ to } \sqrt{7903.1835} = 40.827 \text{ to } 88.900$$

11.71 a. $df = n - 1 = 25 - 1 = 24$

χ^2 for 24 *df* and .005 area in the right tail = 45.559

χ^2 for 24 *df* and .995 area in the right tail = 9.886

The 99% confidence interval for σ^2 is

$$\frac{(n-1)s^2}{\chi^2_{\alpha/2}} \text{ to } \frac{(n-1)s^2}{\chi^2_{1-\alpha/2}} = \frac{(25-1)(.19)}{45.559} \text{ to } \frac{(25-1)(.19)}{9.886} = .1001 \text{ to } .4613$$

The 99% confidence interval for σ is $\sqrt{.1001}$ to $\sqrt{.4613} = .316$ to $.679$

b. Step 1: H_0: $\sigma^2 = .13$, H_1: $\sigma^2 \neq .13$

Step 2: Since this is a test about a population variance, use the chi-square distribution.

Step 3: For $\alpha = .01$ with $df = 24$, the critical values of χ^2 are 9.886 and 45.559.

Step 4: $\chi^2 = (n-1)s^2/\sigma^2 = (25 - 1)(.19)/.13 = 35.077$

Step 5: Do not reject H_0 since $9.886 < 35.077 < 45.559$.

Conclude that the population variance is not different from .13.

11.73 a. $\sum x = 6293$, $\sum x^2 = 4{,}957{,}983$

$$s^2 = \frac{\sum x^2 - \frac{(\sum x)^2}{n}}{n-1} = \frac{4{,}957{,}983 - \frac{(6293)^2}{8}}{8-1} = 1107.4107$$

b. $df = n - 1 = 8 - 1 = 7$

χ^2 for 7 *df* and .025 area in the right tail = 16.013

χ^2 for 7 *df* and .975 area in the right tail = 1.690

The 95% confidence interval for σ^2 is

$$\frac{(n-1)s^2}{\chi^2_{\alpha/2}} \text{ to } \frac{(n-1)s^2}{\chi^2_{1-\alpha/2}} = \frac{(8-1)(1107.4107)}{16.013} \text{ to } \frac{(8-1)(1107.4107)}{1.690} = 484.0989 \text{ to } 4586.9082$$

The 95% confidence interval for σ is $\sqrt{484.0989}$ to $\sqrt{4586.9082}$ = \$22.00 to \$67.73

c. Step 1: H_0: $\sigma^2 = 500$, H_1: $\sigma^2 \neq 500$

Step 2: Since this is a test about a population variance, use the chi-square distribution.

Step 3: For $\alpha = .05$ with $df = 7$, the critical values of χ^2 are 1.690 and 16.013.

Step 4: $\chi^2 = (n-1)s^2/\sigma^2 = (8-1)(1107.4107)/500 = 15.504$

Step 5: Do not reject H_0 since $1.690 < 15.504 < 16.013$.

Conclude that the population variance is not different from 500 square dollars.

11.75 Step 1: H_0: Opinions on disposal site are independent of gender.

H_1: Opinions on disposal site are dependent on gender.

Step 2: Since this is a test of independence, use the chi-square distribution.

Step 3: $df = (R-1)(C-1) = (2-1)(3-1) = 2$

For $\alpha = .05$, the critical value of χ^2 is 5.991.

Step 4: From the given data we can calculate the following:

Total number opposed = 200(.60) = 120

Total number in favor = 200(.32) = 64

Total number undecided = 200(.08) = 16

Number of women opposed = 120(.65) = 78

Number of men opposed = 120 – 78 = 42

Number of men in favor = 64(.625) = 40

Number of women in favor = 64 – 40 = 24

Number of women undecided = 110 – 78 – 24 = 8

Number of men undecided = 16 – 8 = 8

Using these results, we may construct the following contingency table of observations and expected values.

	Opposed	In Favor	Undecided	Total
Women	78	24	8	110
	(66.0)	(35.2)	(8.8)	
Men	42	40	8	90
	(54.0)	(28.8)	(7.2)	
Total	120	64	16	200

$$\chi^2 = \sum \frac{(O-E)^2}{E} = \frac{(78-66.0)^2}{66.0} + \frac{(24-35.2)^2}{35.2} + \frac{(8-8.8)^2}{8.8} + \frac{(42-54.0)^2}{54.0} + \frac{(40-28.8)^2}{28.8} + \frac{(8-7.2)^2}{7.2}$$

$$= 2.182 + 3.564 + .073 + 2.667 + 4.356 + .089 = 12.931$$

Step 5: Reject H_0 since 12.931 > 5.991.

Conclude that opinions on disposal site are dependent on gender.

11.77 Step 1: H_0: The proportions of red and green marbles are the same in all five boxes.

H_1: The proportions of red and green marbles are not the same in all five boxes.

Step 2: Since this is a test of homogeneity, use the chi-square distribution.

Step 3: $df = (R-1)(C-1) = (2-1)(5-1) = 4$

For $\alpha = .05$, the critical value of χ^2 is 9.488.

Step 4: The expected frequencies are given in parentheses below the observed frequencies in the table below.

Box	1	2	3	4	5	Total
Red	20	14	23	30	18	105
	(21)	(21)	(21)	(21)	(21)	
Green	30	36	27	20	32	145
	(29)	(29)	(29)	(29)	(29)	
Total	50	50	50	50	50	250

$$\chi^2 = \sum \frac{(O-E)^2}{E} = \frac{(20-21)^2}{21} + \frac{(14-21)^2}{21} + \frac{(23-21)^2}{21} + \frac{(30-21)^2}{21} + \frac{(18-21)^2}{21} + \frac{(30-29)^2}{29} + \frac{(36-29)^2}{29} + \frac{(27-29)^2}{29} + \frac{(20-29)^2}{29} + \frac{(32-29)^2}{29}$$

$$= .048 + 2.333 + .190 + 3.857 + .429 + .034 + 1.690 + .138 + 2.793 + .310$$

$$= 11.822$$

Step 5: Reject H_0 since 11.822 > 9.488.

Conclude that the proportions of red and green marbles are not the same in all five boxes.

11.79 Step 1: H_0: A normal distribution is an appropriate model for these data.

H_1: A normal distribution is not an appropriate model for these data.

Step 2: Since this is a multinomial experiment, use the chi-square distribution.

Step 3: $k = 10$, $df = k - 1 = 10 - 1 = 9$

For $\alpha = .05$, the critical value of χ^2 is 16.919.

Step 4: We will find the probability values necessary to fill in the table used to calculate the chi-square test statistic using the standard normal distribution table. For example, for the second row, we will find $p = P(-2 < z < -1.5) = P(z < -1.5) - P(z < -2) = .0668 - .0228 = .0440$. Note the symmetry in the standard normal corresponds to the symmetry in the probabilities, and, therefore, in the expected counts.

Category	O	p	$E = np$	$O - E$	$(O - E)^2$	$(O - E)^2/E$
z score below −2	48	.0228	22.8	25.2	635.04	27.853
z score from −2 to −1.5	67	.0440	44.0	23.0	529.00	12.023
z score from −1.5 to −1	146	.0919	91.9	54.1	2926.81	31.848
z score from −1 to −0.5	248	.1498	149.8	98.2	9643.24	64.374
z score from −0.5 to 0	187	.1915	191.5	−4.5	20.25	.106
z score from 0 to 0.5	125	.1915	191.5	−66.5	4422.25	23.093
z score from 0.5 to 1	88	.1498	149.8	−61.8	3819.24	25.496
z score from 1 to 1.5	47	.0919	91.9	−44.9	2016.01	21.937
z score from 1.5 to 2	25	.0440	44.0	−19.0	361.00	8.205
z score of 2 or above	19	.0228	22.8	−3.8	14.44	.633
	$n = 1000$					Sum = 215.568

$\chi^2 = \Sigma (O - E)^2/E = 215.568$

Step 5: Reject H_0 since $215.568 > 16.919$.

Conclude that a normal distribution is not an appropriate model for these data.

11.81 a. Step 1: H_0: All categories are equally likely.

H_1: All categories are not equally likely.

Step 2: Since this is a multinomial experiment, use the chi-square distribution.

Step 3: $k = 4$, $df = k - 1 = 4 - 1 = 3$

For $\alpha = .10$, the largest reasonable significance level, the critical value of χ^2 is 6.251.

Step 4:

Category	O	p	$E = np$	$O - E$	$(O - E)^2$	$(O - E)^2/E$
A	21	.25	25	−4	16	.64
B	26	.25	25	1	1	.04
C	31	.25	25	6	36	1.44
D	22	.25	25	−3	9	.36
	$n = 100$					Sum = 2.48

$\chi^2 = \Sigma (O - E)^2/E = 2.48$

Step 5: Do not reject H_0 since $2.48 < 6.251$.

Conclude that all categories are equally likely. Note that for any $\alpha < .10$, $\chi^2 > 6.251$. Hence, we would not reject H_0 for any reasonable significance level.

b. We must have $\Sigma \mid O - E \mid = 14$ and $\Sigma(O - E) = 0$. Under these constraints, the value of the chi-square test statistic is maximized when the four values for $O - E$ are 0, 0, 7, and −7. In such a case, calculating the chi-square test statistic yields $\chi^2 = \frac{(0)^2}{25} + \frac{(0)^2}{25} + \frac{(7)^2}{25} + \frac{(7)^2}{25} = 3.92$.

Since $3.92 < 6.251$, the p-value $> .10$.

Self-Review Test

1. b **2.** a **3.** c **4.** a **5.** b **6.** b **7.** c **8.** b **9.** a

10. Step 1: H_0: The current distribution of all dietitians' responses does not differ from that for 2004.

H_1: The current distribution of all dietitians' responses differs from that for 2004.

Step 2: Since this is a multinomial experiment, use the chi-square distribution.

Step 3: $k = 5$, $df = k - 1 = 5 - 1 = 4$

For $\alpha = .05$, the critical value of χ^2 is 9.488.

Step 4:

Response	O	p	$E = np$	$O - E$	$(O - E)^2$	$(O - E)^2/E$
Fad diet	48	.27	40.5	7.5	56.25	1.389
Not limiting portions	41	.20	30.0	11.0	121.00	4.033
Inappropriate goals	17	.15	22.5	–5.5	30.25	1.344
Not exercising	12	.13	19.5	–7.5	56.25	2.885
Other	30	.25	37.5	–7.5	56.25	1.500
	$n = 150$					Sum = 11.151

$\chi^2 = \Sigma (O - E)^2/E = 11.151$

Step 5: Reject H_0 since $11.151 > 9.488$.

Conclude that the current distribution of all dietitians' responses differs from that for 2004.

11. Step 1: H_0: Educational level and ever being divorced are independent.

H_1: Educational level and ever being divorced are dependent.

Step 2: Since this is a test of independence, use the chi-square distribution.

Step 3: $df = (R - 1)(C - 1) = (2 - 1)(4 - 1) = 3$

For $\alpha = .01$, the critical value of χ^2 is 11.345.

Step 4: The expected frequencies are given in parentheses below the observed frequencies in the table below.

	Educational Level				
	Less than High school	High school degree	Some college	College degree	Total
Divorced	173 (160.47)	158 (136.04)	95 (98.20)	53 (84.30)	479
Never Divorced	162 (174.54)	126 (147.96)	110 (106.81)	123 (91.70)	521
Total	335	284	205	176	1000

$$\chi^2 = \sum\frac{(O-E)^2}{E} = \frac{(173-160.47)^2}{160.47} + \frac{(158-136.04)^2}{136.04} + \frac{(95-98.20)^2}{98.20} + \frac{(53-84.30)^2}{84.30}$$
$$+\frac{(162-174.54)^2}{174.54} + \frac{(126-147.96)^2}{147.96} + \frac{(110-106.81)^2}{106.81} + \frac{(123-91.70)^2}{91.70}$$
$$= .978 + 3.545 + .104 + 11.621 + .901 + 3.259 + .095 + 10.684 = 31.187$$

Step 5: Reject H_0 since $31.187 > 11.345$.

Conclude that educational level and ever being divorced are dependent.

12. Step 1: H_0: The percentages of people who play the lottery often, sometimes, and never are the same for each income group.

H_1: The percentages of people who play the lottery often, sometimes, and never are not the same for each income group.

Step 2: Since this is a test of homogeneity, use the chi-square distribution.

Step 3: $df = (R-1)(C-1) = (3-1)(3-1) = 4$

For $\alpha = .05$, the critical value of χ^2 is 9.488.

Step 4: The expected frequencies are given in parentheses below the observed frequencies in the table below.

	Income Group			
	Low	Middle	High	Total
Play often	174 (170.80)	163 (142.33)	90 (113.87)	427
Play sometimes	286 (249.20)	217 (207.67)	120 (166.13)	623
Never Play	140 (180.00)	120 (150.00)	190 (120.00)	450
Total	600	500	400	1500

$$\chi^2 = \Sigma\frac{(O-E)^2}{E} = \frac{(174-170.80)^2}{170.80} + \frac{(163-142.33)^2}{142.33} + \frac{(90-113.87)^2}{113.87}$$
$$+ \frac{(286-249.20)^2}{249.20} + \frac{(217-207.67)^2}{207.67} + \frac{(120-166.13)^2}{166.13}$$
$$+ \frac{(140-180.00)^2}{180.00} + \frac{(120-150.00)^2}{150.00} + \frac{(190-120.00)^2}{120.00}$$
$$= .060 + 3.002 + 5.004 + 5.434 + .419 + 12.809 + 8.889 + 6.000 + 40.833$$
$$= 82.450$$

Step 5: Reject H_0 since $82.450 > 9.488$.

Conclude that the percentages of people who play the lottery often, sometimes, and never are not the same for each income group.

13. a. $df = n - 1 = 20 - 1 = 19$

χ^2 for 19 df and .005 area in the right tail = 35.582

χ^2 for 19 df and .995 area in the right tail = 6.844

The 99% confidence interval for the population variance σ^2 is

$$\frac{(n-1)s^2}{\chi^2_{\alpha/2}} \text{ to } \frac{(n-1)s^2}{\chi^2_{1-\alpha/2}} = \frac{(20-1)(.48)}{38.582} \text{ to } \frac{(20-1)(.48)}{6.844} = .2364 \text{ to } 1.3326$$

The 99% confidence interval for σ is $\sqrt{.2364}$ to $\sqrt{1.3326}$ = .486 to 1.154

b. Step 1: H_0: $\sigma^2 = .25$, H_1: $\sigma^2 > .25$

Step 2: Since this is a test about a population variance, use the chi-square distribution.

Step 3: For $\alpha = .01$ with $df = 19$, the critical value of χ^2 is 36.191.

Step 4: $\chi^2 = (n-1)s^2/\sigma^2 = (20 - 1)(.48)/.25 = 36.480$

Step 5: Reject H_0 since $36.480 > 36.191$.

Conclude that the population variance exceeds .25 square ounce.

Chapter Twelve

Section 12.1

12.1 The ***F* distribution** is continuous and skewed to the right. The F distribution has two numbers of degrees of freedom: df for the numerator and df for the denominator. The units of an F distribution, denoted by F, are nonnegative.

12.3 a. $F = 3.73$ b. $F = 3.61$ c. $F = 5.37$

12.5 a. $F = 2.39$ b. $F = 2.27$ c. $F = 3.28$

12.7 a. $F = 9.96$ b. $F = 6.57$

12.9 a. $F = 4.85$ b. $F = 3.22$

Section 12.2

12.11 The following assumptions must hold true to use one-way ANOVA:

1. The populations from which the samples are drawn are normally distributed.
2. The populations from which the samples are drawn have the same variance (or standard deviation).
3. The samples drawn from different populations are random and independent.

12.13 a. For Sample I: $n_1 = 7$, $\Sigma x = 105$, $\Sigma x^2 = 1697$

$\bar{x}_1 = \Sigma x/n_1 = 105/7 = 15$

$$s_1 = \sqrt{\frac{\Sigma x^2 - \frac{(\Sigma x)^2}{n_1}}{n_1 - 1}} = \sqrt{\frac{1697 - \frac{(105)^2}{7}}{7-1}} = 4.50924975$$

For sample II: $n_2 = 7$, $\Sigma x = 77$, and $\Sigma x^2 = 963$

$\bar{x}_2 = \Sigma x/n_2 = 77/7 = 11$

$$s_2 = \sqrt{\frac{\Sigma x^2 - \frac{(\Sigma x)^2}{n_2}}{n_2 - 1}} = \sqrt{\frac{963 - \frac{(77)^2}{7}}{7-1}} = 4.39696865$$

b. Step 1: H_0: $\mu_1 - \mu_2 = 0$, H_1: $\mu_1 - \mu_2 \neq 0$

Step 2: Since σ_1 and σ_2 are unknown but assumed to be equal, the samples are independent, the sample sizes are small, and the populations are normally distributed, use the t distribution.

Step 3: $df = n_1 + n_2 - 2 = 7 + 7 - 2 = 12$

For $\alpha = .05$, the critical values of t are –2.179 and 2.179.

Step 4: $$s_p = \sqrt{\frac{(n_1 - 1)s_1^2 + (n_2 - 1)s_2^2}{n_1 + n_2 - 2}}$$

$$= \sqrt{\frac{(7-1)(4.50924975)^2 + (7-1)(4.39696865)^2}{7+7-2}} = 4.45346307$$

$$s_{\bar{x}_1 - \bar{x}_2} = s_p \sqrt{\frac{1}{n_1} + \frac{1}{n_2}} = 4.45346307\sqrt{\frac{1}{7} + \frac{1}{7}} = 2.38047614$$

$$t = \frac{(\bar{x}_1 - \bar{x}_2) - (\mu_1 - \mu_2)}{s_{\bar{x}_1 - \bar{x}_2}} = \frac{(15-11) - 0}{2.38047614} = 1.680$$

Step 5: Do not reject H_0 since $1.680 < 2.179$.

c. Step 1: H_0: $\mu_1 = \mu_2$, H_1: $\mu_1 \neq \mu_2$

Step 2: Since we are testing for equality of means between two groups, use the F distribution.

Step 3: $k = 2$, $n_1 = 7$, $n_2 = 7$, $n = n_1 + n_2 = 7 + 7 = 14$

df for the numerator $= k - 1 = 2 - 1 = 1$

df for the denominator $= n - k = 14 - 2 = 12$

For $\alpha = .05$, the critical value of F is 4.75.

Step 4: $T_1 = 105$, $T_2 = 77$, $\Sigma x = T_1 + T_2 = 105 + 77 = 182$, $\Sigma x^2 = 2660$

$$\text{SSB} = \left(\frac{T_1^2}{n_1} + \frac{T_2^2}{n_2}\right) - \frac{(\Sigma x)^2}{n} = \left(\frac{(105)^2}{7} + \frac{(77)^2}{7}\right) - \frac{(182)^2}{14} = 2422 - 2366 = 56$$

$$\text{SSW} = \Sigma x^2 - \left(\frac{T_1^2}{n_1} + \frac{T_2^2}{n_2}\right) = 2660 - 2422 = 238$$

MSB = SSB/$(k - 1)$ = 56/1 = 56

MSW = SSW/(n – k) = 238/12 = 19.8333

F = MSB/MSW = 56/19.8333 = 2.82

Step 5: Do not reject H_0 since 2.82 < 4.75.

d. For both parts b and c, conclude that the population means are equal.

12.15 a.

ANOVA TABLE

Source of Variation	Degrees of Freedom	Sum of Squares	Mean Square	Value of the Test Statistic
Between	3	112.5201	37.5067	$F = \frac{37.5067}{9.2154} = 4.07$
Within	15	138.2310	9.2154	
Total	18	250.7511		

b. Step 1: H_0: $\mu_1 = \mu_2 = \mu_3 = \mu_4$, H_1: Not all four population means are equal.

Step 2: Since we are testing for equality of means among four groups, use the F distribution.

Step 3: df for the numerator = 3

df for the denominator = 15

For $\alpha = .05$, the critical value of F is 3.29.

Step 4: From the ANOVA table, $F = 4.07$.

Step 5: Reject H_0 since 4.07 > 3.29.

Conclude that the means of the four populations are not all equal.

12.17 a. H_0: $\mu_1 = \mu_2 = \mu_3$, H_1: Not all three population means are equal.

b. $k = 3$, $n = n_1 + n_2 + n_3 = 10 + 9 + 6 = 25$

df for the numerator = $k - 1 = 3 - 1 = 2$, df for the denominator = $n - k = 25 - 3 = 22$

c. $T_1 = 140$, $T_2 = 86$, $T_3 = 37$, $\Sigma x = T_1 + T_2 + T_3 = 140 + 86 + 37 = 263$, $\Sigma x^2 = 3571$

$$\text{SSB} = \left(\frac{T_1^2}{n_1} + \frac{T_2^2}{n_2} + \frac{T_3^2}{n_3}\right) - \frac{(\Sigma x)^2}{n} = \left(\frac{(140)^2}{10} + \frac{(86)^2}{9} + \frac{(37)^2}{6}\right) - \frac{(263)^2}{25}$$

$$= 3009.9444 - 2766.7600 = 243.1844$$

$$\text{SSW} = \Sigma x^2 - \left(\frac{T_1^2}{n_1} + \frac{T_2^2}{n_2} + \frac{T_3^2}{n_3}\right) = 3571 - 3009.9444 = 561.0556$$

SST = SSB + SSW = 243.1844 + 561.0556 = 804.2400

d. For $\alpha = .01$ and $df = (2,22)$, the critical value of F is 5.72.

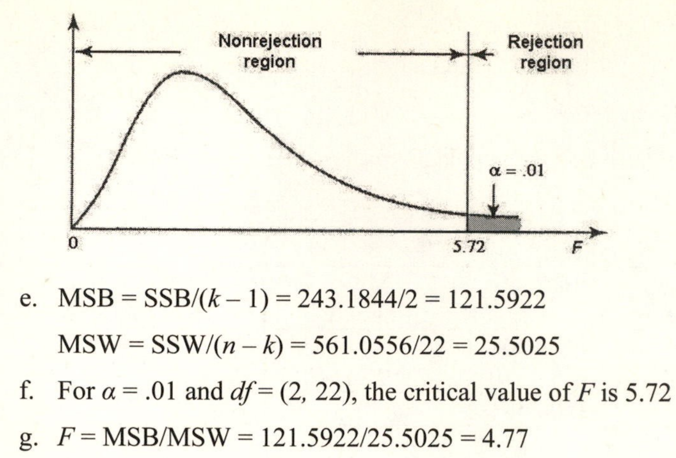

e. MSB = SSB/(k – 1) = 243.1844/2 = 121.5922

MSW = SSW/(n – k) = 561.0556/22 = 25.5025

f. For α = .01 and df = (2, 22), the critical value of F is 5.72

g. F = MSB/MSW = 121.5922/25.5025 = 4.77

h. ANOVA TABLE

Source of Variation	Degrees of Freedom	Sum of Squares	Mean Square	Value of the Test Statistic
Between	2	243.1844	121.5922	$F = \frac{121.5922}{25.5025} = 4.77$
Within	22	561.0556	25.5025	
Total	24	804.2400		

i. Do not reject H_0 since 4.77 < 5.72. Conclude that the mean number of classes missed by all three age groups is the same.

12.19 Step 1: H_0: $\mu_1 = \mu_2 = \mu_3$, H_1: Not all three population means are equal.

Step 2: Since we are testing for equality of means among three groups, use the F distribution.

Step 3: $k = 3$, $n = n_1 + n_2 + n_3 = 8 + 7 + 6 = 21$

df for the numerator = $k - 1 = 3 - 1 = 2$, df for the denominator = $n - k = 21 - 3 = 18$

For α = .05, the critical value of F is 3.55.

Step 4: $T_1 = 408$, $T_2 = 232$, $T_3 = 298$, $\Sigma x = 938$, $\Sigma x^2 = 49{,}322$

$$\text{SSB} = \left(\frac{T_1^2}{n_1}+\frac{T_2^2}{n_2}+\frac{T_3^2}{n_3}\right)-\frac{(\Sigma x)^2}{n} = \left(\frac{(408)^2}{8}+\frac{(232)^2}{7}+\frac{(298)^2}{6}\right)-\frac{(938)^2}{21}$$

= 43,297.8095 – 41,897.3333 = 1400.4762

$$\text{SSW} = \Sigma x^2 - \left(\frac{T_1^2}{n_1}+\frac{T_2^2}{n_2}+\frac{T_3^2}{n_3}\right) = 49{,}322 - 43{,}297.8095 = 6024.1905$$

MSB = SSB/(k – 1) = 1400.4762/2 = 700.2381

MSW = SSW/(n – k) = 6024.1905/18 = 334.6773

F = MSB/MSW = 700.2381/334.6773 = 2.09

Step 5: Do not reject H_0 since 2.09 < 3.55.

Conclude that the mean time taken to find their first job for all 2005 graduates in these three fields is the same.

12.21 Step 1: H_0: $\mu_1 = \mu_2 = \mu_3 = \mu_4$, H_1: Not all four population means are equal.

Step 2: Since we are testing for equality of means among four groups, use the F distribution.

Step 3: $k = 4$, $n = n_1 + n_2 + n_3 + n_4 = 7 + 7 + 7 + 7 = 28$

df for the numerator $= k - 1 = 4 - 1 = 3$, df for the denominator $= n - k = 28 - 4 = 24$

For $\alpha = .025$, the critical value for F is 3.72.

Step 4: $T_1 = 162$, $T_2 = 145$, $T_3 = 172$, $T_4 = 180$, $\Sigma x = 659$, $\Sigma x^2 = 15{,}751$

$$\text{SSB} = \left(\frac{T_1^2}{n_1} + \frac{T_2^2}{n_2} + \frac{T_3^2}{n_3} + \frac{T_4^2}{n_4}\right) - \frac{(\Sigma x)^2}{n}$$

$$= \left(\frac{(162)^2}{7} + \frac{(145)^2}{7} + \frac{(172)^2}{7} + \frac{(180)^2}{7}\right) - \frac{(659)^2}{28}$$

$= 15{,}607.5714 - 15{,}510.0357 = 97.5357$

$$\text{SSW} = \Sigma x^2 - \left(\frac{T_1^2}{n_1} + \frac{T_2^2}{n_2} + \frac{T_3^2}{n_3} + \frac{T_4^2}{n_4}\right) = 15{,}751 - 15{,}607.5714 = 143.4286$$

MSB = SSB/$(k - 1)$ = 97.5357/3 = 32.5119

MSW = SSW/$(n - k)$ = 143.4286/24 = 5.9762

F = MSB/MSW = 32.5119/5.9762 = 5.44

Step 5: Reject H_0 since 5.44 > 3.75.

Conclude that the mean life of bulbs for each of these four brands is not the same.

Supplementary Exercises

12.23 a. Step 1: H_0: $\mu_1 = \mu_2 = \mu_3 = \mu_4$, H_1: Not all four population means are equal.

Step 2: Since we are testing for equality of means among four groups, use the F distribution.

Step 3: $k = 4$, $n = n_1 + n_2 + n_3 + n_4 = 5 + 4 + 5 + 4 = 18$

df for the numerator $= k - 1 = 4 - 1 = 3$,

df for the denominator $= n - k = 18 - 4 = 14$

For $\alpha = .05$, the critical value for F is 3.34.

Step 4: $T_1 = 324$, $T_2 = 210$, $T_3 = 354$, $T_4 = 199$, $\Sigma x = 1087$, $\Sigma x^2 = 68{,}013$

$$SSB = \left(\frac{T_1^2}{n_1}+\frac{T_2^2}{n_2}+\frac{T_3^2}{n_3}+\frac{T_4^2}{n_4}\right)-\frac{(\Sigma x)^2}{n}$$

$$= \left(\frac{(324)^2}{5}+\frac{(210)^2}{4}+\frac{(354)^2}{5}+\frac{(199)^2}{4}\right)-\frac{(1087)^2}{18}$$

$= 66,983.65 - 65,642.7222 = 1340.9278$

$$SSW = \Sigma x^2 - \left(\frac{T_1^2}{n_1}+\frac{T_2^2}{n_2}+\frac{T_3^2}{n_3}+\frac{T_4^2}{n_4}\right) = 68,013 - 66,983.65 = 1029.3500$$

$MSB = SSB/(k-1) = 1340.9278/3 = 446.9759$

$MSW = SSW/(n-k) = 1029.3500/14 = 73.5250$

$F = MSB/MSW = 446.9759/73.5250 = 6.08$

Step 5: Reject H_0 since $6.08 > 3.34$.

Conclude that the mean life of each of the four brands of batteries is not the same.

b. The Type I error would be to conclude that the mean life of all four brands of batteries are not equal when actually they are equal. The probability of making such an error is $\alpha = .05$.

12.25 a. Step 1: H_0: $\mu_1 = \mu_2 = \mu_3$, H_1: Not all three population means are equal.

Step 2: Since we are testing for equality of means among three groups, use the F distribution.

Step 3: $k = 3$, $n = n_1 + n_2 + n_3 = 5 + 5 + 5 = 15$

df for the numerator $= k - 1 = 3 - 1 = 2$,

df for the denominator $= n - k = 15 - 3 = 12$

For $\alpha = .01$, the critical value for F is 6.93.

Step 4: $T_1 = 5.3$, $T_2 = 3.6$, $T_3 = 5.1$, $\Sigma x = 14$, $\Sigma x^2 = 15.08$

$$SSB = \left(\frac{T_1^2}{n_1}+\frac{T_2^2}{n_2}+\frac{T_3^2}{n_3}\right)-\frac{(\Sigma x)^2}{n} = \left(\frac{(5.3)^2}{5}+\frac{(3.6)^2}{5}+\frac{(5.1)^2}{5}\right)-\frac{(14)^2}{15}$$

$= 13.412 - 13.0667 = .3453$

$$SSW = \Sigma x^2 - \left(\frac{T_1^2}{n_1}+\frac{T_2^2}{n_2}+\frac{T_3^2}{n_3}\right) = 15.08 - 13.412 = 1.6680$$

$MSB = SSB/(k-1) = .3453/2 = .1727$

$MSW = SSW/(n-k) = 1.6680/12 = .1390$

$F = MSB/MSW = .1727/.1390 = 1.24$

Step 5: Do not reject H_0 since $1.24 < 6.93$.

Conclude that the mean weight gained by all chickens is the same for each of the three diets.

b. By not rejecting H_0, we may have made a Type II error by concluding that the mean weight gained by all chickens is the same for each of the three diets is the same when in fact it is not.

12.27 a. Step 1: H_0: $\mu_1 = \mu_2 = \mu_3$, H_1: Not all three population means are equal.

Step 2: Since we are testing for equality of means among three groups, use the F distribution.

Step 3: $k = 3$, $n = n_1 + n_2 + n_3 = 5 + 5 + 5 = 15$

df for the numerator $= k - 1 = 3 - 1 = 2$,

df for the denominator $= n - k = 15 - 3 = 12$

For $\alpha = .05$, the critical value for F is 3.89.

Step 4: $T_1 = 521$, $T_2 = 412$, $T_3 = 464$, $\Sigma x = 1397$, $\Sigma x^2 = 132{,}755$

$$SSB = \left(\frac{T_1^2}{n_1} + \frac{T_2^2}{n_2} + \frac{T_3^2}{n_3}\right) - \frac{(\Sigma x)^2}{n} = \left(\frac{(521)^2}{5} + \frac{(412)^2}{5} + \frac{(464)^2}{5}\right) - \frac{(1397)^2}{15}$$

$= 131{,}296.2 - 130{,}107.2667 = 1188.9333$

$$SSW = \Sigma x^2 - \left(\frac{T_1^2}{n_1} + \frac{T_2^2}{n_2} + \frac{T_3^2}{n_3}\right) = 132{,}755 - 131{,}296.2 = 1458.8000$$

$MSB = SSB/(k - 1) = 1188.9333/2 = 594.4667$

$MSW = SSW/(n - k) = 1458.8000/12 = 121.5667$

$F = MSB/MSW = 594.4667/121.5667 = 4.89$

Step 5: Reject H_0 since $4.89 > 3.89$.

Conclude that the mean tips for the three restaurants are not equal.

b. If $\alpha = 0$, there can be no rejection region, and we cannot reject H_0. Therefore, the decision would be "do not reject H_0."

12.29 Step 1: H_0: $\mu_1 = \mu_2 = \mu_3 = \mu_4$, H_1: Not all four population means are equal.

Step 2: Since we are testing for equality of means among four groups, use the F distribution.

Step 3: $k = 4$, $n = n_1 + n_2 + n_3 + n_4 = 5 + 5 + 5 + 5 = 20$

df for the numerator $= k - 1 = 4 - 1 = 3$, df for the denominator $= n - k = 20 - 4 = 16$

For $\alpha = .01$, the critical value for F is 5.29.

Step 4: $n_1 = 5$, $\bar{x}_1 = 295$, $T_1 = 5(295) = 1475$

$n_2 = 5$, $\bar{x}_2 = 380$, $T_2 = 5(380) = 1900$

$n_3 = 5$, $\bar{x}_3 = 405$, $T_3 = 5(405) = 2025$

$n_4 = 5$, $\bar{x}_4 = 345$, $T_4 = 5(345) = 1725$

$\Sigma x = T_1 + T_2 + T_3 + T_4 = 7125$, $\Sigma x^2 = 2{,}890{,}000$

$$\text{SSB} = \left(\frac{T_1^2}{n_1} + \frac{T_2^2}{n_2} + \frac{T_3^2}{n_3} + \frac{T_4^2}{n_4}\right) - \frac{(\Sigma x)^2}{n}$$

$$= \left(\frac{(1475)^2}{5} + \frac{(1900)^2}{5} + \frac{(2025)^2}{5} + \frac{(1725)^2}{5}\right) - \frac{(7125)^2}{20}$$

$$= 2{,}572{,}375 - 2{,}538{,}281.25 = 34{,}093.75$$

$$\text{SSW} = \Sigma x^2 - \left(\frac{T_1^2}{n_1} + \frac{T_2^2}{n_2} + \frac{T_3^2}{n_3} + \frac{T_4^2}{n_4}\right) = 2{,}890{,}000 - 2{,}572{,}375 = 317{,}625$$

MSB = SSB/$(k - 1)$ = 34,093.75/3 = 11,364.5833

MSW = SSW/$(n - k)$ = 317,625/16 = 19,851.5625

F = MSB/MSW = 11,364.5833/19,851.5625 = .57

Step 5: Do not reject H_0 since $.57 < 5.29$.

Conclude that the mean revenue is the same for all four days of the week.

12.31 a. Lay out a route for a test run in a typical city. Let each driver drive each car several times on this route. Make sure that the length of the route is selected in such a way that the total of all tests for each car does not exceed 500 miles. Then, calculate the gas mileage for each of these test runs. Next, compute the mean gas mileage for each of the three cars and compare them in the article.

b. Using the ANOVA procedure and the data collected in part a for three cars, test the null hypothesis that the mean gas mileage is the same for all three cars against the alternative hypothesis that all three means are not equal. Select your own significance level. Then, write a report explaining and interpreting these results.

12.33 a. Since the *df* for the between source of variation is equal to $k - 1 = 4$, there are 5 groups. The within degrees of freedom are equal to $n - k = 45$, so $n - 5 = 45$ gives $n = 50$. Dividing these observations equally between the 5 groups yields $n_1 = n_2 = n_3 = n_4 = n_5 = 10$.

b. Technology is used to solve this problem, as the maximum *df* for the denominator in the F distribution table for $\alpha = .05$ is 100. If we try $n = 105$ so that *df* for the denominator = 100, the value of the test statistic is calculated as follows: MSB = 200/4 = 50, MSW = 3547/100 = 35.47, and F = MSB/MSW = 50/35.47 = 1.41. This is not significant at the 5% significance level since the critical value of F with $df = (4,100) = 2.46$. Hence, each group size must be larger than 21. Utilizing Excel to calculate p-values for group sizes from 30 to 40 yielded the following table. Note that MSB does not change as we change the sample size.

Group size	*df* for denominator	MSW	*F*	*p*-value
30	145	24.4621	2.04	.091
31	150	23.6467	2.11	.082
32	155	22.8839	2.18	.073
33	160	22.1688	2.26	.065
34	165	21.4970	2.33	.059
35	170	20.8647	2.40	.052
36	175	20.2686	2.47	.047
37	180	19.7056	2.54	.042
38	185	19.1730	2.61	.037
39	190	18.6684	2.68	.033
40	195	18.1897	2.75	.030

From the above table, we see that the *p*-value drops below .05 when *df* for the denominator = 175. Hence, $n = 180$, and $n_1 = n_2 = n_3 = n_4 = n_5 = 36$.

Self-Review Test

1. a **2.** b **3.** c **4.** a **5.** a **6.** a **7.** b **8.** a

9. See solution to Exercise 12.11.

10. a. Step 1: H_0: $\mu_1 = \mu_2 = \mu_3 = \mu_4$, H_1: Not all four population means are equal.

Step 2: Since we are testing for equality of means among four groups, use the *F* distribution.

Sep 3: $k = 4$, $n = n_1 + n_2 + n_3 + n_4 = 6 + 6 + 6 + 6 = 24$

df for the numerator $= k - 1 = 4 - 1 = 3$, *df* for the denominator $= n - k = 24 - 4 = 20$

For $\alpha = .05$, the critical value for *F* is 3.10.

Step 4: $T_1 = 124.1$, $T_2 = 152.1$ $T_3 = 156.2$, $T_4 = 152$, $\Sigma x = 584.4$, $\Sigma x^2 = 14{,}503.26$

$$\text{SSB} = \left(\frac{T_1^2}{n_1} + \frac{T_2^2}{n_2} + \frac{T_3^2}{n_3} + \frac{T_4^2}{n_4}\right) - \frac{(\Sigma x)^2}{n}$$

$$= \left(\frac{(124.1)^2}{6} + \frac{(152.1)^2}{6} + \frac{(156.2)^2}{6} + \frac{(152)^2}{6}\right) - \frac{(584.4)^2}{24}$$

$$= 14{,}339.61 - 14{,}230.14 = 109.4700$$

$$\text{SSW} = \Sigma x^2 - \left(\frac{T_1^2}{n_1} + \frac{T_2^2}{n_2} + \frac{T_3^2}{n_3} + \frac{T_4^2}{n_4}\right) = 14{,}503.26 - 14{,}339.61 = 163.6500$$

$\text{MSB} = \text{SSB}/(k - 1) = 109.4700/3 = 36.4900$

$\text{MSW} = \text{SSW}/(n - k) = 163.6500/20 = 8.1825$

$F = \text{MSB}/\text{MSW} = 36.4900/8.18250 = 4.46$

Step 5: Reject H_0 since $4.46 > 3.10$.

Conclude that the mean prices for all four pizza parlors are not the same.

b. By rejecting H_0, we may have committed a Type I error.

Chapter Thirteen

Section 13.1 – 13.2

13.1 A regression model that includes only two variables, one independent and one dependent, is called a **simple regression model**. The **dependent variable** is the one being explained, and the **independent variable** is the one used to explain the variation in the dependent variable. A (simple) regression model that gives a straight–line relationship between two variables is called a (simple) **linear regression model**.

13.3 In an **exact relationship**, the value of the dependent variable y is determined exactly by the independent variable x; that is, for a given value of x there is a unique value of y. In a **nonexact relationship**, there are many (perhaps infinitely many) values of y for a given value of x.

13.5 A **simple regression model** has only one independent variable, while a **multiple regression model** has more than one independent variable. Both models have just one dependent variable.

13.7 The **random error term** ϵ is included in a regression model to capture the effect of all the **missing or omitted variables** which have not been included in the model and to represent **random variation**.

13.9 **SSE** denotes the **error sum of squares**, which is the sum of squared differences between the actual and predicted values of y, SSE = $\Sigma(y - \hat{y})^2$. SSE represents the portion of the variation in y that is not explained by the regression model.

13.11 When x and y have a **positive linear relationship**, y increases as x increases.

13.13
a. A regression line obtained by using the population data is called the **population regression line**. It gives the true values of A and B and is written as $\mu_{y|x} = A + Bx$.
b. A **sample regression line** is obtained from sample data. It uses estimated values, a and b, and is written as $\hat{y} = a + bx$. Here, a is an estimate of A, and b is an estimate of B.

c. The **true values of *A* and *B*** are the values obtained from the population regression line. They are the **population parameters.**

d. The **estimated values of *A* and *B*** are the values in a regression model that is obtained by using the sample data. Such an estimated model is written as $\hat{y} = a + bx$, where a is an estimate of A, and b is an estimate of B.

13.15 a.

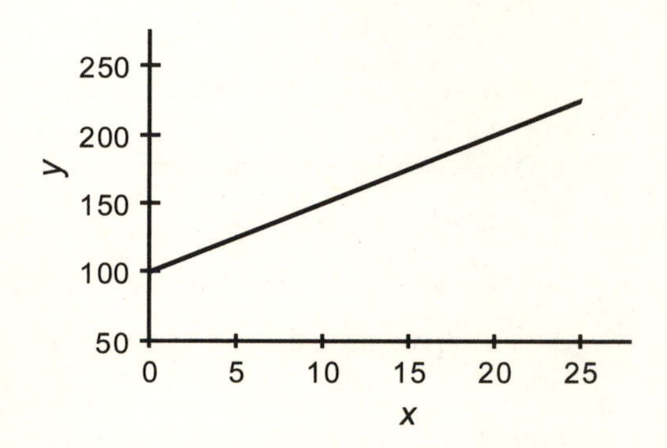

The y–intercept is 100, which is the value for y when $x = 0$. The slope is 5, which means that for a one unit increase in x, there will be a five unit increase in y. Since the slope has the positive value of 5, there is a positive relationship between x and y. **(Note that the vertical axis in the graph is truncated as it starts at 50. This will be true of almost all graphs in this chapter.)**

b.

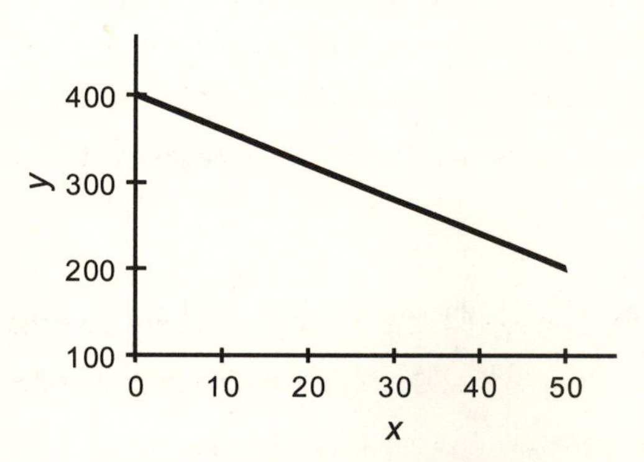

The y–intercept is 400, which is the value for y when $x = 0$. The slope is –4, which means that for a one unit increase in x, there will be a four unit decrease in y. Since the slope has the negative value of –4, there is a negative relationship between x and y.

13.17 $SS_{xx} = \sum x^2 - \frac{(\sum x)^2}{N} = 485{,}870 - \frac{(9880)^2}{250} = 95{,}412.4$

$SS_{xy} = \sum xy - \frac{(\Sigma x)(\Sigma y)}{N} = 85{,}080 - \frac{(9880)(1456)}{250} = 27{,}538.88$

$\mu_x = \Sigma x/N = 9880/250 = 39.52$, $\mu_y = \Sigma y/N = 1456/250 = 5.824$

$B = SS_{xy}/SS_{xx} = 27{,}538.88/95{,}412.4 = .2886$

$A = \mu_y - B\mu_x = 5.824 - (.2886)(39.52) = -5.5815$

$\mu_{y|x} = -5.5815 + .2886\,x$

13.19 $SS_{xx} = \sum x^2 - \frac{(\sum x)^2}{n} = 1140 - \frac{(100)^2}{10} = 140$

$SS_{xy} = \sum xy - \frac{(\Sigma x)(\Sigma y)}{n} = 3680 - \frac{(100)(220)}{10} = 1480$

$\bar{x} = \Sigma x/n = 100/10 = 10$, $\bar{y} = \Sigma y/n = 220/10 = 22$

$b = SS_{xy}/SS_{xx} = 1480/140 = 10.5714$

$a = \bar{y} - b\bar{x} = 22 - (10.5714)(10) = -83.7140$

$\hat{y} = -83.7140 + 10.5714x$

13.21
a. For $x = 100$, $y = 40 + .20(100) = 60$
b. Every person who rents a car from this agency for one day and drives it 100 miles will pay the same amount, \$60. This is due to the fact that for any value x, the equation $y = 40 + .20x$ yields a unique value of y.
c. The relationship is exact.

13.23
a. For $x = 2$, $\hat{y} = 3.6 + 11.75(2) = 27.1$
b. The four companies that spent \$2 million each on advertising would not have the same actual gross sales for 2005. The \$27.1 million obtained in part a is the mean gross sales for companies spending \$2 million on advertising. The actual gross sales would differ due to the influence of variables not included in the model and random variation.
c. The relationship is nonexact.

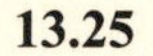

13.25 Let x = age of a car (in years) and y = price of a car (in hundreds of dollars).

a. & d.

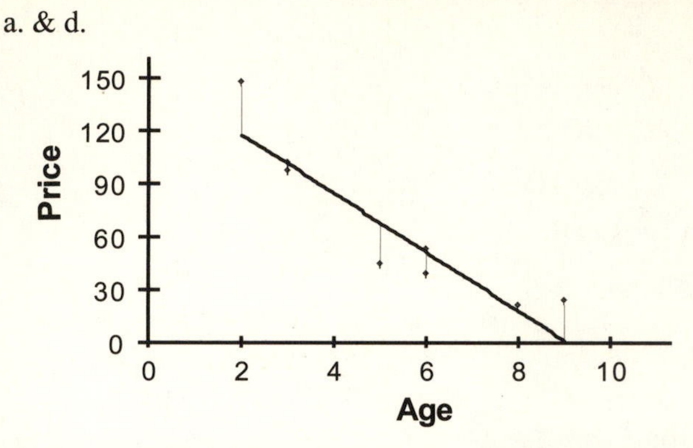

The scatter diagram exhibits a linear relationship between ages and prices of cars.

b. $n = 8$, $\Sigma x = 42$, $\Sigma y = 505$, $\Sigma x^2 = 264$, $\Sigma xy = 1928$

$\bar{x} = \Sigma x/n = 42/8 = 5.25$, $\bar{y} = \Sigma y/n = 505/8 = 63.125$

$$SS_{xx} = \sum x^2 - \frac{(\sum x)^2}{n} = 264 - \frac{(42)^2}{8} = 43.5$$

$$SS_{xy} = \sum xy - \frac{(\Sigma x)(\Sigma y)}{n} = 1928 - \frac{(42)(505)}{8} = -723.25$$

$b = SS_{xy}/SS_{xx} = -723.25/43.5 = -16.6264$

$a = \bar{y} - b\bar{x} = 63.125 - (-16.6264)(5.25) = 150.4136$

$\hat{y} = 150.4136 - 16.6264x$

c. The value of $a = 150.4136$ represents the price of a new car (in hundreds of dollars). The value of $b = -16.6264$ means that, on average, for every one year increase in the age of a car, its price decreases by \$1663.

e. For $x = 7$, $\hat{y} = 150.4136 - 16.6264(7) = 34.0288$. Thus, the price of a 7-year-old car is \$3403.

f. For $x = 18$, $\hat{y} = 150.4136 - 16.6264(18) = -148.8616$. The negative price makes no sense. The regression line is based on data for cars from 2 to 8 years in age. Since $x = 18$ is outside this range, the estimate is invalid.

13.27 Let x = annual income (in thousands of dollars) and y = amount of life insurance policy (in thousands of dollars).

a. & d.

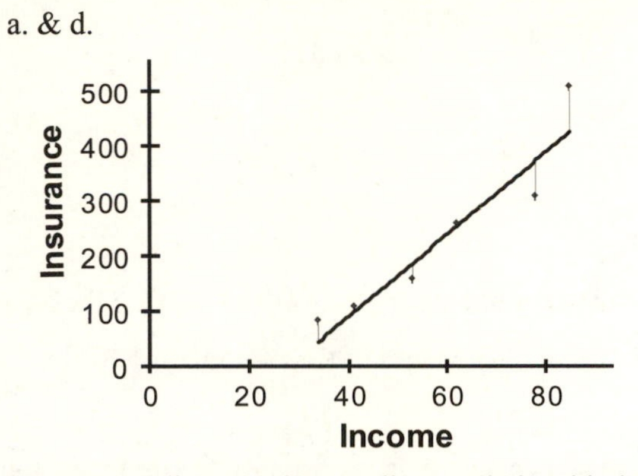

The scatter diagram shows a linear relationship between the annual incomes and amounts of life insurance.

b. $n = 6$, $\Sigma x = 353$, $\Sigma y = 1375$, $\Sigma x^2 = 22{,}799$, $\Sigma xy = 96{,}000$

$\bar{x} = \Sigma x/n = 353/6 = 58.8333$, $\bar{y} = \Sigma y/n = 1375/6 = 229.1667$

$$SS_{xx} = \sum x^2 - \frac{(\sum x)^2}{n} = 22{,}799 - \frac{(353)^2}{6} = 2030.8333$$

$$SS_{xy} = \sum xy - \frac{(\Sigma x)(\Sigma y)}{n} = 96{,}000 - \frac{(353)(1375)}{6} = 15{,}104.1667$$

$b = SS_{xy}/SS_{xx} = 15{,}104.1667/2030.8333 = 7.4374$

$a = \bar{y} - b\bar{x} = 229.1667 - 7.4374(58.8333) = -208.4001$

$\hat{y} = -208.4001 + 7.4374x$

c. The value of $a = -208.4001$ represents the amount of life insurance for a person with a zero income. This negative dollar amount makes no sense. The value $x = 0$ is outside the range of the data ($34,000 to $85,000). The value of $b = 7.4374$ means that, on average, the amount of life insurance increases by $7437 for every $1000 increase in the annual income of a person.

e. For $x = 55$, $\hat{y} = -208.4001 + 7.4374(55) = 200.6569$ or $200,656.90. Thus, the estimated value of life insurance for a person with an annual income of $55,000 is $200,656.90.

f. For $x = 78$, $\hat{y} = -208.4001 + 7.4374(78) = 371.7171$ or $371,717.10.

$e = y - \hat{y} = \$300{,}000 - \$371{,}717.10 = -\$71{,}717.10$

13.29 Let x = total payroll (in millions of dollars) and y = percentage of games won.

a. $N = 16$, $\Sigma x = 1084$, $\Sigma y = 799.3$, $\Sigma x^2 = 81{,}938$, $\Sigma xy = 55{,}640.6$

$\mu_x = \Sigma x/N = 1084/16 = 67.75$, $\mu_y = \Sigma y/N = 799.3/16 = 49.9563$

$$SS_{xx} = \sum x^2 - \frac{(\sum x)^2}{N} = 81{,}938 - \frac{(1084)^2}{16} = 8497$$

$$SS_{xy} = \sum xy - \frac{(\Sigma x)(\Sigma y)}{N} = 55{,}640.6 - \frac{(1084)(799.3)}{16} = 1488.025$$

$B = SS_{xy}/SS_{xx} = 1488.025/8497 = .1751$

$A = \mu_y - B\mu_x = 49.9563 - (.1751)(67.75) = 38.0933$

$\mu_{y|x} = 38.0933 + .1751x$

b. The regression line obtained in part a is the population regression line because the data are given for all National League baseball teams. The values of the y–intercept and slope obtained above are those of A and B.

c. The value of $A = 38.0933$ represents the percentage of games won by a team with a total payroll of zero dollars. The value of $B = .1751$ means that, on average, the percentage of games won increases by 17.51% for every \$1 million increase in payroll of a National League baseball team.

d. For $x = 55$, $\mu_{y|x} = 38.0933 + .1751(55) = 47.7238$. Thus, a team with a total payroll of \$55 million is expected to win about 47.72% of its games.

Sections 13.3 – 13.4

13.31 For a simple linear regression model, $df = n - 2$.

13.33 **SST** is the sum of squared differences between the actual y values and $\bar{y}$, SST = $\Sigma(y - \bar{y})^2$. **SSR** is the portion of SST that is explained by the use of the regression model.

13.35 From the solution to Exercise 13.18: $SS_{xy} = 3987.3913$ and $B = .2636$.

$$SS_{yy} = \sum y^2 - \frac{(\sum y)^2}{N} = 39{,}347 - \frac{(2650)^2}{460} = 24{,}080.6957$$

$$\sigma_\epsilon = \sqrt{\frac{SS_{yy} - B(SS_{xy})}{N}} = \sqrt{\frac{24{,}080.6957 - (.2636)(3987.3913)}{460}} = 7.0756$$

$\rho^2 = BSS_{xy}/SS_{yy} = (.2636)(3987.3913)/24{,}080.6957 = .04$

13.37 From the solution to Exercise 13.20: $SS_{xy} = -990$, $SS_{xx} = 33$, and $b = -30$.

$$SS_{yy} = \sum y^2 - \frac{(\sum y)^2}{n} = 58{,}734 - \frac{(588)^2}{12} = 29{,}922$$

$$s_e = \sqrt{\frac{SS_{yy} - b(SS_{xy})}{n-2}} = \sqrt{\frac{29{,}922 - (-30)(-990)}{12-2}} = 4.7117$$

$r^2 = bSS_{xy}/SS_{yy} = (-30)(-990)/29{,}922 = .99$

13.39 Let x = fat consumption (in grams) per day and y = cholesterol level (in milligrams per hundred milliliters).

a. $n = 8$, $\Sigma x = 421$, $\Sigma y = 1514$, $\Sigma x^2 = 23{,}743$, $\Sigma y^2 = 292{,}116$, $\Sigma xy = 82{,}517$

$\bar{x} = \Sigma x/n = 421/8 = 52.625$, $\bar{y} = \Sigma y/n = 1514/8 = 189.25$

$$SS_{xx} = \sum x^2 - \frac{(\sum x)^2}{n} = 23{,}743 - \frac{(421)^2}{8} = 1587.875$$

$$SS_{yy} = \sum y^2 - \frac{(\sum y)^2}{n} = 292{,}116 - \frac{(1514)^2}{8} = 5591.5$$

$$SS_{xy} = \sum xy - \frac{(\Sigma x)(\Sigma y)}{n} = 82{,}517 - \frac{(421)(1514)}{8} = 2842.75$$

b. $b = SS_{xy}/SS_{xx} = 2842.75/1587.875 = 1.7903$

$$s_e = \sqrt{\frac{SS_{yy} - b(SS_{xy})}{n-2}} = \sqrt{\frac{5591.5 - (1.7903)(2842.75)}{8-2}} = 9.1481$$

c. $a = \bar{y} - b\bar{x} = 189.25 - (1.7903)(52.625) = 95.0355$

$\hat{y} = 95.0355 - 1.7903x$

x	y	$\hat{y} = 95.0355 - 1.7903x$	$e = y - \hat{y}$	e^2
55	180	193.5020	−13.5020	182.3040
68	215	216.7759	−1.7759	3.1538
50	195	184.5505	10.4495	109.1921
34	165	155.9057	9.0943	82.7063
43	170	172.0184	−2.0184	4.0739
58	204	198.8729	5.1271	26.2872
77	235	232.8886	2.1114	4.4580
36	150	159.4863	−9.4863	89.9899
				$\Sigma e^2 = 502.1652$

SST = SS_{yy} = 5591.5, SSE = Σe^2 = 502.1652

SSR = SST – SSE = 5591.5 – 502.1652 = 5089.3348

d. $r^2 = bSS_{xy}/SS_{yy} = (1.7903)(2842.75)/5591.5 = .91$

13.41 Let x = lowest temperature and y = number of calls.

a. From the solution to Exercise 13.26: $n = 7$, $\Sigma y = 118$, $SS_{xy} = -857.1429$, and $b = -.5249$.

$\Sigma y^2 = 2506$

$$SS_{yy} = \sum y^2 - \frac{(\sum y)^2}{n} = 2506 - \frac{(118)^2}{7} = 516.8571$$

$$s_e = \sqrt{\frac{SS_{yy} - b(SS_{xy})}{n-2}} = \sqrt{\frac{516.8571-(-.5249)(-857.1429)}{7-2}} = 3.659$$

b. $r^2 = bSS_{xy}/SS_{yy} = (-.5249)(-857.1429)/516.8571 = .87$

Thus, 87% of the total squared errors (SST) is explained by the regression model.

13.43 Let x = size of a house (in hundreds of square feet) and y = monthly rent (in dollars).

a. From the solution to Exercise 13.28, $n = 6$, $\Sigma y = 7450$, $SS_{xy} = 13{,}311.6667$, and $b = 51.8300$.

$\Sigma y^2 = 9{,}975{,}300$

$$SS_{yy} = \sum y^2 - \frac{(\sum y)^2}{n} = 9{,}975{,}300 - \frac{(7450)^2}{6} = 724{,}883.3333$$

$$s_e = \sqrt{\frac{SS_{yy} - b(SS_{xy})}{n-2}} = \sqrt{\frac{724{,}883.3333-(51.8300)(13{,}311.6667)}{6-2}} = 93.4608$$

b. $r^2 = bSS_{xy}/SS_{yy} = (51.8300)(13{,}311.6667)/724{,}883.3333 = .95$

Thus, 95% of the variation in monthly rents is explained by the sizes of the houses, and 5% is not explained.

13.45 Let x = total payroll (in millions of dollars) and y = percentage of games won.

a. From the solution to Exercise 13.30: $N = 14$, $\Sigma y = 700.4$, $SS_{xy} = 2746.5714$, and $B = .1215$.

$\Sigma y^2 = 35{,}930.9$

$$SS_{yy} = \sum y^2 - \frac{(\sum y)^2}{N} = 35{,}930.9 - \frac{(700.4)^2}{14} = 890.8886$$

$$\sigma_\epsilon = \sqrt{\frac{SS_{yy} - B(SS_{xy})}{N}} = \sqrt{\frac{890.8886-(.1215)(2746.5714)}{14}} = 6.3086$$

b. $\rho^2 = BSS_{xy}/SS_{yy} = .1215(2746.5714)/890.8886 = .37$

Section 13.5

13.47 a. $b = 6.32$, $s_b = s_e/\sqrt{SS_{xx}} = 1.951/\sqrt{340.700} = .1057$, $df = n - 2 = 16 - 2 = 14$

The 99% confidence interval for B is

$b \pm ts_b = 6.32 \pm (2.977)(.1057) = 6.32 \pm .31 = 6.01$ to 6.63

b. Step 1: H_0: $B = 0$, H_1: $B > 0$

Step 2: Since σ_ϵ is unknown, use the t distribution.

Step 3: For $\alpha = .05$ with $df = 14$, the critical value of t is 2.145.

Step 4: $t = (b - B)/s_b = (6.32 - 0)/.1057 = 59.792$

Step 5: Reject H_0 since $59.792 > 2.145$.

Conclude that B is positive.

c. Step 1: H_0: $B = 0$, H_1: $B \neq 0$

Step 2: Since σ_ϵ is unknown, use the t distribution.

Step 3: For $\alpha = .01$ with $df = 14$, the critical values of t are –2.977 and 2.977.

Step 4: $t = (b - B)/s_b = (6.32 - 0)/.1057 = 59.792$

Step 5: Reject H_0 since $59.792 > 2.977$.

Conclude that B is different from zero.

d. Step 1: H_0: $B = 4.50$, H_1: $B \neq 4.50$

Step 2: Since σ_ϵ is unknown, use the t distribution.

Step 3: For $\alpha = .02$ with $df = 14$, the critical values of t are –2.624 and 2.624.

Step 4: $t = (b - B)/s_b = (6.32 - 4.50)/.1057 = 17.219$

Step 5: Reject H_0 since $17.219 > 2.624$.

Conclude that B is different from 4.50.

13.49 a. $b = 2.50$, $s_b = s_e / \sqrt{SS_{xx}} = 1.464/\sqrt{524.884} = .0639$

Since $n = 100$ is very large, we use the normal distribution to approximate the t distribution.

The 98% confidence interval for B is

$b \pm z s_b = 2.50 \pm (2.33)(.0639) = 2.50 \pm .15 = 2.35$ to 2.65

b. Step 1: H_0: $B = 0$, H_1: $B > 0$

Step 2: Since σ_ϵ is unknown, use the t distribution. However, since $n = 100$ is very large, we use the normal distribution to approximate the t distribution.

Step 3: For $\alpha = .02$, the critical value of z is 2.05.

Step 4: $z = (b - B)/s_b = (2.50 - 0)/.0639 = 39.12$

Step 5: Reject H_0 since $39.12 > 2.05$.

Conclude that B is positive.

c. Step 1: H_0: $B = 0$, H_1: $B \neq 0$

Step 2: Since σ_ϵ is unknown, use the t distribution. However, since $n = 100$ is very large, we use the normal distribution to approximate the t distribution.

Step 3: For $\alpha = .01$, the critical values of z are –2.58 and 2.58.

Step 4: $z = (b - B)/s_b = (2.50 - 0)/.0639 = 39.12$

Step 5: Reject H_0 since $39.12 > 2.58$.

Conclude that B is different from zero.

d. Step 1: H_0: $B = 1.75$, H_1: $B > 1.75$

Step 2: Since σ_ϵ is unknown, use the t distribution. However, since $n = 100$ is very large, we use the normal distribution to approximate the t distribution.

Step 3: For $\alpha = .01$, the critical value of z is 2.33.

Step 4: $z = (b - B)/s_b = (2.50 - 1.75)/.0639 = 11.74$

Step 5: Reject H_0 since $11.74 > 2.33$.

Conclude that B is greater than 1.75.

13.51 Let x = age of a car (in years) and y = price of a car (in hundreds of dollars).

From the solutions to Exercises 13.25 and 13.40: $n = 8$, $SS_{xx} = 43.5$, $b = -16.6264$, and $s_e = 18.6361$.

$s_b = s_e / \sqrt{SS_{xx}} = 18.6361/\sqrt{43.5} = 2.8256$

a. $df = n - 2 = 8 - 2 = 6$

The 95% confidence interval for B is

$b \pm ts_b = -16.6264 \pm (2.447)(2.8256) = -16.6264 \pm 6.9142 = -23.5406$ to -9.7122

b. Step 1: H_0: $B = 0$, H_1: $B < 0$

Step 2: Since σ_ϵ is unknown, use the t distribution.

Step 3: For $\alpha = .05$ with $df = 6$, the critical value of t is -1.943.

Step 4: $t = (b - B)/s_b = (-16.6264 - 0)/2.8256 = -5.884$

Step 5: Reject H_0 since $-5.884 < -1.943$.

Conclude that B is negative.

13.53 Let x = years of experience and y = monthly salary.

$n = 9$, $\Sigma x = 80$, $\Sigma y = 318$, $\Sigma x^2 = 968$, $\Sigma y^2 = 11{,}710$, $\Sigma xy = 3162$

$\bar{x} = \Sigma x/n = 80/9 = 8.8889$, $\bar{y} = \Sigma y/n = 318/9 = 35.3333$

$$SS_{xx} = \sum x^2 - \frac{(\sum x)^2}{n} = 968 - \frac{(80)^2}{9} = 256.8889$$

$$SS_{yy} = \sum y^2 - \frac{(\sum y)^2}{n} = 11{,}710 - \frac{(318)^2}{9} = 474$$

$$SS_{xy} = \sum xy - \frac{(\Sigma x)(\Sigma y)}{n} = 3162 - \frac{(80)(318)}{9} = 335.3333$$

a. $b = SS_{xy}/SS_{xx} = 335.3333/256.8889 = 1.3054$

$a = \bar{y} - b\bar{x} = 35.3333 - (1.3054)(8.8889) = 23.7297$

$\hat{y} = 23.7297 + 1.3054x$

b. $s_e = \sqrt{\dfrac{SS_{yy} - b(SS_{xy})}{n-2}} = \sqrt{\dfrac{474 - (1.3054)(335.3333)}{9-2}} = 2.2758$

$s_b = s_e / \sqrt{SS_{xx}} = 2.2758 / \sqrt{256.8889} = .1420$

$df = n - 2 = 9 - 2 = 7$

The 98% confidence interval for B is

$b \pm ts_b = 1.3054 \pm (2.998)(.1420) = 1.3054 \pm .4257 = .8797$ to 1.7311

c. Step 1: H_0: $B = 0$, H_1: $B > 0$

Step 2: Since σ_ϵ is unknown, use the t distribution.

Step 3: For $\alpha = .025$ with $df = 7$, the critical value of t is 2.365.

Step 4: $t = (b - B)/s_b = (1.3054 - 0)/.1420 = 9.193$

Step 5: Reject H_0 since $9.193 > 2.365$.

Conclude that B is greater than zero.

13.55 Let x = annual income (in thousands of dollars) and y = amount of life insurance policy (in thousands of dollars).

From the solutions to Exercises 13.27 and 13.42: $n = 6$, $SS_{xx} = 2030.8333$, $b = 7.4374$, and $s_e = 57.4132$.

$s_b = s_e / \sqrt{SS_{xx}} = 57.4132 / \sqrt{2030.8333} = 1.2740$

a. $df = n - 2 = 6 - 2 = 4$

The 99% confidence interval for B is

$b \pm ts_b = 7.4374 \pm 4.604(1.2740) = 7.4374 \pm 5.8655 = 1.5719$ to 13.3029

b. Step 1: H_0: $B = 0$, H_1: $B \neq 0$

Step 2: Since σ_ϵ is unknown, use the t distribution.

Step 3: For $\alpha = .01$ with $df = 4$, the critical values of t are –4.604 and 4.604.

Step 4: $t = (b - B)/s_b = (7.4374 - 0)/1.2740 = 5.838$

Step 5: Reject H_0 since $5.838 > 4.604$.

Conclude that B is different from zero.

13.57 Let x = hours worked per week and y = GPA.

From the solution to Exercise 13.38: $n = 7$, $SS_{xx} = 181.4286$, $a = 4.4948$, $b = -.1019$, and $s_e = .3615$.

$s_b = s_e / \sqrt{SS_{xx}} = .3615/\sqrt{181.4286} = .0268$

a. $\hat{y} = 4.4948 - .1019x$

b. $df = n - 2 = 7 - 2 = 5$

The 95% confidence interval for B is

$b \pm ts_b = -.1019 \pm 2.571(.0268) = -.1019 \pm .0689 = -.1708$ to $-.0330$

c. Step 1: H_0: $B = -.04$, H_1: $B < -.04$

Step 2: Since σ_ϵ is unknown, use the t distribution.

Step 3: For $\alpha = .05$ with $df = 5$, the critical value of t is -2.015.

Step 4: $t = (b - B)/s_b = [-.1019 - (-.04)]/.0268 = -2.310$

Step 5: Reject H_0 since $-2.310 < -2.015$.

Conclude that B is less than $-.04$.

Section 13.6

13.59 The **linear correlation coefficient** measures the strength of the linear association between two variables. Its value always lies in the range –1 to 1.

13.61
a. **Perfect positive linear correlation** occurs when all the points in the scatter diagram lie on a straight line with positive slope. In this case, $r = 1$. See Figure 13–18(a) in the text.
b. **Perfect negative linear correlation** occurs when all the points in the scatter diagram lie on a straight line with negative slope. In this case, $r = -1$. See Figure 13–18(b) in the text.
c. If the correlation between two variables is positive and close to 1, they are said to have a **strong positive correlation**. See Figure 13–19(a) in the text.
d. If the correlation between two variables is negative and close to –1, they are said to have a **strong negative correlation**. See Figure 13–19(c) in the text.
e. If the correlation between two variables is positive and close to zero, they are said to have a **weak positive correlation**. See Figure 13–19(b) in the text.
f. If the correlation between two variables is negative and close to zero, they are said to have a **weak negative correlation**. See Figure 13–19(d) in the text.
g. If the data points are scattered all over the diagram (hence, r is close to zero) there is **no linear correlation** between the variables. See Figure 13–18(c) in the text.

13.63 The answer is **a**, because r and b always have the same sign for a given sample.

13.65 The linear correlation coefficient r measures only linear relationships. Thus, r may be zero and the variables might still have a nonlinear relationship.

13.67
a. Positive
b. Postive
c. Positive
d. Negative
e. Zero

13.69 From the solution to Exercise 13.35: $SS_{xx} = 15{,}124.7826$, $SS_{yy} = 24{,}080.6957$, and $SS_{xy} =$ 3987.3913.

$$\rho = \frac{SS_{xy}}{\sqrt{SS_{xx}SS_{yy}}} = \frac{3987.3913}{\sqrt{(15{,}124.7826)(24{,}080.6957)}} = .21$$

13.71
a. From the solution to Exercise 13.37: $SS_{xx} = 33$, $SS_{yy} = 29{,}922$, and $SS_{xy} = -990$.

$$r = \frac{SS_{xy}}{\sqrt{SS_{xx}SS_{yy}}} = \frac{-990}{\sqrt{(33)(29{,}922)}} = -.996$$

b. Step 1: H_0: $\rho = 0$, H_1: $\rho < 0$

Step 2: Assuming the population distributions for both variables are normally distributed, use the t distribution.

Step 3: $df = n - 2 = 12 - 2 = 10$

For $\alpha = .01$, the critical value of t is -2.764.

Step 4: $t = r\sqrt{\dfrac{n-2}{1-r^2}} = -.996\sqrt{\dfrac{12-2}{1-(-.996)^2}} = -35.249$

Step 5: Reject H_0 since $-35.249 < -2.764$.

Conclude that ρ is negative.

13.73 Let x = years of experience and y = monthly salary.

a. We expect experience and monthly salaries to be positively related because, on average, more experienced secretaries command higher salaries.

b. From the solution to Exercise 13.53: $SS_{xx} = 256.8889$, $SS_{yy} = 474$, and $SS_{xy} = 335.3333$.

$$r = \frac{SS_{xy}}{\sqrt{SS_{xx}SS_{yy}}} = \frac{335.3333}{\sqrt{(256.8889)(474)}} = .96$$

c. Step 1: H_0: $\rho = 0$, H_1: $\rho > 0$

Step 2: Assuming the population distributions for both variables are normally distributed, use the t distribution.

Step 3: $df = n - 2 = 9 - 2 = 7$

For $\alpha = .05$, the critical value of t is 1.895.

Step 4: $t = r\sqrt{\dfrac{n-2}{1-r^2}} = .96\sqrt{\dfrac{9-2}{1-(.96)^2}} = 9.071$

Step 5: Reject H_0 since 9.071 > 1.895.

Conclude that ρ is positive.

13.75 Let x = husband's age and y = wife's age.

a. We expect the ages of husbands and wives to be positively correlated because, on average, a younger husband will have a younger wife and an older husband will have an older wife.

b.

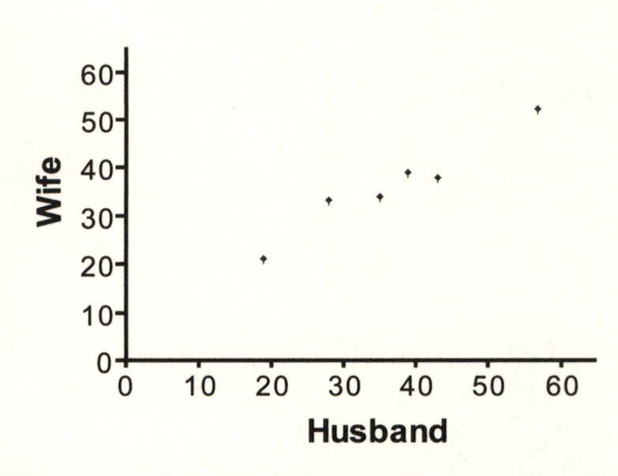

We expect the correlation coefficient to be close to 1 because the points in the scatter diagram show a very strong positive correlation.

c. $n = 6$, $\Sigma x = 221$, $\Sigma y = 211$, $\Sigma x^2 = 8989$, $\Sigma y^2 = 7927$, $\Sigma xy = 8411$

$$SS_{xx} = \sum x^2 - \frac{(\sum x)^2}{n} = 8989 - \frac{(221)^2}{6} = 848.8333$$

$$SS_{yy} = \sum y^2 - \frac{(\sum y)^2}{n} = 7927 - \frac{(211)^2}{6} = 506.8333$$

$$SS_{xy} = \sum xy - \frac{(\Sigma x)(\Sigma y)}{n} = 8411 - \frac{(221)(211)}{6} = 639.1667$$

$$r = \frac{SS_{xy}}{\sqrt{SS_{xx}SS_{yy}}} = \frac{639.1667}{\sqrt{(848.8333)(506.8333)}} = .97$$

This value of r is consistent with what we expected in parts a and b.

d. Step 1: H_0: $\rho = 0$, H_1: $\rho \neq 0$

Step 2: Assuming the population distributions for both variables are normally distributed, use the t distribution.

Step 3: $df = n - 2 = 6 - 2 = 4$

For $\alpha = .05$, the critical values of t are –2.776 and 2.776.

Step 4: $t = r\sqrt{\dfrac{n-2}{1-r^2}} = .97\sqrt{\dfrac{6-2}{1-(.97)^2}} = 7.980$

Step 5: Reject H_0 since 7.980 > 2.776.

Conclude that ρ is different from zero.

13.77 Let x = fat consumption (in grams) per day and y = cholesterol level (in milligrams per hundred milliliters).

a. From the solution to Exercise 13.39: $SS_{xx} = 1587.875$, $SS_{yy} = 5591.5$, and $SS_{xy} = 2842.75$.

$$r = \frac{SS_{xy}}{\sqrt{SS_{xx}SS_{yy}}} = \frac{2842.75}{\sqrt{(1587.875)(5591.5)}} = .95$$

The sign of b calculated in Exercise 13.58 is also positive.

b. Step 1: H_0: $\rho = 0$, H_1: $\rho \neq 0$

Step 2: Assuming the population distributions for both variables are normally distributed, use the t distribution.

Step 3: $df = n - 2 = 8 - 2 = 6$

For $\alpha = .01$, the critical values of t are –3.707 and 3.707.

Step 4: $t = r\sqrt{\dfrac{n-2}{1-r^2}} = .95\sqrt{\dfrac{8-2}{1-(.95)^2}} = 7.452$

Step 5: Reject H_0 since 7.452 > 3.707.

Conclude that ρ is different from zero.

13.79 Let x = total payroll (in millions of dollars) and y = percentage of games won.

From the solutions to Exercises 13.30 and 13.45: $SS_{xx} = 22{,}605.4286$, $SS_{yy} = 890.8886$, and $SS_{xy} = 2746.5714$.

$$\rho = \frac{SS_{xy}}{\sqrt{SS_{xx}SS_{yy}}} = \frac{2746.5714}{\sqrt{(22{,}605.4286)(890.8886)}} = .61$$

Section 13.7

13.81 Let x = age of man and y = cholesterol level.

a. $n = 10$, $\Sigma x = 512$, $\Sigma y = 1896$, $\Sigma x^2 = 28{,}110$, $\Sigma y^2 = 364{,}280$, $\Sigma xy = 98{,}307$

$\bar{x} = \Sigma x/n = 512/10 = 51.2$, $\bar{y} = \Sigma y/n = 1896/10 = 189.6$

$$SS_{xx} = \sum x^2 - \frac{(\sum x)^2}{n} = 28{,}110 - \frac{(512)^2}{10} = 1895.6$$

$$SS_{yy} = \sum y^2 - \frac{(\sum y)^2}{n} = 364{,}280 - \frac{(1896)^2}{10} = 4798.4$$

$$SS_{xy} = \sum xy - \frac{(\Sigma x)(\Sigma y)}{n} = 98{,}307 - \frac{(512)(1896)}{10} = 1231.8$$

b. $b = SS_{xy}/SS_{xx} = 1231.8/1895.6 = .6498$

$a = \bar{y} - b\bar{x} = 189.6 - (.6498)(51.2) = 156.3302$

$\hat{y} = 156.3302 + .6498x$

c. The value of $a = 156.3302$ represents the cholesterol level of a man with an age of zero years. The value of $b = .6498$ means that, on average, the cholesterol level of a man increases by .6498 for every one year increase in age.

d. $$r = \frac{SS_{xy}}{\sqrt{SS_{xx}SS_{yy}}} = \frac{1231.8}{\sqrt{(1895.6)(4798.4)}} = .41$$

$r^2 = bSS_{xy}/SS_{yy} = (.6498)(1231.8)/4798.4 = .17$

The value of $r = .41$ indicates that the two variables have a positive correlation but they are not strongly related. The value of $r^2 = .17$ means that only 17% of the total squared errors (SST) is explained by our regression model.

e.

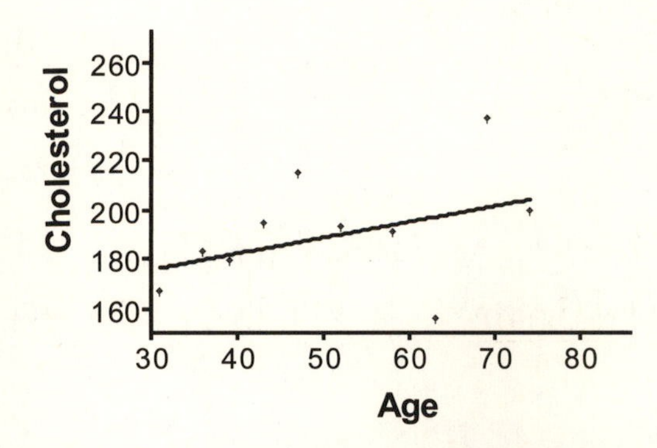

f. For $x = 60$, $\hat{y} = 156.3302 + .6498(60) = 195.3182$

Thus, a 60 year old man is expected to have a cholesterol level of about 195.

g. $$s_e = \sqrt{\frac{SS_{yy} - b(SS_{xy})}{n-2}} = \sqrt{\frac{4798.4 - (.6498)(1231.8)}{10-2}} = 22.3550$$

h. $s_b = s_e / \sqrt{SS_{xx}} = 22.3550/\sqrt{1895.6} = .5135$

$df = n - 2 = 10 - 2 = 8$

The 95% confidence interval for B is

$b \pm ts_b$ = .6498 ± 2.306(.5135) = .6498 ± 1.1841= –.5343 to 1.8339

i. Step 1: H_0: $B = 0$, H_1: $B > 0$

Step 2: Since σ_ϵ is unknown, use the t distribution.

Step 3: For $\alpha = .05$ with $df = 8$, the critical value of t is 1.860.

Step 4: $t = (b - B)/s_b = (.6498–0)/.5135 = 1.265$

Step 5: Do not reject H_0 since 1.265 < 1.860.

Conclude that B is not positive.

j. Step 1: H_0: $\rho = 0$, H_1: $\rho > 0$

Step 2: Assuming the population distributions for both variables are normally distributed, use the t distribution.

Step 3: For $\alpha = .025$ with $df = 8$, the critical value of t is 2.306.

Step 4: $$t = r\sqrt{\frac{n-2}{1-r^2}} = .41\sqrt{\frac{10-2}{1-(.41)^2}} = 1.271$$

Step 5: Do not reject H_0 since 1.271 < 2.306.

Conclude that ρ is not positive.

13.83 Let x = income and y = charitable contributions.

a. $n = 10$, $\Sigma x = 641$, $\Sigma y = 141$, $\Sigma x^2 = 45{,}349$, $\Sigma y^2 = 2927$, $\Sigma xy = 10{,}934$

$\bar{x} = \Sigma x/n = 641/10 = 64.1$, $\bar{y} = \Sigma y/n = 141/10 = 14.1$

$$SS_{xx} = \sum x^2 - \frac{(\sum x)^2}{n} = 45{,}349 - \frac{(641)^2}{10} = 4260.9$$

$$SS_{yy} = \sum y^2 - \frac{(\sum y)^2}{n} = 2927 - \frac{(141)^2}{10} = 938.9$$

$$SS_{xy} = \sum xy - \frac{(\Sigma x)(\Sigma y)}{n} = 10{,}934 - \frac{(641)(141)}{10} = 1895.9$$

b. $b = SS_{xy}/SS_{xx} = 1895.9/4260.9 = .4450$

$a = \bar{y} - b\bar{x} = 14.1 - (.4450)(64.1) = -14.4245$

$\hat{y} = -14.4245 + .4450x$

c. Although $a = -14.4245$ represents the charitable contributions of a household with no income, the negative value makes no sense. This is because incomes in the sample, x, varied from \$36,000 to \$102,000, but $x = 0$ is far outside that range. The value of $b = .4450$ means that, on average, charitable contributions increase by \$44.50 for every \$1000 increase in a household's income.

d. $r = \dfrac{SS_{xy}}{\sqrt{SS_{xx}SS_{yy}}} = \dfrac{1895.9000}{\sqrt{(4260.9000)(938.9000)}} = .95$

$r^2 = SS_{xy}/SS_{yy} = (.4450)(1895.9000)/938.9000 = .90$

The value of $r = .95$ indicates that the two variables have a very strong positive linear correlation. The value of $r^2 = .90$ means that 90% of the total squared errors (SST) is explained by the regression model.

e. $s_e = \sqrt{\dfrac{SS_{yy} - b(SS_{xy})}{n-2}} = \sqrt{\dfrac{938.9000-(.4950)(1895.9000)}{10-2}} = 3.4501$

f. $s_b = s_e / \sqrt{SS_{xx}} = 3.4501/\sqrt{4260.9000} = .0529$

$df = n - 2 = 10 - 2 = 8$

The 99% confidence interval for B is

$b \pm ts_b = .4450 \pm 3.355(.0529) = .4450 \pm .1775 = .2675$ to $.6225$

g. Step 1: H_0: $B = 0$, H_1: $B > 0$

Step 2: Since σ_ϵ is unknown, use the t distribution.

Step 3: For $\alpha = .01$ with $df = 8$, the critical value of t is 2.896.

Step 4: $t = (b - B)/s_b = (.4450 - 0)/.0529 = 8.412$

Step 5: Reject H_0 since $8.412 > 2.896$.

Conclude that B is positive.

h. Step 1: H_0: $\rho = 0$, H_1: $\rho \neq 0$

Step 2: Assuming the population distributions for both variables are normally distributed, use the t distribution.

Step 3: For $\alpha = .01$ with $df = 8$, the critical values of t are –3.355 and 3.355.

Step 4: $t = r\sqrt{\dfrac{n-2}{1-r^2}} = .95\sqrt{\dfrac{10-2}{1-(.95)^2}} = 8.605$

Step 5: Reject H_0 since $8.605 > 3.355$.

Conclude that ρ is different from zero.

13.85 a. Let x = GPA and y = starting salary (in thousands of dollars).

$n = 7$, $\Sigma x = 20.57$, $\Sigma y = 277$, $\Sigma x^2 = 63.8111$, $\Sigma y^2 = 11{,}247$, $\Sigma xy = 843.18$

$\bar{x} = \Sigma x/n = 20.57/7 = 2.9386$, $\bar{y} = \Sigma y/n = 277/7 = 39.5714$

$$SS_{xx} = \sum x^2 - \frac{(\sum x)^2}{n} = 63.8111 - \frac{(20.57)^2}{7} = 3.3647$$

$$SS_{yy} = \sum y^2 - \frac{(\sum y)^2}{n} = 11{,}247 - \frac{(277)^2}{7} = 285.7143$$

$$SS_{xy} = \sum xy - \frac{(\Sigma x)(\Sigma y)}{n} = 843.18 - \frac{(20.57)(277)}{7} = 29.1957$$

b. $b = SS_{xy}/SS_{xx} = 29.1957/3.3647 = 8.6771$

$a = \bar{y} - b\bar{x} = 39.5714 - (8.6771)(2.9386) = 14.0729$

$\hat{y} = 14.0729 + 8.6771x$

c. The value of $a = 14.0729$ represents the starting salary (about $14,073) for a college graduate with a GPA of zero. The value of $b = 8.6771$ means that, on average, the starting salary of a college graduate increases by $8677 for every one point increase in GPA.

d. $$r = \frac{SS_{xy}}{\sqrt{SS_{xx}SS_{yy}}} = \frac{29.1957}{\sqrt{(3.3647)(285.7143)}} = .94$$

$r^2 = bSS_{xy}/SS_{yy} = (8.6771)(29.1957)/285.7143 = .89$

The value of $r = .94$ indicates that the two variables have a very strong positive linear correlation. The value of $r^2 = .89$ means that 89% of the total squared errors (SST) is explained by the regression model.

e. $$s_e = \sqrt{\frac{SS_{yy} - b(SS_{xy})}{n-2}} = \sqrt{\frac{285.7143 - (8.6771)(29.1957)}{7-2}} = 2.5448$$

f. $s_b = s_e / \sqrt{SS_{xx}} = 2.5448/\sqrt{3.3647} = 1.3873$

$df = n - 2 = 7 - 2 = 5$

The 95% confidence interval for B is

$b \pm ts_b = 8.6771 \pm 2.571(1.3873) = 8.6771 \pm 3.5667 = 5.1104$ to 12.2438

g. Step 1: H_0: $B = 0$, H_1: $B \neq 0$

Step 2: Since σ_ϵ is unknown, use the t distribution.

Step 3: For $\alpha = .01$ with $df = 5$, the critical values of t are -4.032 and 4.032.

Step 4: $t = (b - B)/s_b = (8.6771 - 0)/1.3873 = 6.255$

Step 5: Reject H_0 since $6.255 > 4.032$.

Conclude that B is different from zero.

h. Step 1: H_0: $\rho = 0$, H_1: $\rho > 0$

Step 2: Assuming the population distributions for both variables are normally distributed, use the t distribution.

Step 3: For $\alpha = .01$ with $df = 5$, the critical value of t is 3.365.

Step 4: $t = r\sqrt{\frac{n-2}{1-r^2}} = .94\sqrt{\frac{7-2}{1-(.94)^2}} = 6.161$

Step 5: Reject H_0 since 6.161 > 3.365.

Conclude that ρ is positive.

Sections 13.8 – 13.9

13.87 a. For $x = 15$, $\hat{y} = 3.25 + .80(15) = 15.25$

$$s_{\hat{y}_m} = s_e\sqrt{\frac{1}{n}+\frac{(x_0-\bar{x})^2}{SS_{xx}}} = (.954)\sqrt{\frac{1}{10}+\frac{(15-18.52)^2}{144.65}} = .4111$$

$df = n - 2 = 10 - 2 = 8$

The 99% confidence interval for $\mu_{y|15}$ is

$\hat{y} \pm ts_{\hat{y}_m} = 15.25 \pm 3.355(.4111) = 15.25 \pm 1.3792 = 13.8708$ to 16.6292

$$s_{\hat{y}_p} = s_e\sqrt{1+\frac{1}{n}+\frac{(x_0-\bar{x})^2}{SS_{xx}}} = (.954)\sqrt{1+\frac{1}{10}+\frac{(15-18.52)^2}{144.65}} = 1.0388$$

The 99% prediction interval for y_p for $x = 15$ is

$\hat{y} \pm ts_{\hat{y}_p} = 15.25 \pm 3.355(1.0388) = 15.25 \pm 3.4852 = 11.7648$ to 18.7352

b. For $x = 12$, $\hat{y} = -27 + 7.67(12) = 65.04$

$df = n - 2 = 10 - 2 = 8$

$$s_{\hat{y}_m} = s_e\sqrt{\frac{1}{n}+\frac{(x_0-\bar{x})^2}{SS_{xx}}} = (2.46)\sqrt{\frac{1}{10}+\frac{(12-13.43)^2}{369.77}} = .7991$$

The 99% confidence interval for $\mu_{y|12}$ is

$\hat{y} \pm ts_{\hat{y}_m} = 65.04 \pm 3.355(.7991) = 65.04 \pm 2.6810 = 62.3590$ to 67.7210

$$s_{\hat{y}_p} = s_e\sqrt{1+\frac{1}{n}+\frac{(x_0-\bar{x})^2}{SS_{xx}}} = (2.46)\sqrt{1+\frac{1}{10}+\frac{(12-13.43)^2}{369.77}} = 2.5865$$

The 99% prediction interval for y_p for $x = 12$ is

$\hat{y} \pm ts_{\hat{y}_p} = 65.04 \pm 3.355(2.5865) = 65.04 \pm 8.6777 = 56.3623$ to 73.7177

13.89 From the solution to Exercise 13.53: $n = 9$, $\bar{x} = 8.8889$, $SS_{xx} = 256.8889$, and $s_e = 2.2758$

For $x = 10$, $\hat{y} = 23.7297 + 1.3054(10) = 36.7837$

$df = n - 2 = 9 - 2 = 7$

$$s_{\hat{y}_m} = s_e\sqrt{\frac{1}{n}+\frac{(x_0-\bar{x})^2}{SS_{xx}}} = (2.2758)\sqrt{\frac{1}{9}+\frac{(10-8.8889)^2}{256.8889}} = .7748$$

The 90% confidence interval for $\mu_{y|10}$ is

$\hat{y} \pm ts_{\hat{y}_m} = 36.7837 \pm 1.895(.7748) = 36.7837 \pm 1.4682 = 35.3155$ to 38.2519

$$s_{\hat{y}_p} = s_e\sqrt{1+\frac{1}{n}+\frac{(x_0-\bar{x})^2}{SS_{xx}}} = (2.2758)\sqrt{1+\frac{1}{9}+\frac{(10-8.8889)^2}{256.8889}} = 2.4041$$

The 90% prediction interval for y_p for $x = 10$ is:

$\hat{y} \pm ts_{\hat{y}_p} = 36.7837 \pm 1.895(2.4041) = 36.7837 \pm 4.5558 = 32.2279$ to 41.3395

13.91 From the solution to Exercise 13.82: $n = 7$, $\bar{x} = 91.8571$, $SS_{xx} = 2104.8571$, and $s_e = 3.6221$

For $x = 90$, $\hat{y} = 37.2235 + .9027(90) = 118.4665$

$df = n - 2 = 7 - 2 = 5$

$$s_{\hat{y}_m} = s_e\sqrt{\frac{1}{n}+\frac{(x_0-\bar{x})^2}{SS_{xx}}} = (3.6221)\sqrt{\frac{1}{7}+\frac{(90-91.8571)^2}{2104.8571}} = 1.3769$$

The 99% confidence interval for $\mu_{y|90}$ is

$\hat{y} \pm ts_{\hat{y}_m} = 118.4665 \pm 4.032(1.3769) = 118.4665 \pm 5.5517 = 112.9148$ to 124.0182

$$s_{\hat{y}_p} = s_e\sqrt{1+\frac{1}{n}+\frac{(x_0-\bar{x})^2}{SS_{xx}}} = (3.6221)\sqrt{1+\frac{1}{7}+\frac{(90-91.8571)^2}{2104.8571}} = 3.8750$$

The 99% prediction interval for y_p for $x = 90$ is

$\hat{y} \pm ts_{\hat{y}_p} = 118.4665 \pm 4.032(3.8750) = 118.4665 \pm 15.6240 = 102.8425$ to 134.0905

13.93 From the solution to Exercise 13.83: $n = 10$, $\bar{x} = 64.1$, $SS_{xx} = 4260.9000$, and $s_e = 3.4501$

For $x = 64$, $\hat{y} = -14.4245 + .4450(64) = 14.0555$

$df = n - 2 = 10 - 2 = 8$

$$s_{\hat{y}_m} = s_e\sqrt{\frac{1}{n}+\frac{(x_0-\bar{x})^2}{SS_{xx}}} = (3.4501)\sqrt{\frac{1}{10}+\frac{(64-64.1)^2}{4260.9}} = 1.0910$$

The 95% confidence interval for $\mu_{y|64}$ is

$\hat{y} \pm t s_{\hat{y}_m} = 14.0555 \pm 2.306(1.0910) = 14.0555 \pm 2.5158 = 11.5397$ to 16.5713

$$s_{\hat{y}_p} = s_e\sqrt{1+\frac{1}{n}+\frac{(x_0-\bar{x})^2}{SS_{xx}}} = (3.4501)\sqrt{1+\frac{1}{10}+\frac{(64-64.1)^2}{4260.9}} = 3.6185$$

The 95% prediction interval for y_p for $x = 64$ is

$\hat{y} \pm t s_{\hat{y}_p} = 14.0555 \pm 2.306(3.6185) = 14.0555 \pm 8.3443 = 5.7112$ to 22.3998

Supplementary Exercises

13.95 Let x = age (in years) of a machine and y = the number of breakdowns.

a. As the age of a machine increases (that is, the machine becomes older), the number of breakdowns is expected to increase. Hence, we expect a positive relationship between these two variables. Consequently, B is expected to be positive.

b. $n = 7$, $\Sigma x = 55$, $\Sigma y = 41$, $\Sigma x^2 = 527$, $\Sigma y^2 = 339$, $\Sigma xy = 416$

$\bar{x} = \Sigma x/n = 55/7 = 7.8571$, $\bar{y} = \Sigma y/n = 41/7 = 5.8571$

$$SS_{xx} = \sum x^2 - \frac{(\sum x)^2}{n} = 527 - \frac{(55)^2}{7} = 94.8571$$

$$SS_{yy} = \sum y^2 - \frac{(\sum y)^2}{n} = 339 - \frac{(41)^2}{7} = 98.8571$$

$$SS_{xy} = \sum xy - \frac{(\Sigma x)(\Sigma y)}{n} = 416 - \frac{(55)(41)}{7} = 93.8571$$

$b = SS_{xy}/SS_{xx} = 93.8571/94.8571 = .9895$

$a = \bar{y} - b\bar{x} = 5.8571 - (.9895)(7.8571) = -1.9175$

$\hat{y} = -1.9175 + .9895x$

The sign of $b = .9895$ is positive, which is consistent with what we expected.

c. The value of $a = -1.9175$ represents the number of breakdowns per month for a new machine (age = 0). The value of $b = .9895$ means that the average number of breakdowns per month increases by about .9895 for every one year increase in the age of such a machine.

d. $$r = \frac{SS_{xy}}{\sqrt{SS_{xx}SS_{yy}}} = \frac{93.8571}{\sqrt{(94.8571)(98.8571)}} = .97$$

$r^2 = bSS_{xy}/SS_{yy} = (.9895)(93.8571)/98.571 = .94$

The value of $r = .97$ indicates that the two variables have a very strong positive correlation.

The value of $r^2 = .94$ means that 94% of the total squared errors (SST) is explained by our regression model.

e. $s_e = \sqrt{\frac{SS_{yy} - bSS_{xy}}{n-2}} = \sqrt{\frac{98.8571-(.9895)(93.8571)}{7-2}} = 1.0941$

f. $s_b = s_e / \sqrt{SS_{xx}} = 1.0941/\sqrt{94.8571} = .1123$

$df = n - 2 = 7 - 2 = 5$

The 99% confidence interval for B is

$b \pm ts_b = .9895 \pm 4.032(.1123) = .9895 \pm .4528 = .5367$ to 1.4423

g. Step 1: H_0: $B = 0$, H_1: $B > 0$

Step 2: Since σ_ϵ is unknown, use the t distribution.

Step 3: For $\alpha = .025$ with $df = 5$, the critical value of t is 2.571.

Step 4: $t = (b - B)/s_b = (.9895 - 0)/.1123 = 8.811$

Step 5: Reject H_0 since $8.811 > 2.571$.

Conclude that B is positive.

h. Step 1: H_0: $\rho = 0$, H_1: $\rho > 0$

Step 2: Assuming the population distributions for both variables are normally distributed, use the t distribution.

Step 3: For $\alpha = .025$ with $df = 5$, the critical value of t is 2.571.

Step 4: $t = r\sqrt{\frac{n-2}{1-r^2}} = .97\sqrt{\frac{7-2}{1-(.97)^2}} = 8.922$

Step 5: Reject H_0 since $8.922 > 2.571$.

Conclude that ρ is positive. The conclusion is the same as that of part g (reject H_0).

13.97 Let $x =$ number of promotions per day and $y =$ number of units (in hundreds) sold per day.

a. We would expect an increase in the number of promotions to yield increased sales, implying a positive relationship between the two variables. Consequently, we expect B to be positive.

b. $n = 7$, $\Sigma x = 177$, $\Sigma y = 144$, $\Sigma x^2 = 5285$, $\Sigma y^2 = 3224$, $\Sigma xy = 4049$

$\bar{x} = \Sigma x/n = 177/7 = 25.2857$, $\bar{y} = \Sigma y/n = 144/7 = 20.5714$

$$SS_{xx} = \sum x^2 - \frac{(\sum x)^2}{n} = 5285 - \frac{(177)^2}{7} = 809.5286$$

$$SS_{yy} = \sum y^2 - \frac{(\sum y)^2}{n} = 3224 - \frac{(144)^2}{7} = 261.7143$$

$$SS_{xy} = \sum xy - \frac{(\Sigma x)(\Sigma y)}{n} = 4049 - \frac{(177)(144)}{7} = 407.8571$$

$b = SS_{xy}/SS_{xx} = 407.8571/809.5286 = .5039$

$a = \bar{y} - b\bar{x} = 20.5714 - (.5039)(25.2857) = 7.8299$

$\hat{y} = 7.8299 + .5039x$

The sign of $b = 13.6123$ is positive, which is consistent with what we expected.

c. The value of $a = 7.8299$ represents the number of units (in hundreds) sold if there are no promotions. The value of $b = .5039$ means that the sales are expected to increase by about 50 units per day for each additional promotion.

d. $$r = \frac{SS_{xy}}{\sqrt{SS_{xx}SS_{yy}}} = \frac{407.8571}{\sqrt{(809.4286)(261.7143)}} = .89$$

$r^2 = bSS_{xy}/SS_{yy} = (.5039)(407.8571)/261.7143 = .79$

The value of $r = .89$ indicates that the two variables have a strong positive correlation.

The value of $r^2 = .78$ means that 78% of the total squared errors (SST) is explained by our regression model.

e. For $x = 35$, $\hat{y} = 7.8299 + .5039(35) = 25.4664$

Thus, we expect sales of about 2547 units in a day with 35 promotions.

f. $$s_e = \sqrt{\frac{SS_{yy} - bSS_{xy}}{n-2}} = \sqrt{\frac{261.7143 - (.5039)(407.8571)}{7-2}} = 3.3525$$

g. $s_b = s_e / \sqrt{SS_{xx}} = 3.3525/\sqrt{809.4286} = .1178$

$df = n - 2 = 7 - 2 = 5$

The 98% confidence interval for B is

$b \pm ts_b = .5039 \pm 3.365(.1178) = .5039 \pm .3964 = .1075$ to $.9003$

h. Step 1: H_0: $B = 0$, H_1: $B > 0$

Step 2: Since σ_ϵ is unknown, use the t distribution.

Step 3: For $\alpha = .01$ with $df = 5$, the critical value of t is 3.365.

Step 4: $t = (b - B)/s_b = (.5039 - 0)/.1178 = 4.278$

Step 5: Reject H_0 since $4.278 > 3.365$.

Conclude that B is positive.

i. Step 1: H_0: $\rho = 0$, H_1: $\rho \neq 0$

Step 2: Assuming the population distributions for both variables are normally distributed, use the t distribution.

Step 3: For $\alpha = .02$ with $df = 5$, the critical values of t are –3.365 and 3.365.

Step 4: $t = r\sqrt{\dfrac{n-2}{1-r^2}} = .89\sqrt{\dfrac{7-2}{1-(.89)^2}} = 4.365$

Step 5: Reject H_0 since $4.365 > 3.365$.

Conclude that ρ is different from zero.

13.99 Let x = time and y = number of Americans who took cruises (in millions).

a.

x	0	1	2	3	4	5	6	7	8	9
y	4.4	4.7	5.1	5.4	5.9	6.9	6.9	7.6	8.2	9.0

b. $n = 10$, $\Sigma x = 45$, $\Sigma y = 64.1$, $\Sigma x^2 = 285$, $\Sigma y^2 = 432.65$, $\Sigma xy = 330.4$

$\bar{x} = \Sigma x/n = 45/10 = 4.5$, $\bar{y} = \Sigma y/n = 64.1/10 = 6.41$

$$SS_{xx} = \sum x^2 - \frac{(\sum x)^2}{n} = 285 - \frac{(45)^2}{10} = 82.5$$

$$SS_{yy} = \sum y^2 - \frac{(\sum y)^2}{n} = 432.65 - \frac{(64.1)^2}{10} = 21.769$$

$$SS_{xy} = \sum xy - \frac{(\Sigma x)(\Sigma y)}{n} = 330.4 - \frac{(45)(64.1)}{10} = 41.95$$

c.

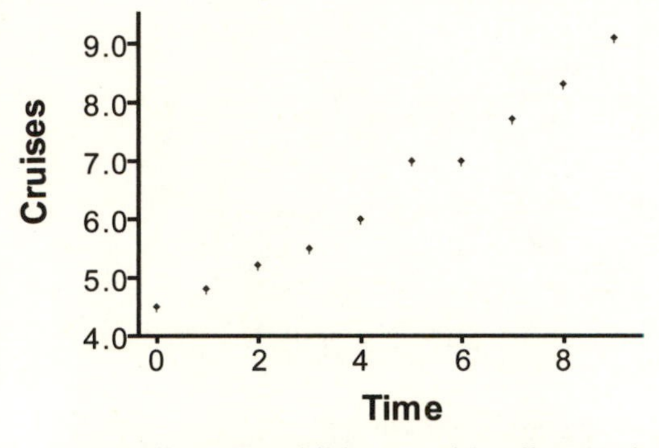

The scatter diagram exhibits a positive linear relationship between time and number of Americans who took cruises (in millions).

d. $b = SS_{xy}/SS_{xx} = 41.95/82.5 = .5085$

$a = \bar{y} - b\bar{x} = 6.41 - (.5085)(4.5) = 4.1218$

$\hat{y} = 4.1218 + .5085x$

e. The value of $a = 4.1218$ represents the number of Americans (in millions) who took cruises at time zero. The value of $b = 3.2158$ means that, on average, the number of Americans (in millions) who took cruises increased by .5085 million (508,500) per year from 1995 to 2004.

f. $r = \frac{SS_{xy}}{\sqrt{SS_{xx}SS_{yy}}} = \frac{41.95}{\sqrt{(82.5)(21.769)}} = .99$

g. For $x = 14$, $\hat{y} = 4.1218 + .5085(14) = 11.2408$

Thus, the predicted number of Americans who will take cruises in 2009 ($x = 14$) is 11,240,800. Note that this prediction is based on the regression equation derived from data for 1995 through 2004. This prediction assumes that the same linear relationship will continue for 5 or more years into the future, a questionable assumption.

13.101 Let x = year and y = time (in seconds)

a.

x	0	1	2	3	4	5	6	7	8
y	58.6	55.7	54.8	55.9	54.9	54.6	54.5	53.8	53.8

b. $n = 9$, $\Sigma x = 36$, $\Sigma y = 496.6$, $\Sigma x^2 = 204$, $\Sigma y^2 = 27{,}418.6$, $\Sigma xy = 1959.6$

$\bar{x} = \Sigma x/n = 36/9 = 4$, $\bar{y} = \Sigma y/n = 496.6/9 = 55.1778$

$$SS_{xx} = \sum x^2 - \frac{(\sum x)^2}{n} = 204 - \frac{(36)^2}{9} = 60$$

$$SS_{yy} = \sum y^2 - \frac{(\sum y)^2}{n} = 27{,}418.6 - \frac{(496.6)^2}{9} = 17.3156$$

$$SS_{xy} = \sum xy - \frac{(\Sigma x)(\Sigma y)}{n} = 1959.6 - \frac{(36)(496.6)}{9} = -26.8$$

c.

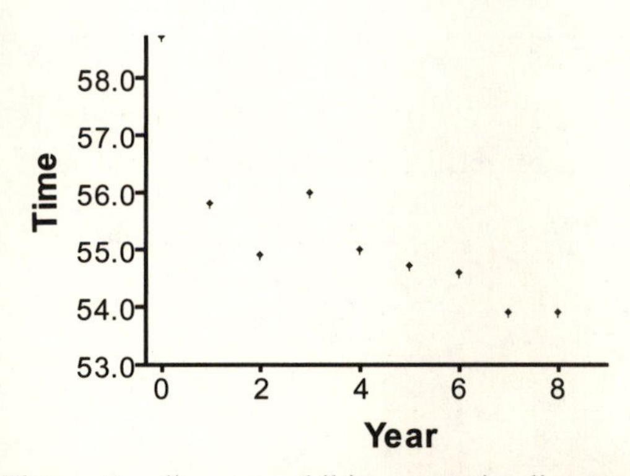

The scatter diagram exhibits a negative linear relationship between year and winning time in the women's 100-meter freestyle.

d. $b = SS_{xy}/SS_{xx} = -26.8/60 = -.4667$

$a = \bar{y} - b\bar{x} = 55.1778 - (-.4667)(4) = 57.0446$

$\hat{y} = 57.0446 - .4667x$

e. The value of $a = 57.0446$ represents the winning time in the women's 100-meter freestyle at year zero. The value of $b = -.4667$ means that, on average, the winning time decreases by .4667 seconds per time period (four years).

f. $r = \dfrac{SS_{xy}}{\sqrt{SS_{xx}SS_{yy}}} = \dfrac{-26.8}{\sqrt{(60)(17.3156)}} = -.83$

g. For $x = 10$, $\hat{y} = 57.0446 - .4667(10) = 52.3776$

Thus, the predicted winning time in the women's 100-meter freestyle event is 52.3776 seconds in the 2012 Olympics. Note that this prediction is based on the regression equation derived from data for 1972 through 2004. This prediction assumes that the same linear relationship will continue for 8 or more years into the future, a questionable assumption.

13.103 From the solution to Exercise 13.96: $n = 7$, $\bar{x} = 5.4429$, $SS_{xx} = 17.3771$, and $s_e = 13.4087$

For $x = 7$, $\hat{y} = -16.2333 + 13.6123(7) = 79.0528$

$df = n - 2 = 7 - 2 = 5$

$$s_{\hat{y}_m} = s_e\sqrt{\frac{1}{n} + \frac{(x_0 - \bar{x})^2}{SS_{xx}}} = (13.4087)\sqrt{\frac{1}{7} + \frac{(7 - 5.4429)^2}{17.3771}} = 7.1253$$

The 95% confidence interval for $\mu_{y|7}$ is

$\hat{y} \pm ts_{\hat{y}_m} = 79.0528 \pm 2.571(7.1253) = 79.0528 \pm 18.3191 = 60.7337$ to 97.3719

$$s_{\hat{y}_p} = s_e\sqrt{1 + \frac{1}{n} + \frac{(x_0 - \bar{x})^2}{SS_{xx}}} = (13.4087)\sqrt{1 + \frac{1}{7} + \frac{(7 - 5.4429)^2}{17.3771}} = 15.1843$$

The 95% prediction interval for y_p for $x = 7$ is

$\hat{y} \pm ts_{\hat{y}_p} = 79.0528 \pm 2.571(15.1843) = 79.0528 \pm 39.0388 = 40.0140$ to 118.0916

13.105 From the solution to Exercise 13.98: $n = 8$, $\bar{x} = 88.8750$, $SS_{xx} = 522.8750$, and $s_e = 11.7635$

For $x = 95$, $\hat{y} = -61.3189 + 6.1583(95) = 223.7196$

$df = n - 2 = 8 - 2 = 6$

$$s_{\hat{y}_m} = s_e\sqrt{\frac{1}{n} + \frac{(x_0 - \bar{x})^2}{SS_{xx}}} = (11.7635)\sqrt{\frac{1}{8} + \frac{(95 - 88.8750)^2}{522.8750}} = 5.2179$$

The 98% confidence interval for $\mu_{y|95}$ is:

$\hat{y} \pm ts_{\hat{y}_m} = 223.7196 \pm 3.143(5.2179) = 223.7196 \pm 16.3999 = 207.3197$ to 240.1195

$$s_{\hat{y}_p} = s_e\sqrt{1+\frac{1}{n}+\frac{(x_0-\bar{x})^2}{SS_{xx}}} = (11.7635)\sqrt{1+\frac{1}{8}+\frac{(95-88.8750)^2}{522.8750}} = 12.8688$$

The 98% prediction interval for y_p for $x = 95$ is

$\hat{y} \pm ts_{\hat{y}_p} = 223.7196 \pm 3.143(12.8688) = 223.7196 \pm 40.4466 = 183.2730$ to 264.1662

13.107 a. $\hat{y} = -432 + 7.7x$, $s_e = 28.17$, $SS_{xx} = 607$, and $\bar{x} = 87.5$

$b = 7.7$, $s_b = s_e / \sqrt{SS_{xx}} = 28.17/\sqrt{607} = 1.1434$

Step 1: H_0: $B = 0$, H_1: $B > 0$

Step 2: Since σ_ϵ is unknown, use the t distribution.

Step 3: $df = n - 2 = 20 - 2 = 18$

For $\alpha = .05$, the critical value of t is 1.734.

Step 4: $t = (b - B)/s_b = (7.7 - 0)/1.1434 = 6.734$

Step 5: Reject H_0 since $6.734 > 1.734$.

Conclude that B is positive. The maximum temperature and bowling activity between twelve noon and 6:00 pm have a positive association.

b. For $x = 90$, $\hat{y} = -432 + 7.7(90) = 261$

$$s_{\hat{y}_m} = s_e\sqrt{\frac{1}{n}+\frac{(x_0-\bar{x})^2}{SS_{xx}}} = (28.17)\sqrt{\frac{1}{20}+\frac{(90-87.5)^2}{607}} = 6.9172$$

The 95% confidence interval for $\mu_{y|90}$ is:

$\hat{y} \pm ts_{\hat{y}_m} = 261 \pm 2.101(6.9172) = 261 \pm 14.5330 = 246.4670$ to 275.5330 lines.

c. $$s_{\hat{y}_p} = s_e\sqrt{1+\frac{1}{n}+\frac{(x_0-\bar{x})^2}{SS_{xx}}} = (28.17)\sqrt{1+\frac{1}{20}+\frac{(90-87.5)^2}{607}} = 29.0068$$

The 95% prediction interval for y_p for $x = 90$ is

$\hat{y} \pm ts_{\hat{y}_p} = 261 \pm 2.101(29.0068) = 261 \pm 60.9433 = 200.0567$ to 321.9433 lines.

d. The mean value $\mu_{y|90}$ could be at either extreme of the interval in part b. Given a particular mean, the individual data points for this mean will have a certain variation, hence the prediction interval for y_p must be larger than the prediction interval for $\mu_{y|x}$.

e. $y = -432 + 7.7(100) = 338$ lines

Our regression line is only valid for the range of x values in our sample ($77°$ to $95°$ Fahrenheit).

We should interpret this estimate very cautiously and not attach too much value to it.

13.109 Burton's logic is faulty. The correlation coefficient merely describes the quantitative relationship between the two variables (frequency of mowing the lawn and size of corn ears). The high correlation does not prove that there is a cause–and–effect relation between the two variables. In this case, the correlation is due to the effect of other variables, such as amounts of sunshine and rain, and fertility of the soil. In years in which there are favorable amounts of sun and rain (and perhaps when Burton applies optimal amounts of fertilizers to both lawn and garden) the corn grows larger and the grass grows faster, thus requiring more frequent mowing. Thus, each of these other variables (amount of sunshine, amount of rain, and amount of fertilizer) is highly correlated with the size of the corn ears. Each of them is also highly correlated with the growth rate of the grass, (and therefore with the frequency of mowing). To obtain larger corn ears next year, Burton should be sure to plant the corn in a sunny part of his garden, water the corn during periods of dry weather, and apply fertilizer consistently.

13.111 Let $x =$ executive's test score and $y =$ executive's salary.

a. We are given that $\bar{x} = 44$ and $\bar{y} = 200{,}000$.

For U.S. executives, a loss of $16,836 for every five points scored above average on the test is equivalent to a loss of $3367.20 for each point scored above average. Thus, based on the given information, $b = -3367.20$.

$a = \bar{y} - b\bar{x} = 200{,}000 - (-3367.20)(44) = 348{,}156.80$

Thus, the regression equation is $\hat{y} = 348{,}156.80 - 3367.20x$

b. Nothing is said about the salaries of U.S. executives who scored below average, so the equation may not be valid for values of x below 44. It is also given that the maximum possible score on the test is 60. Thus, the equation is valid for the values of x from 44 to 60.

13.113 For 13.53: The value $b = 1.3054$ means that, on average, secretaries' salaries increase by $130.54 per month for each one year increase in experience. The value $a = 23.7297$ means that the starting salary for secretaries with no experience is expected to be $2372.97 per month. This value is logical.

For 13.54: The value of $b = -.5249$ means that, on average, the number of calls decreases by .5249 for every one degree increase in the temperature. The value of $a = 24.6556$ represents the number of calls when the temperature is at zero degrees. This value is logical.

For 13.55: The value of $b = 7.4374$ means that, on average, the amount of life insurance increases by \$7437 for every \$1000 increase in the annual income of a person. The value of $a = -208.4001$ represents the amount of life insurance for a person with a zero income. This value is not logical, as a negative amount of life insurance does not make sense. The value of $x = 0$ is outside the range of the data (\$34,000 to \$85,000).

For 13.56: The value of $b = 51.8300$ means that, on average, the rent of a house increases by \$51.83 for every 100 square feet increase in the size of the house. The value of $a = 58.2168$ represents the rent for a house with an area of zero square feet. This is not logical, as a house with no square footage does not make sense. The value of $x = 0$ is outside the range of the data (1300 to 3400 square feet).

Self-Review Test

1. d	**2.** a	**3.** b	**4.** a	**5.** b
6. b	**7.** True	**8.** True	**9.** a	**10.** b

11. See solution to Exercise 13.7.

12. The values of *A* and *B* for a regression model are obtained by using the population data. On the other hand, if a regression model is estimated by using the sample data, then we obtain the values of *a* and *b*.

13. The following are the assumptions of the regression model:

1. The random error term ϵ has a mean equal to zero for each *x*.
2. The errors associated with different observations are independent.
3. For any given *x*, the distribution of errors is normal.
4. The distribution of population errors for each *x* has the same (constant) standard deviation, which is denoted by σ_{ϵ}.

14. A regression line obtained by using the population data is called the population regression line. It gives values of *A* and *B* and is written as $\mu_{y|x} = A + Bx$. A regression line obtained by using the sample data is called the sample regression line. It gives the estimated values of *A* and *B*, which are denoted by *a* and *b*. The sample regression line is written as $\hat{y} = a + bx$.

15. a. The attendance depends on temperature. With a higher temperature more people attend the minor league baseball game. Hence, a higher temperature is expected to draw bigger crowds.

b. As mentioned in part a, a higher temperature is expected to bring in more ticket buyers on average. Consequently, we expect B to be positive.

c.

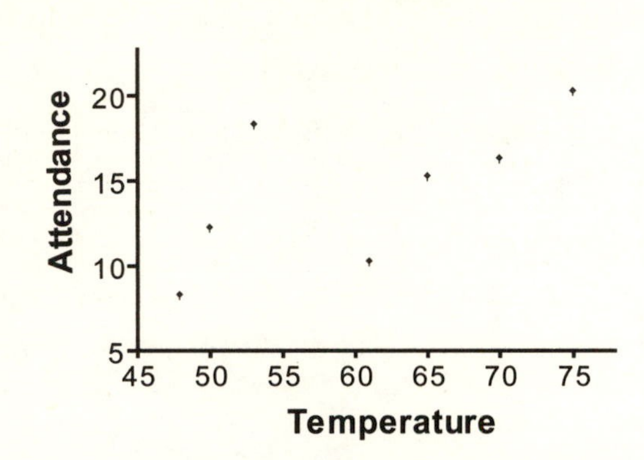

The scatter diagram exhibits a linear relationship between temperature and the attendance at a minor league baseball game but this relationship does not seem to be strong.

d. Let $x =$ temperature (in degrees) and $y =$ attendance (in hundreds).

$n = 7, \Sigma x = 422, \Sigma y = 99, \Sigma x^2 = 26{,}084, \Sigma y^2 = 1513, \Sigma xy = 6143$

$\bar{x} = \Sigma x/n = 422/7 = 60.2857, \ \bar{y} = \Sigma y/n = 99/7 = 14.1429$

$$SS_{xx} = \sum x^2 - \frac{(\sum x)^2}{n} = 26{,}084 - \frac{(422)^2}{7} = 643.4286$$

$$SS_{yy} = \sum y^2 - \frac{(\sum y)^2}{n} = 1513 - \frac{(99)^2}{7} = 112.8571$$

$$SS_{xy} = \sum xy - \frac{(\Sigma x)(\Sigma y)}{n} = 6143 - \frac{(422)(99)}{7} = 174.7143$$

$b = SS_{xy}/SS_{xx} = 174.7143/643.4286 = .2715$

$a = \bar{y} - b\bar{x} = 14.1429 - (.2715)(60.2857) = -2.2247$

$\hat{y} = -2.2247 + .2715x$

The sign of b is consistent with what we expected in part b.

e. The value of $a = -2.2247$ represents the number of people attending a minor league game when the temperature is zero. The value of $b = .2715$ means that, on average, the people attending a minor league games increases by about .27 for every one degree increase in temperature.

f. $$r = \frac{SS_{xy}}{\sqrt{SS_{xx}SS_{yy}}} = \frac{174.7143}{\sqrt{(643.4286)(112.8571)}} = .65$$

$r^2 = bSS_{xy}/SS_{yy} = (.2715)(174.7143)/112.8571 = .42$

The value of r = .65 indicates that the two variables have a positive correlation, which is not very strong. The value of r^2 = .42 means that 42% of the total squared errors (SST) is explained by our regression model.

g. For $x = 60$, $\hat{y} = -2.2247 + .2715(60) = 14.0653$

Thus, with a sixty degree temperature the minor league game is expected to sell about 1407 tickets.

h. $$s_e = \sqrt{\frac{SS_{yy} - bSS_{xy}}{n-2}} = \sqrt{\frac{112.8571 - .2715(174.7143)}{7-2}} = 3.6172$$

i. $s_b = s_e / \sqrt{SS_{xx}} = 3.6172 / \sqrt{643.4286} = .1426$

$df = n - 2 = 7 - 2 = 5$

The 99% confidence interval for B is

$b \pm ts_b = .2715 \pm 4.032(.1426) = .2715 \pm .5750 = -.3035$ to $.8464$

j. Step 1: H_0: $B = 0$, H_1: $B > 0$

Step 2: Since σ_ϵ is unknown, use the t distribution.

Step 3: For $\alpha = .01$ with $df = 5$, the critical value of t is 3.365.

Step 4: $t = (b - B)/s_b = (.2715 - 0)/.1426 = 1.904$

Step 5: Do not reject H_0 since $1.904 < 1.904$.

Conclude that B is not positive.

k. For $x = 60$, $\hat{y} = -2.2247 + .2715(60) = 14.0653$

$$s_{\hat{y}_m} = s_e\sqrt{\frac{1}{n} + \frac{(x_0 - \bar{x})^2}{SS_{xx}}} = (3.6172)\sqrt{\frac{1}{7} + \frac{(60 - 60.2857)^2}{643.4286}} = 1.3678$$

The 95% confidence interval for $\mu_{y|60}$ is

$\hat{y} \pm ts_{\hat{y}_m} = 14.0653 \pm 2.571(1.3678) = 14.0653 \pm 3.5166 = 10.5487$ to 17.5819

l. The standard deviation of $\hat{y}$ for predicting y for $x = 60$ is:

$$s_{\hat{y}_p} = s_e\sqrt{1 + \frac{1}{n} + \frac{(x_0 - \bar{x})^2}{SS_{xx}}} = (3.6172)\sqrt{1 + \frac{1}{7} + \frac{(60 - 60.2857)^2}{643.4286}}$$

The 95% prediction interval for y_p for $x = 60$ is

$\hat{y} \pm ts_{\hat{y}_p} = 14.0653 \pm 2.571(3.8672) = 14.0653 \pm 9.9426 = 4.1227$ to 24.0079

m. Step 1: H_0: $\rho = 0$, H_1: $\rho > 0$

Step 2: Assuming the population distributions for both variables are normally distributed, use the t distribution.

Step 3: For $\alpha = .01$ with $df = 5$, the critical value of t is 3.365.

Step 4: $t = r\sqrt{\frac{n-2}{1-r^2}} = .65\sqrt{\frac{7-2}{1-(.65)^2}} = 1.913$

Step 5: Do not reject H_0 since 1.913 < 3.365.

Conclude that ρ is different from zero.

Appendix A

A.1 Data sources can be divided into three categories: 1) internal sources, 2) external sources, and 3) surveys and experiments. The data sources such as a company's own personal files or accounting records are called **internal sources**. Sources of data from outside the company are called **external sources**. Data obtained from external sources may be primary or secondary data. **Primary data** are the data obtained from the organization which originally collected them. **Secondary data** are data obtained from a source which did not originally collect them. Surveys or experiments may be conducted when the necessary data is not available from the company's internal or external sources. In a **survey**, data are collected from the members of a population or sample with no particular control over the factors that may affect the characteristic of interest or the results of the survey. In an **experiment**, data are collected from members of a population or sample with some control over the factors that may affect the characteristic of interest or the results of the experiment.

A.3 A **census** is a survey that includes every member of the population. The technique of collecting information from a portion of the population is called a **sample survey**. A sample survey is preferred over a census for the following reasons:

1. Conducting a census is very time consuming because the size of the population is usually very large.
2. Conducting a census is very expensive.
3. In many cases it is almost impossible to identify and access every member of the target population.

A.5

a. A sample drawn in such a way that each element of the population has some chance of being included in the sample is called a **random sample**.

b. A sample in which some members of the population may have no chance of being selected is called a **nonrandom sample**.

c. A **convenience sample** is a sample in which the most accessible members of the population are selected.

d. A **judgment sample** is a sample in which members of a population are selected based on the judgment and prior knowledge of an expert.

e. A **quota sample** is a sample selected in such a way that each group or subpopulation is represented in the sample in exactly the same proportion as in the target population.

A.7 Simple random sample

A.9
a. This is a nonrandom sample. The students are not picked randomly but are chosen according to the professor's preferences.
b. This is a judgment sample since the professor uses his knowledge and expertise to determine which students to include in the sample.
c. This sample is subject to selection error since only those students the professor considers appropriate are included in the sample.

A.11
a. This is a random sample since the sampling frame is the entire class.
b. This is a simple random sample since the software package gives each sample of 20 students an equal chance of being chosen.
c. There should be no systematic error, since the sampling frame is the entire population and the software package would give each sample of 20 students an equal chance of being selected.

A.13
a. This is a non-random sample. Only readers of the magazine were able to answer the survey.
b. This sample is subject to voluntary response error, since only those who feel strongly enough about the issues to complete the questionnaire will respond. It also suffers from selection error since only the magazine's readers are included in the sampling frame.

A.15 This survey is subject to response error since some parents may be reluctant to give honest answers to an interviewer's questions about sensitive family matters.

A.17
a. This is a designed experiment since the doctors controlled the assignment of volunteers to the treatment and control groups.
b. There is not enough information to determine if this is a double-blind study. We would need to know if the women and/or the doctors were aware of which women were assigned to the treatment group and which were assigned to the control group.

A.19
a. This is a designed experiment since the doctors controlled the assignment of people to the treatment and control groups.
b. The study is double-blind since neither the patients nor the doctors knew who was given the aspirin and who was given the placebo.

A.21 This is a designed experiment since the researchers selected participants randomly from the entire population of families on welfare and then controlled which families received the treatment (job training) and which did not.

A.23 If the data showed that the percentage of families who got off welfare was higher in the group that received job training, the conclusion is justified. Since families were randomly assigned to treatment and control groups, the two groups should have been similar, and the difference in outcomes should be due to treatment (job training).

A.25

a. Since the study relies on volunteers, it may not be representative of the entire population of people suffering from compulsive behavior. Furthermore, the doctors used their own judgment to form the treatment and control groups. Thus, subjective factors may have influenced them, and the two groups may not be comparable. As a result, the effect of the medicine on compulsive behavior may be confounded with other variables. Therefore, the conclusion is not justified.

b. Although this study technically satisfies the criteria for a designed experiment (experimenters controlled the assignment of people to treatment groups), it suffers from the weaknesses of an observational study, as pointed out in part a.

c. The study is not double-blind since the physicians knew who received the treatment.

A.27

a. This is a designed experiment, since the doctors controlled the assignment of patients to the treatment and control groups.

b. The study is double-blind since neither patients nor doctors know who was receiving the medicine.

A.29

a. These 10 pigs represent a convenience sample since the first ten (easiest to catch) pigs comprise the sample. Convenience samples are nonrandom samples.

b. From part a we know these 10 pigs comprise a nonrandom sample. Therefore, they are not likely to be representative of the entire population. Faster pigs, for example, are not as likely to be included in the sample.

c. They form a convenience sample.

d. Answers will vary, but one better procedure is as follows. Assign numbers 1 through 40 to the pigs, and write the numbers 1 through 40 on separate pieces of paper, put them in a hat, mix them, and then draw 10 numbers. Pick the pigs whose numbers were drawn.

A.31

Seventy-eight percent of members of Health Maintenance Organizations (HMOs) responding to a recent survey reported that they had experienced denial of claims by their HMOs. Of those who had suffered such denials, 25% had unable to resolve the problem to their satisfaction in at least one instance. The survey was based on questionnaires sent to 5000 randomly chosen HMO members, of which 1200 actually completed their questionnaires and returned them.

The results of this survey should be interpreted with caution, because the percentage may not be representative of all HMO members. The most likely source of bias is *nonresponse error*, since only 1200 of the 5000 questionnaires were actually returned. Members who have had a claim denied may be angry and consequently have more motivation to return their questionnaires. Thus, there is likely to be a higher percentage of denied claims among the 1200 members who actually responded. Therefore, 78% may be an overestimate of the true percentage of HMO members who have experienced denied claims. Similarly, those who were unable to resolve their problem would be even more strongly motivated to respond to the survey, so 25% is likely to overestimate the corresponding percentage for the whole group.

A.33 a. We would expect $61,200 to be a biased estimate of the current mean annual income for all 5432 alumni because only 1240 of the 5432 alumni answered the income question. These 1240 are unlikely to be representative of the entire group of 5432.

b. The following types of bias are likely to be present.

Nonresponse error: Alumni with low incomes may be ashamed to respond. Thus, the 1240 who actually returned their questionnaires and answered the income question would tend to have higher than average incomes.

Response error: Some of those who answered the income question may give a value that is higher than their actual income in order to appear more successful.

c. We would expect the estimate of $61,200 to be above the current mean annual income of all 5432 alumni, for the given reasons in part b.